THE GREATEST POWER ON EARTH

By the same author:

THE BIRTH OF THE BOMB

THE RISE OF THE BOFFINS

TIZARD

EINSTEIN: THE LIFE AND TIMES

THE MAN WHO BROKE PURPLE

THE HUXLEYS

THE LIFE OF BERTRAND RUSSELL

FREUD: THE MAN AND THE CAUSE

RONALD W. CLARK

The Greatest Power on Earth

The Story of Nuclear Fission

Foreword by
LORD ZUCKERMAN

SIDGWICK & JACKSON
LONDON

First published in Great Britain by
Sidgwick and Jackson Limited in 1980

ISBN 0 283 98715 4

Typeset by Computacomp (UK) Ltd,
Fort William, Scotland.
Printed in Great Britain by
The Garden City Press
Letchworth, Hertfordshire
for Sidgwick and Jackson Limited
1 Tavistock Chambers, Bloomsbury Way
London WC1A 2SG

Acknowledgments

The author emphasizes that the opinions expressed in *The Greatest Power on Earth* are his own unless the contrary is made clear in the text. Nevertheless, he wishes to thank members of the U.K. Atomic Energy Authority and Richard G. Hewlett, official historian of the U.S. Atomic Energy Commission, for help and advice on a number of points. He thanks Imperial Chemical Industries for permission to quote the letter from Lord McGowan on pages 128–9 and the English-Speaking Union for permission to quote Lord Cherwell's letters on pages 270 and 273. Crown Copyright Records, Public Records Office, appear by permission of the Controller of Her Majesty's Stationery Office.

Contents

Part Three

ON TO THE DARKLING PLAIN

Illustrations

Foreword by Lord Zuckerman

The keystones of our century's history are those scientific and technological advances – radio, the internal combustion engine, antibiotics – that have determined the pattern of the life which half the world now leads, and for which the other half strives. And a handful of scientific treatises on nuclear physics which, in the years before the Second World War, embodied the discoveries of Einstein, of Planck, Bohr, J. J. Thomson, Rutherford, Joliot Curie, and Szilard – to name a few of the small band who transformed earlier conceptions of the nature of energy and the structure of matter – have proved more critical to the worldwide political and social upheavals of our own times than all the thoughts and ambitions of the prime ministers, presidents and dictators who have exercised power in this century. The intellectual achievements of these experimental and mathematical physicists have been described countless times, while volume after volume has tried to explain the practical consequences of their search for scientific truth. There are scores of books that tell of the need for nuclear energy, and of the dangers of nuclear power stations; books on nuclear weapons and the dangers of nuclear war; on the medical and scientific value of man-made radio-isotopes. There are technical works and official histories for the specialist, and popular expositions for the layman. But for most of us the subject still remains shrouded in mystery, mystery that generates fear and despair; fear and despair which, in turn, generate both a sense of doom and a hostility to the scientific process and to the scientists who brought these things about.

Starting with *The Birth of the Bomb* in 1961 and *The Rise of the Boffins* in 1962, Ronald Clark has written many works that focus on the impact of new scientific discovery on national affairs. Since the publication of his first book on nuclear weapons, a mass of new information on the subject has been published not only in the United Kingdom and the United States, but in France, Germany and the U.S.S.R., as well as in other countries. The United Nations organization has contributed its share. All this new material Ronald Clark has sifted in writing his new book, *The Greatest Power on Earth*.

It begins with an account of the development of our understanding of the principles of nuclear physics, written in terms that should be understandable to the non-technical reader. He goes on to tell the story of the first efforts that were made to harness the energy that resides in the nucleus of the atom. We read how in the days before the war scientists of all countries were freely exchanging ideas about how this might be done, and how the relatively enormous technical problems concerned were overcome for the first time when Fermi, an Italian refugee, constructed a nuclear pile in Chicago in 1942. We read about the secrecy that then supervened. Ronald Clark tells how the minds of scientists turned to the concept of devising a nuclear bomb, and of the feverish work that was started in the United Kingdom to devise a weapon once two refugee scientists, Peierls and Frisch, had provided the outlines of the necessary recipe. He describes the birth of the United States Manhattan nuclear bomb Project, and explains why in 1943 the United Kingdom abandoned its own efforts in the face of the vastly greater American programme. He reminds us that the very small number of politicians and military men who were in the know supposed that what was being done was no more than the development of some super-explosive, on a par with, but immensely more powerful than, the conventional ones which had been known for centuries. And he relates how only a few scientists – among them, and outstandingly, Niels Bohr – realized the political implications of the development of nuclear weapons; how Bohr and a few other scientists who thought like him could already see that the emergence of the bomb, a weapon which at a stroke could eliminate any great city, was a total break with the strategic past, the beginning of a new and dangerous stage in man's social evolution. He tells the story of the mutual suspicions of the U.S.A. and the U.S.S.R., and of the determination of the United Kingdom and France to be independent of the U.S.A. in the development of nuclear weapons, and of the widespread urge of other countries to become nuclear powers – in the quest for prestige – because of the belief that nuclear weapons mean military might. Once it was known that the atom could be split, once it was known that a hydrogen bomb could be made, it was inevitable that a nuclear arms race would follow.

In his characteristically readable way, Mr Clark tells the history of the technological steps that have led the world to its present dangerous pass. Let us hope that he will write yet another book, a book which will describe how reason ultimately prevailed among nations because of the realization that nuclear weapons are not weapons of war, but weapons of mutual annihilation; that they can deter, but that they cannot defend.

September 1980

PART ONE

Prologue to a Discovery

CHAPTER 1

From the Greeks to the Edwardians

The autumn of 1919 was a dangerous time for the human race. The war to end war had ended but in the United States the Senate had by its resolution on article 10 of the League of Nations Covenant virtually rejected the Peace Treaty. In Germany and the tattered rump of what had been the Austro-Hungarian Empire, inflation and starvation were steadily eroding civilized life. In Bavaria German troops had only with difficulty recaptured the great city of Munich from a self-appointed Communist Government. Across the bleak wastes of Russia ill-organized armies struggled to maintain, or to overthrow, an impressive tyranny, the forerunner of a formidable machine which would soon ensure that a population of 160,000,000 were 'by a gigantic system of governmental pressure hermetically sealed against the invasion of unwelcome truth'.* Even in Britain, a country where revolution does not come easily, troops returning from the wars had at places taken the law into their own hands, set up soviets in their demobilization camps and had in one place burned down the town hall as a warning of the wrath to come.

Yet this record of battery and barricade, which added a bloody touch to the first autumn of peace, was of infinitely less importance to the future of the human race than three events in the worlds of astronomy and physics. At the time, all appeared to emphasize the remoteness of science from everyday life and to imply that scientists themselves, however brilliant their intellects might be, lived in a tiny self-contained world of their own. Many of them still believed this to be true.

Of the events in the autumn of 1919 which were to help shape the world of the 1980s, the confirmation of Einstein's theory of general

* Full references to quoted material begin on p. 305.

relativity, by the British expeditions which had gone to Brazil and West Africa to observe an eclipse of the sun, was celebrated throughout the world. This was natural since there was something encouraging in the fact that at the height of the war, two years earlier, the internationalism of science had so overridden national feeling that the British Government had spent a substantial sum on a two-part expedition to test the theory of a German Jew. One might even be returning to the days of comparative civilization as when, during the Napoleonic Wars, a body of British scientists was allowed safe passage to and from Paris to attend an international meeting.

When the results of Arthur Eddington's 'eclipse' expeditions were announced in London on 6 November 1919 they astounded the world for a far more significant reason. Einstein was apparently right in his general theory. As Eddington put it in the last lines of a parody on the *Rubáiyát*, 'One thing is certain, and the rest debate / Light-rays, when near the Sun, do not go straight.' After 250 years the Newtonian framework on which so much of science relied, now had to be revised. The fabric of the universe was not quite what men had believed it to be.

The support which the eclipse expeditions gave to Einstein's theory of general relativity did not only make him the international figure who was fêted on his triumphant visit to the United States two years later. It also added substance to his earlier theory of special relativity and his famous equation that followed from it, the $E = mc^2$ equation expressing the relationship between energy and mass, the pointer towards the day when man might be able to release the greatest power on earth.

The famous equation was not, as is often believed, contained in the revolutionary paper which outlined the special theory to an astonished scientific world in 1905. It was, in fact, something of an afterthought, tossed off in a brief paper published later in the same year. Einstein had already maintained that space and time, previously thought to be absolute, depended on relative motion. But there was a third yardstick used to measure the nature of the physical world. This was mass, still also regarded as absolute, and he began to consider whether this, as well as space and time, might not be linked with the speed of light. 'This thought is amusing and infectious,' he wrote to an old friend, Conrad Habicht, 'but I cannot possibly know whether the good Lord does not laugh at it and has led me up the garden path.'

Einstein continued to brood on the amusing and infectious idea, decided that the good Lord had not misled him, and put together what was almost a footnote to his earlier paper. His conclusion was embodied in the famous equation: the energy locked up in matter is equal to its mass multiplied by the square of the velocity of light. If the mass were measured in grams, and the velocity of light in centimetres per second, the units customary in Einstein's day, then the energy would be given in ergs. Today the mass is often expressed in kilograms, the velocity of light in metres per second, and the energy in joules. The velocity of light being what it is – 186,281 miles, or 300 million metres, per second – no mathematical expertise is needed to appreciate that a very small mass is equivalent to a very large amount of energy.

In 1905 the mass-energy equation was of little more than academic importance. It helped to simplify some of the experiments being carried out across the world on the nature of the atom but it seemed of little practical value. Mass might indeed be, as Einstein inferred, merely congealed energy; all energy might be merely liberated matter. But it seemed unlikely that the curious proposition would have any effect on the lives of ordinary men and women.

The comparative isolation of Einstein's theoretical physics from the practicalities of everyday life dissolved only slowly during the first two decades of the twentieth century. Scientists would occasionally let their imagination play with the possibilities of transforming matter into energy, but to preserve their reputations assured the world that they were only joking. Poets and authors had a little more latitude, being able to contemplate, without accusations of irresponsibility, Pope's view of 'Atoms or systems into ruin hurled / And now a bubble burst and now a world.' The change from imaginative figment to distant possibility came at different times in the minds of different men. Yet the year in which the ideological hump was passed can be given as 1919, the very year in which Sir J. J. Thomson, the man more responsible than most for conceiving matter itself as a series of electrical charges, compared Einstein's theories, as now underpinned by the eclipse expeditions, not with 'the discovery of an outlying island but [with that] of a whole continent of new scientific ideas'. For in 1919, Sir Ernest Rutherford, the bluff booming-voiced New Zealander often described as the father of the nuclear age, had transformed the atoms of one element into the atoms of another, the dream of the medieval alchemists. And in Cambridge Francis Aston, a gentleman-physicist whose leisurely investigations had been rudely interrupted by the war, produced the third of the trio of events which were to change the world: the calculation of numerical values for the energy locked within the atom.

The two streams of thought which had produced Einstein's relativity and the atomic theories of Rutherford and Aston, two streams which began to flow more nearly parallel from 1919 on, were of course different branches of one river. But while Einstein's theories had sprung very largely from his contemplation, as he put it, of 'how God created this world', the work of physicists such as Rutherford and Aston owed more to man's age-old curiosity about the actual materials God had used, the materials of the physical world. And it was an end-product of this curiosity, in whose creation Einstein's theories had been used as an essential tool, which eventually led into the nuclear age.

To understand how this is so, it is necessary to consider how men had considered brute matter in earlier years. A basic subject for argument was whether matter itself could be divided indefinitely into smaller and ever smaller pieces, or whether there came a point when these could no longer be divisible; was it like a jelly – which could, if one had the right sort of tools, apparently be divided for ever without ceasing to be jelly – or was it like a bushel of peas whose possible division ceased after the last two peas had been put into separate piles?

The Greek philosopher Democritus settled for the second of the two theories, coining the word atom, which means 'that which cannot be cut', and maintaining: 'The only existing things are atoms and empty space; all else is mere opinion.' This view gave a superficially plausible explanation of the different forms of certain materials, such as water which could be solid as in ice, liquid, or gaseous as when it was boiled and was turned into steam. In each successive state, it was argued, all that happened was that the separate atoms moved farther apart.

The idea that a bar of metal or a bucketful of water was composed not only of matter but of an intervening succession of empty spaces was an offence to the common sense of ordinary men. The theories of Democritus and the Atomists of his school of thought were thus overwhelmed by those of Aristotle to whom the world appeared as combinations of primeval elements – earth, water, air and fire.

Aristotle's theory was maintained in various forms for something like a millenium and a half, and it was only in the seventeenth century that first Robert Boyle and then Isaac Newton began to revive the atomic theory. 'All these things being consider'd,' Newton wrote, 'it seems probable to me, that God in the Beginning, form'd Matter in solid, massy, hard, impenetrable, moveable Particles, of such Sizes and Figures, and with such other Properties, and in such Proportion to Space, as most conduced to the End for which he form'd them; and that these primitive Particles being Solids, are incomparably harder than any pourous Bodies compounded of them; even so very hard, as never to wear or break in pieces; no ordinary Power being able to divide what God himself made one in the first Creation.'

Antoine Lavoisier, and Joseph Prout, who had speculated that two elements could combine only in the ratio of whole numbers – 6 : 1, for instance, but never 5·9 : 1 – were others whose work led back towards Democritus's particulate theory of matter. However, basic scientific observation was still lacking and it was only with the rise of chemistry in the eighteenth and nineteenth centuries that the way was cleared for the next step forward. It was taken by John Dalton, the Manchester chemist, poor and colour-blind, who saw that the recently discovered chemical laws could be explained if the elements, of which he named twenty, consisted of invisible and indivisible particles for which he used Democritus's name of atoms. The atoms of any one element were always of the same weight, but the atoms making up different elements were of different weights. Furthermore, when atoms of different elements combined to form molecules of a compound they always combined in fixed and simple ratios. As for the atom itself, Dalton saw this as a solid sphere, a hard billiard-ball of matter.

Dalton's revolutionary ideas were outlined in 1803, and were developed five years later in his *New System of Chemical Philosophy*. They were to hold the stage until the later years of the nineteenth century. But critics remained. Most of them stressed the fact that there was no experimental evidence for such views, and even at the end of the nineteenth century eminent scientists such as Mach and Ostwald did not believe in the physical existence of atoms at all. Instead, they regarded Dalton's atomic theory 'more as a visualizing symbol than as knowledge concerning the factual construction of matter'. The change came only in 1905 when Einstein's explanation of the Brownian motion, first observed seventy years previously, provided decisive evidence for the reality of atoms, evidence confirmed experimentally by Jean Perrin in Paris three years later.

More general than total disbelief in atoms during the last decades of the nineteenth century was the view that although they might possibly exist, Dalton's explanation was unsatisfactory. Thus the French chemist, Marcelin Berthelot, would admit the existence of atoms, but appeared to believe that another century would be needed before men had unravelled their complexities – an almost inspired guess. Only after another hundred years, he maintained in 1869, would the physical and chemical sciences have allowed man to 'know of what the atom is constituted and ... be able, at will, to moderate, extinguish, and light up the sun as if it were a gas-lamp.' The physiologist Claude Bernard with equal imagination, forecast at the same time a future in which 'man would be so completely the master of organic law that he would create life in competition with God.' To which the Goncourt brothers, foreshadowing the critics of the nuclear age, commented: 'To all this we raised no objection, but we have the feeling that when this time comes in science, God with His white beard will come down to earth, swinging a bunch of keys, and will say to humanity, the way they say at five o'clock at the Salon, "Closing time, gentlemen." '

Dalton's theory of the billiard-ball atom was an article of faith among most physicists until the closing years of the nineteenth century although the work of chemists during its last third made acceptance of other ideas increasingly possible. Important among such work was Dmitri Ivanovich Mendeleev's notion of the periodic table which he expounded in 1869 and 1871. This was an arrangement of the chemical elements which showed that their properties were dependent on their atomic weights. The table, which did much to remove the confusion surrounding the elements – and from which Mendeleev was able to forecast the existence and properties of undiscovered elements – was an important step towards recognition of the basis of atomic structure.

But belief in Dalton clung on. 'I was brought up to look at the atom as a nice hard fellow, red or gray in colour, according to taste,' said the great Ernest Rutherford, born in 1871. The remark is typical of Rutherford, the man who with the Dane Niels Bohr dominated atomic research for the first third of the present century. Rutherford could produce a homely simile for most things and always maintained that a good scientific theory should be explicable to a barmaid. His essential beliefs are suggested by the remark to a young student who spoke of nuclear reactions as 'my reactions': 'Are you God that you call them *my* reactions?' Yet he had the light touch many physicists lacked, proposing a vote of thanks to Heisenberg with the words: 'We are all much obliged for your exposition of a lot of interesting nonsense,' and commenting on a lecture by Bohr on the uncertainty principle: 'You know, Bohr, your conclusions seem to me as uncertain as the premises upon which they are built.'

Rutherford's estimate of what his work on the atom would really mean is still something of a mystery, as later events will show. It is no doubt due in part to the fact that his scientific life began in a world where Dalton's simple and undemanding concept of the atom was only slowly giving way to a realization that these billiard-balls of matter might contain an almost inconceivable reservoir of energy, available to anyone who discovered how to turn on the tap. Rutherford, in his old-fashioned wisdom, was not quite sure that that was what God intended.

The Daltonian view of the atom disintegrated following work in France, Holland, Denmark, Germany, and Britain by Becquerel, the Curies, Lorentz, Bohr, Röntgen, Planck, J. J. Thomson and Rutherford, to name only the more important leaders of the revolution which ended the classical age of physics. Their reports and papers circulated across frontiers in a climate where the word 'security' was irrelevant and where the main aim was to explore the new world now seen in the distance. Most scientists still believed in the ideal that Rutherford was to put into words between the two world wars: '... science is international, and I trust will ever remain so.'

As early as 1875 William Crookes had shown that when a cathode, or negative electrode, was placed under strong electric potential within a vacuum tube, there came from it an emission of what he christened cathode rays. At first the rays were thought to be a form of electromagnetic radiation, but Crookes finally saw them as streams of electrically charged particles, an idea which to most scientists seemed perverse. However, two years later the Dutchman Hendrik Lorentz suggested that large numbers of small elementary electrical charges, both positive and negative, existed inside all ordinary matter. Unlikely as it seemed, this hypothesis, which claimed that whatever atoms were they could not be the hard massy impenetrable objects of Newton and Dalton, helped provide an explanation of discoveries which came within a few months of each other two decades later. First was Röntgen's demonstration in 1895 that a new kind of ray, designated 'X' because of its mystery, could apparently pass through solid matter, 'the first strong breeze of discovery which blew the science of physics out of the doldrums in which it had lingered for many years.' A few months later Antoine Henri Becquerel observed in Paris that photographic plates could be fogged by streams of particles thrown out by the metal uranium and having characteristics somewhat like those of X-rays. Almost simultaneously, Marie Curie and her husband isolated the elements radium and thorium and discovered, to general scientific consternation, that both produced, as Max Born was later vividly to describe it, 'enormous quantities of energy, apparently from nothing'. These awkward revelations were compounded in 1897 when J. J. Thomson in Cambridge demonstrated that Crookes's cathode rays were in fact streams of negatively charged particles which he now christened electrons. Each electron had a mass roughly 1/1,837th of the lightest known atom, that of hydrogen; but, more important, it could have come from nowhere but the metallic cathode.

Becquerel and the Curies now began to discover the characteristics of the rays emitted by uranium, radium and thorium, the radioactive elements as they were soon called. Two different kinds were eventually identified: alpha rays, found to be nuclei, or the central cores, of helium atoms, and beta rays which had far greater radiation, soon referred to as gamma rays and found to be essentially very high-energy X-rays without mass or electric charge.

All this made the first years of the twentieth century an exciting time for scientists, particularly for those having what Hobbes called the 'lust of the mind', an insatiable curiosity to know more. Pierre Curie was one of the few who openly voiced doubts.

> One may also imagine that in criminal hands radium might become very dangerous and here we may ask ourselves if humanity had anything to gain by learning the secrets of nature, if it is ripe enough to profit by them, or if this knowledge is not harmful. The example of Nobel's discoveries is characteristic: powerful explosives have permitted men to perform admirable work. They are also a terrible means of destruction in the hands of the great criminals who lead the peoples towards war.
>
> I am among those who think, with Nobel, that humanity will obtain more good than evil from the new discoveries.

However, one thing was certain. The concept of the billiard-ball atom had gone for ever. In its place there came, for a while, the idea of the raisin-cake atom proposed by Thomson. In this the atom was still solid; but in Thomson's view it was a sphere of positively charged matter, studded with just enough negatively charged particles – the raisins in the cake – to neutralize the positive charge.

The proposition was to hold the ring for only a short while. During the first years of the twentieth century it was replaced by Rutherford, who after a short while with J. J. Thomson in the Cavendish Laboratory, Cambridge, had moved to McGill University, Montreal, and begun work with one of his colleagues, Frederick Soddy. Rutherford's theory was revolutionary in the accurate meaning of the word and, like most revolutions, it aroused both opposition and scorn.

This was understandable since the nub of his explanation, encompassing the discoveries of the past decade, rested on the theory that radioactivity was the spontaneous transformation of radioactive elements into different kinds of matter. The proposition had its own particular dangers, and Rutherford said to Soddy: 'Don't call it *transmutation*. They'll have our heads off as alchemists.'

But whatever precautions Rutherford took against the accusation, there was no doubt in his mind that some elements were, without external aid, constantly changing themselves into something different. The so-called rays given off by these radioactive substances were, Rutherford confirmed, of three different kinds: the positively charged alpha rays; the beta rays, which were streams of electrons – the cathode rays of Crookes; and the gamma rays, electromagnetic waves of very short wavelength. At first Rutherford detected his particles by 'catching' them on a screen coated with zinc sulphide on which they showed up as scintillations; later by a counter, invented by one of his pupils, Hans Geiger, in which was recorded the particle's electrical discharge.

Rutherford's developing picture of the atom led him to make an intriguing suggestion.

> The difference between the energy originally possessed by the matter, which has undergone the change, and the final inactive products which arise, is a measure of the total amount of energy released.
>
> There seems to be no reason to suppose that the atomic energy of all the elements is not of a similar order of magnitude. ... If it were ever found possible to control at will the rate of disintegration of the radio-elements, an enormous amount of energy could be obtained from a small quantity of matter.

Just how enormous the amount might be was not yet clear; but in a paper on radioactive change, Rutherford and Soddy pointed out that the energy released when an atom of one element was turned into the atom of another must be at least 20,000 times, and might be even a million times, more than that released when a molecule, made up of different atoms, is transformed by chemical change into a molecule of another substance. Here of course was the fundamental difference between the energy of the pre-nuclear and the post-nuclear worlds. When oil is burned in air, energy is produced by a rearrangement of the unaltered atoms of hydrogen, carbon and oxygen; the 'burning' of uranium in a bomb or a nuclear power station results from rearrangements within the atomic nucleus itself.

Einstein with his $E=mc^2$ had done no more than quantify a figure for circumstances that were, as he believed then and continued to believe for a considerable while, never likely to exist. Nevertheless, Rutherford had already made what a correspondent, Sir William Dampier-Whetham, called a 'playful suggestion that, could a proper detonator be found, it was just conceivable that a wave of atomic disintegration might be started through matter, which would indeed make this old world vanish in smoke.' Indeed, almost half a century later, as scientists from the Manhattan Project waited in the chill dawn for the test of the world's first nuclear weapon, a few still uneasily wondered if that might not happen.

In the first years of the century all this was still very speculative. Theoretical deductions followed theories that might or might not be correct. Even if correct they might be greatly modified by facts from the vast black unknown surrounding the small area of atomic knowledge which the physicists had so far been able to illuminate. In addition there seemed not the remotest prospect of being able to control the natural disintegration that had been noted by Becquerel, the Curies and others. As for causing the disintegration of non-radioactive atoms, not even Rutherford had yet proposed that this would ever be possible.

Rutherford and Soddy developed their extraordinary idea that some of the world's elements were constantly turning into different elements without any help from the alchemists, those medieval figures who had so unsuccessfully tried to transmute base metals into gold, or even anything at all into anything else. In 1905 Einstein erupted into a surprised scientific world, a small friendly employee of the Swiss patent office who in a single paper destroyed the certainties of time and space which had been taken for granted since time began, and who showed to an almost incredulous humanity that one man's 'now' was another man's 'then'. Einstein's $E=mc^2$, pulled from the intellectual stratosphere, was complementary to Rutherford's conception of radioactivity as atomic transformation involving the release of energy, yet few men linked the two. Among those who did was Max Planck, the German physicist who, wandering in Berlin's Grunewald five years earlier, had turned to his son with the words: 'Today I have made a discovery as important as that of Newton.' The discovery was the quantum theory which assumed that energy was emitted not in the continuous flow that common sense suggested but in discontinuous bursts or 'quanta' – the Latin for 'how much'. It was to become second only to Einstein's relativity in changing man's view of the world in which he lived. Mulling over the implications of the mass-energy equation in 1908, Planck commented of the atom's latent energy that 'though the actual production of such a "radical" process might have appeared extremely small only a decade ago, it is now in the range of the possible, through the discovery of radio-active elements and their transmutation ...'

Even this was something of an extravagance, since in 1908 knowledge of the atom itself consisted largely of the negative fact that it was not the solid if minuscule sphere that men had for almost a century believed it to be. Radioactive atoms appeared to turn themselves into others, and both positively and negatively charged particles appeared to be thrown out from certain atoms under certain conditions; but as yet there existed only the most speculative ideas about how this was done or could be done.

The situation was radically altered in 1908 when Rutherford, by this time back in Britain as Langworthy Professor of Physics in the University of Manchester, began to direct the series of experiments which revealed the basic structure of the atom as it is known today. Under his direction Hans Geiger and Ernest Marsden began the bombardment of various metal foils with the positively charged alpha particles. The crucial phase of the experiment came when the 'target' was a thin gold foil, only one fifty-thousandth of an inch thick but containing no less than 2,000 layers of gold atoms. It was expected that as the particles passed close to the gold atoms they might be deflected off beam by a few degrees, but not more. That is what took place in most cases, although from the record of the particles on the photographic plate behind the gold foil it appeared, astonishingly, as though the individual atom must consist mainly of empty space. But a very few particles – one in every hundred thousand or so – were deflected considerably more than the rest and Rutherford therefore asked Marsden to carry out an experiment to see if any of the particles were deflected by more than ninety degrees. The results were astounding; a few of the particles were not only deflected through more than ninety degrees but appeared almost to bounce back from the target. It was this last extraordinary behaviour, as surprising, Rutherford later said, 'as if you had fired a fifteen-inch shell at a sheet of tissue paper, and it had come back and hit you', which provided the vital clue to the structure of the atom. For while the passage of most alpha particles indicated the huge empty spaces within the atom, those which bounced back showed that there must somewhere be areas of electrical charge so strong that, like repelling like, the positively charged alpha particles were bounced back towards their source.

From this single series of experiments Rutherford was able to build up a picture of the atom not so very different from that known today. At its centre, he maintained, there is a nucleus containing one or more positively charged particles or protons. Circling the nucleus at a relatively immense distance are the negatively charged electrons, one for each proton under normal conditions. The numbers of protons and electrons in each element are different, and the chemical characteristics of each element depend on these differing numbers of charged particles.

The atom itself was soon known to be almost indescribably small, something of the order of a hundred millionth of an inch across, and various efforts were made to indicate what such dimensions meant. The time taken for the whole population of the world to count the number of atoms in a single bubble of soda-water gas was almost laughably high; the size of an atom in a single fibre that had been magnified to tree-trunk size was almost laughably small. Rutherford would have liked the illustration that if a minute cube were cut from a cigarette paper so that its breadth, height and width were all equal, then a cube no bigger than a speck of dust would contain about a thousand million million atoms. Today it is known that the nucleus also contains electrically neutral particles called neutrons, while several hundred other kinds of particles that do not normally occur in atoms have been discovered. Nevertheless Rutherford's idea of the atom as a miniature solar system was essentially correct and was to be the basis for most of the atomic investigation that was to follow.

In the early days there had been one barrier to acceptance of the Rutherford model of the atom, quite apart from the difficulty of accepting that matter consisted largely of empty space: according to classical physics, the electrons orbiting the nucleus would lose energy by radiation and then spiral into the nucleus with the inevitability of a cricket-ball falling to the ground. But apparently this did not happen with electrons.

It was Niels Bohr who explained this contradiction when he began to investigate why a hot gas radiates light only at specific frequencies, which depend on the elements in the gas, thus producing the line spectrum used, for example, to determine the composition of stars. Two suppositions were necessary. The first was that atoms existed only in well defined stationary states and that in each state the electrons circle the nucleus in specific 'allowed' orbits. The second supposition was that when an electron jumped, for whatever reason, from one of its 'allowed' orbits into another nearer to the nucleus, then radiation was emitted; by contrast, when an atom absorbed radiation, one or more of its orbiting electrons jumped from its 'allowed' orbit to another farther from the nucleus. Thus in one stroke of supreme genius Bohr had reconciled Planck's conception of radiation in discontinuous surges of energy with Rutherford's startling picture of the nuclear atom.

Bohr's explanation came in 1913, thus tidying up the situation in time for the outbreak of the First World War which was to divert scientific attention from such erudite matters on to practical development of the most efficient weapons. That the atom itself might become such a weapon appears to have occurred at this date to only one man; yet it was not a scientist but one of Arthur O'Shaughnessy's 'dreamers of dreams ... the movers and shakers of the world for ever, it seems', who first saw the possibilities.

While Planck's guarded warning had been ignored, even by Einstein, H. G. Wells, already famous as an author of science fiction, had been fired by Soddy's book, *The Interpretation of Radium*, first published in 1909. He had, as he wrote to his friend A. T. Simmons, editor of *The Times Educational Supplement*, suddenly broken out into one of the good old scientific romances again. Men were supposed to find out how to set up atomic degeneration in the heavy elements just as they found out long ago how to set up burning in coal. Hence limitless energy. Perhaps Simmons would tell him what books he should consult.

With the books and a bundle of ideas, Wells set off for Switzerland where, in the summer retreat of Elizabeth von Arnim's Chalet Soleil near Montana-sur-Sierre, he wrote *The World Set Free*. Published early in 1914, a 'porridge composed of Mr Wells's vivid imagination, his discontents and his Utopian aspirations', as *The Times Literary Supplement* called it, the book was dedicated to Soddy's book and, said the author, owed 'long passages to the eleventh chapter' of it. Wells forecast the release of nuclear energy and, after a brief period of prosperity, the outbreak of nuclear war between the Allies and the Central Powers. The new weapons left areas of bubbling radioactivity, uninhabitable for years, and in other ways described future events with considerable foresight. The patent problems that were to bedevil Anglo-U.S. relations in the 1940s appeared in the shape of the Bengali

Mr Das with his patent claims, and the breaching of the Dutch dykes with conventional bombs in the last months of the Second World War was graphically attributed by Wells to nuclear weapons.

Few reviewers troubled to describe the release of nuclear energy as nonsense. Reasonably so, since it was quite outside most men's perceptions. 'The nightmare will not come true,' *Blackwood's Magazine* comfortingly commented, 'and we shall go on living a life of reasonable happiness, committing follies and paying for them, and doing our duty in the old fashion, indifferent whether it is a horse we ride behind or a steam-engine that wafts us on our way ... even the atomic energy imagined by Mr Wells would be powerless always against the unbroken traditions of the human race.'

Within a year of Wells's book, as the long line of the 1914 defences hardened up between the Channel coast and the Swiss frontier, these traditions began to be tested as rarely before. And, as this happened, the possibilities of nuclear weapons could no longer be entirely ignored. Speaking on 'Radiations from Radium' in 1916, Rutherford told his audience that it had 'to be borne in mind that in releasing such energy at such a rate as we might desire, it would be possible from one pound of the material to obtain as much energy practically as from one hundred million pounds of coal.' Fortunately, he went on, a method had not yet been found of so dealing with these forces, and personally he was very hopeful it would not be found until Man was living at peace with his neighbours.

Like more than one other physicist who worked in the age of what was still nuclear innocence, Rutherford at times revealed an almost religious feeling that atomic exploitations of matter must have been ruled out by God. Professor Lindemann, better known as Lord Cherwell, Churchill's *eminence grise* of the Second World War, was so repelled by the idea that 'he could scarcely believe that the universe was constructed in this way'. And Sir Henry Tizard, who was to help co-ordinate Britain's first probings towards the bomb, showed much the same feeling when he asked a colleague: 'Do you really think that the universe was made in this way?'

The fact that it might be made in such a way appeared rather more likely as the First World War was ending. Rutherford, by this time deeply involved in work for the Services, had continued his scientific investigations in Manchester and one day arrived late for an important meeting of a Government committee. He excused himself on the grounds that he had just found definite evidence that it might be possible to disintegrate atoms at will, and that if this proved to be true, it was something far more important than the war.

In Manchester he had been using the alpha rays from radioactive radium to bombard a target of nitrogen. When one of these particles, or helium bullets, penetrated the nucleus of a nitrogen atom with its seven protons, one of the original protons was ejected and replaced by the two protons of the helium nucleus. The result was a nucleus with eight protons – that of the oxygen atom. The energy released by this transformation was greater than that of the helium 'bullet': but only about one in every million of the bombarding helium nuclei hit a target, the rest passing through the relatively immense spaces between the nitrogen nuclei and their encircling electrons. The process, as Einstein was to say of another nuclear experiment, was rather like trying to shoot birds in the dark in a country where there were not many birds in the sky. There was thus a net loss of energy rather than any gain and if Rutherford's experiment achieved anything – other than adding to nuclear knowledge – it was to underline the practical problems of getting more energy from the nuclear stockpot than was put into it.

Nevertheless, there were the optimists. One of them was Sir Oliver Lodge, the British physicist who had played a significant role in the development of wireless during the first years of the century and who spoke out with genial optimism soon after the armistice of 1918. During the war the physicists of the belligerent powers had been fully occupied with the development of aircraft, of wireless interception, of artillery ranging, and with various side-issues to the art of killing without being killed. Rutherford and the great French physicist Paul Langevin helped to devise anti-submarine measures. Madame Curie drove Red Cross ambulances while even Einstein, in time off from his scientific labours, tried to design a new airfoil which would combine maximum lift with minimum drag. The interesting matter of the power within the atom had been given a rest.

Lodge was one of those men constitutionally built to look on the bright side, and when, in September 1919, he came to speak in the city of Birmingham on the centenary of James Watt's death, he could not help painting an enthusiastic picture of the future that lay in the peace ahead. According to the report in *The Times*, Lodge claimed that if Watt were living in 1919, he would be looking for alternative sources of energy and there was little doubt he would be considering atomic energy 'which, if it could be utilized on an extensive scale, would, he

[Lodge] believed, greatly ameliorate the conditions of factory life. There would be no smoke due to imperfect combustion and no dirt due to the transit of coal or ashes, while the power would be very compact and clean. Possibly, there might occasionally be explosions due to the liberation of power more quickly than it was wanted,' but in general he was optimistic. 'If you were able,' he added, 'to use mechanically the energy contained in an ounce of matter, and you reckon the foot tons thereby obtained, you will find enough energy to raise the German Navy and pile it on the top of a Scottish mountain.'

Only later did a small doubt appear to cross Lodge's mind. 'And if ever the human race get hold of a means of tapping even a small fraction of the Energy contained in the atoms of their own planet,' he warned, 'the consequences will be beneficent or destructive according to the state of civilization at that time attained, and the beneficence or malevolence of their spiritual development.'

In December 1919 he expanded on the theme in the Trueman Wood Lecture to the Royal Society of Arts in London, pointing out that the atomic energy in the piece of chalk he was using would be sufficient to lift a hundred thousand tons some three thousand feet. *The Times* reported the lecture:

> He felt that we were on the brink of making a discovery with regard to the utilization of this source of energy. He did not know whether it would come tomorrow or take a century. But he did not believe that our descendants would be consuming stored material, such as coal, using chemical energy and burning up air when they wished to drive machinery. They would be taking the energy out of an ounce or two of matter instead of out of a thousand tons of coal.

Lodge's general optimism, which except for its qualification about the possibility of worrying little explosions might be tailor-made advocacy for today's nuclear-power enthusiasts, rested on something more than an extrapolation from Rutherford's first nuclear transformations. A few weeks before Lodge spoke in Birmingham, Francis Aston, whose investigations in Cambridge had during the war been abandoned for work at the Royal Aircraft Establishment, had for the first time given with some exactitude figures for the amount of energy locked within the atom.

Aston had learned much of his science as assistant in a brewery, but was genuinely dedicated to better things. On being left a small fortune while still a young man he decided to devote his life to the purest of pure physics and in 1919 was happy to return to his work in Cambridge, devoting his energies to one of the inexplicable oddities which he had helped to reveal before the war.

Rutherford's nuclear atom, with its negatively charged electrons for ever circling a tight positively charged central nucleus, had at least one thing in common with Dalton's billiard-ball picture. In both, all the atoms of one element had the same mass, an apparently incontrovertible fact of scientific life to which physicists clung with the near-desperation of a rock-climber clutching a jug-handle hold in a difficult place. The reason why an atom of the gas neon had a mass ten times that of an atom of hydrogen might not be entirely clear, but unless some curious fact of nature had been overlooked then all atoms of neon would also have the same mass.

But it was becoming disconcertingly apparent that this was not always so. An early hint of this had been given years previously when Crookes, speaking in 1886 at the Birmingham meeting of the British Association for the Advancement of Science, had put forward what he called 'an audacious speculation'. 'I conceive, therefore,' he said, 'that when we say the atomic weight of, for instance, Calcium is 40, we really express the fact that, while the majority of calcium atoms have an actual atomic weight of 40, there are not a few which are represented by 39 or 41, a less number by 38 or 42, and so on.' The first outline of what Crookes's 'audacious speculation' might really mean had been made before the war when Frederick Soddy, working with Rutherford in Canada, found that at least some radioactive elements appeared to exist in two forms: these, although chemically identical, had different atomic masses. The differences were not great but they were great enough, and certain enough, to underline the fact that there were some areas of atomic knowledge where ignorance still reigned supreme.

Bearing in mind that his awkward elements had the same chemical properties, Soddy christened them iso-topes from the Greek for 'same place' – the same place in the periodic table of the elements, that is – and for a while attributed their behaviour to some hitherto unexplained aspect of their radioactivity. This idea was first punctured by Thomson and Aston. In 1912 Thomson discovered two isotopic forms of the gas neon and shortly afterwards Aston found a means of separating them. His method foreshadowed things to come in a rather bizarre way since it involved diffusing the gas through pipeclay, a first primitive example of the process used some thirty years later at Oak Ridge, Tennessee, to separate the rare uranium isotope needed for the world's first nuclear weapons.

Returned to his isotopes in 1919, Aston tended to be a lone worker, although for a year he had the help of K. T. Bainbridge, then a young student from Harvard. 'Aston's laboratory,' wrote a colleague, 'was a dingy darkened room, in a corner of the ground floor [of the Cavendish laboratory] its wall covered with discarded apparatus and with samples of rare gases in tubes over mercury, which he had separated by fractionating residues from liquid air.'

Here he set about the task of confirming that the atomic masses of the two isotopes of neon were respectively 20 and 22 times the atomic mass of hydrogen; then of finding out if other non-radioactive materials also existed in different isotopic forms. For the work, he developed the mass spectrograph, a device which, by subjecting streams of particles to magnetic and electric fields, separated those of different masses and recorded their presence at different places on a photographic plate. In Aston's own words, this 'by its peculiar sequence of electric and magnetic fields eliminated the effect of varying velocity and gave a spectrum dependent upon mass alone.' The result can be compared to what happens when two cars, of different weights but travelling at the same speed, are forced to change direction. The light car will follow a tighter curve than the heavy car. In the same way the charged atoms of different masses followed different paths and hit the detector placed in their path at different places.

Aston was not alone. Bainbridge back in the United States at the Franklin Institute and Dempster in the University of Chicago were working along similar lines. But just as it was the Americans whose technological expertise was to make nuclear weapons possible a quarter of a century later so was it Aston in Britain – a genius of an experimenter who made much of his own equipment – who gave figures for the energy locked within the atom.

After discovering the atomic masses of the two isotopes of neon he had, within a few weeks, assigned different atomic masses to various isotopes of chlorine and of mercury, and had soon done the same for argon and helium. It was quickly becoming clear that many if not most elements existed in two or more isotopic forms. This explained how the atomic masses previously measured had actually been averages of the various isotopic forms of each element, and this seemed to explain why they were not exact multiples of the atomic mass of hydrogen.

Aston's first mass spectrograph was capable of measuring with an accuracy of one part in 1,000. Before long an improved version raised the accuracy to one part in 10,000. But once an accuracy of this order had been reached it was impossible to ignore one important conclusion that had to be drawn from certain of the results.

It was already reasonably certain that the hydrogen atom, consisting of a single proton and its encircling electron, was the building block of all other atoms; and, specifically, that an atom of helium consisted of four atoms of hydrogen, fused together in the immense heat of some cosmic past. The only logical deduction was that the mass of the helium atom would be four times that of the hydrogen atom. But however carefully the experiments were conducted, this was not so: the weight of the atom of helium was persistently less than that of four atoms of hydrogen. For a variety of reasons the figures found by Aston vary slightly from those of today, but their implications are the same. Nowadays atomic masses are measured in terms of the atomic mass unit, whose symbol is u. The modern value for the atomic weight of hydrogen is 1·00797 u, but even this is an average figure, for in 1932 Harold Urey, at Columbia University, New York, finally demonstrated that hydrogen occurs naturally in two isotopic forms: by far the most abundant is protium, with atomic mass 1·00782 u. Only 0·015 per cent of naturally occurring hydrogen is the other isotope, deuterium, which has an atomic mass of 2·0140 u, accounting for its popular name of 'heavy hydrogen'.

Using present-day figures, four atoms of protium should make a single atom of helium and so helium should have an atomic mass of 4·03128 u. Yet the mass of the helium atom is only 4·0026 u, or about 0·7 per cent less than it should be.

The reasons for the mass defect, as it is called, were somewhat obscure in Aston's day and are not entirely understood even now. Yet some facts were indisputably clear: that in the fusion process which had created the helium, mass had been lost: that if Einstein's belief in the interchangeability of mass and energy were correct, this loss of mass would be accompanied by the release of energy: and if Einstein's $E = mc^2$ was correct, then the amount of energy would be immense.

As a result of Aston's experiments it now became possible to give a figure for nuclear energy – 'nuclear' rather than 'atomic' since it was from the nucleus that the energy came. Aston himself pointed out that changing the hydrogen in a tumbler of water into helium would liberate sufficient energy to drive the *Mauretania* across the Atlantic and back at full speed. As yet there was no indication that this energy could ever be liberated. It seemed unlikely to most scientists that it ever would be possible because of the almost inconceivably high temperatures necessary for fusion of the lighter elements. It seemed even more unlikely that it would be possible to split any of the heavier elements, even were that to produce comparable results.

Aston, almost alone among physicists, thought otherwise. 'I have little doubt myself,' he said, as early as 1925, 'that man will one day be able to liberate and control this tremendous force, and I am optimist enough to believe that he will not devote it entirely to blowing his neighbours to pieces.' For the next two decades, through 'the long week-end' between the two world wars, this possibility of utilizing nuclear energy remained a thought at the back of many physicists' minds. But it was an idea to play with, a subject for Senior Common Room debate rather than one which had to be thrust on to the political or international stage. Wells had audaciously shown the way and between the wars other writers were to follow. But for most of the period the subject was too remote from reality for scientists to do more than speculate idly on what might just conceivably happen in the rather distant future. Politicians, who in any case had a poor view of scientists in general – although there might be exceptions such as Rutherford – usually had their time occupied with other matters.

Only in the mid-1930s, as Germany flexed her muscles and the hopes of permanent peace grew dim, did an ominous question begin to exercise a few men's minds: if the energy locked within the atom were ever unleashed, would its masters be the leaders of the new Germany, or those who had fought her for more than four bitter years? And would control fall into the hands of the military, of the politicians, or of the scientists? If Clemenceau had been right with his aphorism about war and the military, might not this new and extraordinary power be too serious a subject for anything less than a new and commanding international body?

But neither scientists nor politicians thought about that. With the League of Nations being increasingly ignored, and its effectiveness approaching zero, it is difficult to blame them.

CHAPTER 2

Towards the Greatest Power on Earth

As scientists took up once again the work which had been interrupted by four years of war, a mixed bag of warnings and prophecies about the potentials of nuclear energy started to titillate popular expectations. The earliest came from Russia where in *Red Star* the writer Aleksandr Bogdanov envisaged Martians visiting the planet earth in a spaceship powered by some unidentified radioactive material. Two years later, Arthur Eddington told the British Association that the vast reservoirs of energy upon which the stars were continually drawing must be subatomic energy which existed in all matter, and added: 'We sometimes dream that man will one day learn how to release it and use it for his service.' Professor Lindemann, who philosophically drew away from the implications of nuclear energy as its prospects increased, speaking at the Royal Society on isotopes in 1921, described the energy which 'it is hoped will one day be available'.

The dangers of this prospect were noted in the same year by Hans Thirring, the Austrian scientist who in 1945 made the first public forecast of the hydrogen bomb. It took one's breath away, he wrote in 1921, 'to think of what might happen in a town, if the dormant energy of a single brick were to be set free, say in the form of an explosion. It would suffice to raze a city with a million inhabitants to the ground.' But in the same year Einstein, approached in Prague by a young man who had ideas for a weapon based on the mass-energy equation, was given a brusque reception. 'You haven't lost anything if I don't discuss your work with you in detail,' he was told. 'Its foolishness is evident at first glance. You cannot learn anything more from a longer discussion.'

Different reasons for caution were advanced by Aston who, in a lecture to the Franklin Institute, Philadelphia, in March 1922 told his listeners:

Should the research worker of the future discover some means of releasing this energy [from the hydrogen atom] in a form which could be employed, the human race will have at its command powers beyond the dreams of scientific fiction; but the remote possibility must always be considered that the energy once liberated will be completely uncontrollable and by its intense violence detonate all neighboring substances. In this event, the whole of the hydrogen on the earth might be transformed at once, and the success of the experiment published at large to the universe as a new star.

There was also the voice of Winston Churchill for whom the awful prospects of Hans Thirring presented an opportunity to be grasped without delay. 'May there not be methods of using explosive energy incomparably more intense than anything heretofore discovered?' he asked in a 1924 article called 'Shall we all commit suicide?'

Might not a bomb no bigger than an orange be found to possess a secret power to destroy a whole block of buildings – nay, to concentrate the force of a thousand tons of cordite and blast a township at a stroke? Could not explosives even of the existing type be guided automatically in flying machines by wireless or other rays, without a human pilot, in ceaseless procession upon a hostile city, arsenal, camp, or dockyard?

Yet despite the optimistic – although extremely generalized – prophecies of men such as Eddington and Lindemann, few physicists close to nuclear research considered the release of nuclear energy to be a practical proposition in the foreseeable future. The reasons for this were obvious enough. It seemed clear that the fusion of light elements would require temperatures at least equal to those within the sun, temperatures which appeared quite incapable of creation on earth, even if any method of utilizing them could be found. It was true that if the atoms of heavier elements could be split in two – the converse of fusing the lighter ones – then a comparable release of energy might be brought about. But the nuclear transformations produced by Rutherford were 'chipping' rather than 'splitting', and it seemed inconceivable that any way would ever be found of dealing with the exceptionally strong forces which bound together the particles in the nucleus of the heavier elements.

Quite as important in damping down any optimistic thought that limitless energy might be round the corner were the awkward figures involved in Rutherford's nuclear transformations – figures which seemed likely to be similar should any other way of doing the same thing ever be found. About one in every million of the particles ejected by Rutherford's radium penetrated a nitrogen nucleus and transmuted it into the nucleus of an oxygen atom. The energy released was greater than that of the helium 'bullet'; but since most of the bullets missed their targets and passed through the relatively vast spaces between the targets and their encircling electrons, the final result was a considerable loss of energy.

This awkward fact of nuclear research was clearest to those closest to the problem and as late as 1937 Rutherford himself was to emphasize the net loss of energy and to add that consequently the 'outlook for gaining useful energy from the atoms by artificial processes of transformation does not look promising.' The following year – only a few months before the key to the problem was to be discovered – Ernest Lawrence, who already had his atom-smashing cyclotron operating in Berkeley, California, was to say: 'The fact is, at this time, that although we now know that matter can be converted into energy, we are aware of no greater prospect of destroying nuclear matter for power purposes than of cooling the ocean ... and extracting the heat for profitable work.' Yet within four years, the world's first nuclear fire was to be lit in Chicago by Enrico Fermi and his colleagues, an exultant Arthur Compton was to inform Dr Conant that 'the Italian navigator has just landed in the New World', and the nuclear age had begun.

The New World was thus entered with comparatively little warning; science fiction had become science with a speed which sprang solely from the demands of war. Only the most cursory thought had been given to what nuclear energy could mean, in war or in peace, and the dangers of radioactivity, so far limited to those unfortunate enough to be early and innocent victims of X-rays or radium, were given scant attention. As to the genetic dangers, no one had paused to consider them in the rush to be first with 'the bomb'.

Yet the period of almost two decades between the end of the First World War and the discouraging remarks of Rutherford and Lawrence in 1937 and 1938 had contained increasing hints of things to come. There are many parallels. George Cayley, the Lilienthal brothers and Octave Chanute prepared the way for the Wright brothers' appointment with destiny at Kittyhawk; Marconi, Taylor and Young in the United States, and Heinrich Lowy in Vienna, paved the way for the radar of Britain's early-warning chain. In the same way there were pioneers whose steady accumulation of evidence on the nucleus helped fill the gap between Rutherford's 'alchemy' and the fission of Hahn and Strassman which in December 1938 burst upon a surprised world in the wake of Munich and as the world prepared for war.

Although each man made his own contribution it would be almost misleading to claim that each worked independently in these inter-war years, when secrecy was still anathema in science. The order of the day was publication as quickly as possible. And while scientists in any one field might vie with each other for results, each worked in some ways as member of a relay team, catching the discovery as it was thrown from one laboratory and using it until, weeks, months or years later, an improved version was passed on to someone else. Thus it is not possible entirely to isolate the separate steps which within a few years made the release of nuclear energy less unlikely – even though few physicists realized it at the time.

Not until 1932 did there arrive the *annus mirabilis*, the year when three separate advances in knowledge of the nucleus each brought the release of nuclear energy nearer. Yet in the preceding years there was played out a prologue quite as important, a theoretical argument which owed as much to philosophy as to physics and whose decisions were to provide the foundations for the theoretical and experimental physics which is alive today, and without which the crucial steps towards the release of nuclear energy would not have been taken.

The subject which was to produce a cascade of new ideas within a few years was Einstein's belief that light must be regarded sometimes as a wave, sometimes as a particle. This was taken up by Louis de Broglie, a young Frenchman who had come to physics by the unusual route of medieval history. 'I had a sudden inspiration,' he later wrote. 'Einstein's wave-particle dualism was an absolutely general phenomenon extending to all physical nature, and, that being the case, the motion of all particles, photons, electrons, protons, or any others, must be associated with the propagation of a wave.' The startling essence of de Broglie's perception – which, according to Einstein, 'lifted a corner of the great veil' – was that particles such as electrons were guided by what were soon to be called 'de Broglie waves' or 'matter waves'. The theory was next expanded by Erwin Schrödinger, a young Viennese who within four months developed the basis of what became known as wave mechanics. Here, the emphasis of the de Broglie waves on the electron particle was taken a step further. The particle itself now gave way to what was, in effect, a standing electron wave; instead of being a wave-controlled corpuscle it became a corpuscular wave.

This Franco-Austrian theory appeared to contradict not only common sense, which was already suspicious of the subatomic world, but also rational logic. Could anything really be one thing, and its opposite, at one and the same time? Niels Bohr answered the question with a yes, provided by his principle of complementarity: in the subatomic world, reality is a function of the experimental conditions.

Even while de Broglie, Schrödinger and Bohr were thus writing a more plausible, even if more complicated, scenario for what went on inside the atom, the subject was being tackled from a totally different direction by Werner Heisenberg. Starting from an existing assumption that theories should be based on physically verifiable phenomena, he seized upon the spectral lines that were the fingerprints, as it were, of the atoms forming the different elements. The wavelengths for these could be determined by the use of a mathematical system called quantum mechanics, and it was soon clear that Heisenberg's purely mathematical approach supported the de Broglie–Schrödinger concept.

The surge of intellectual outpourings among the physicists in the later 1920s was to produce Heisenberg's famous indeterminacy principle, end determinacy in subatomic physics and, since at this level all appeared to be governed not by causality but by chance, arouse Einstein's agonized cry: 'I, at any rate, am convinced that He does not throw dice.' It also provided the clue which led to the first disintegration of the atomic nucleus not by radioactive emissions but by humanly speeded-up particles.

'Artificial' disintegration was one of the problems being studied in the Cavendish Laboratory under Rutherford's direction by John Cockcroft. It seemed essential that if artificially speeded-up streams of protons were to be used as 'bullets' rather than the alpha particles ejected by radium, then the protons would have to be subjected to voltages counted in millions. In the 1920s this was impossible. The situation was changed when in 1929 George Gamow, a Russian theoretical physicist who had been working with Bohr in Copenhagen, visited the Cavendish. In Denmark Gamow had been applying wave mechanics to the theory of nuclear collisions and transmutations and had come to a surprising conclusion. 'It appeared,' Cockcroft later said, 'that the mountainous barriers of electrostatic potential guarding the nucleus did not have to be scaled by our artificially produced hydrogen projectiles – they could, it appeared, wriggle through the mountain by reason of their wave-like properties.' The result was that voltages of only a few hundred thousand instead of many millions would be required.

Cockcroft's first efforts, made in 1930 and 1931, were unsuccessful.

Then, in the middle of the year, after being joined by a 27-year-old Dubliner, E. T. S. Walton, he was forced by reorganization of the laboratory to dismantle his apparatus and move to a much larger room, one with a high ceiling which had been used as a lecture theatre. Here, in the second half of 1931, the two men began to assemble the apparatus which would transmute matter, with the help not of naturally occurring radium but of an assembly of man-made equipment. Judged by today's standards the equipment looks remarkably ramshackle; yet its cost, a mere £500, was reported to be more than the laboratory had ever spent on a single piece of apparatus. Against one wall there stood a large black transformer, frequently leaking oil, of the kind seen at a generating station. Electricity was fed in from the mains and then stepped up to a high voltage before being led into a tall column of glass cylinders in the middle of the room. These were columns from the old-style petrol pumps then used in Britain. They had been evacuated and were the units which not only further stepped up the current to about 300,000 volts but also rectified it to flow in one direction. Finally the current was stored in condensers.

High voltages could thus be applied, as necessary, to the second part of the equipment, one of J. J. Thomson's positive ray tubes, or hydrogen discharge tubes, from which the necessary stream of protons could be produced. Below this, a second stack of glass cylinders formed the accelerator tube and below this again there was a place for the lithium 'target' plate. Mark Oliphant, also a member of the Cavendish, has described some of the problems which the primitive apparatus raised.

There were frequent vacuum troubles due to heat softening the plasticine, or to puncturing of one of the glass cylinders by a spark. Cockcroft and Walton spent a large part of their time perched on ladders, locating and repairing such leaks, or just rubbing over every plasticine joint with their fingers in the hope that they would eventually make the system vacuum tight again.

The first three months of 1932 were used in removing teething troubles, and it was only on 13 April that the first crucial experiments were made. Walton has described them:

When the voltage and the current of protons reached a reasonably high value, I decided to have a look for scintillations. So I left the control table while the apparatus was running and I crawled over to the hut under the accelerating tube. Immediately I saw scintillations on the screen. I then went back to the control table and switched off the power to the proton source. On returning to the hut, no scintillations could be seen. After a few more repetitions of this kind of thing, I became convinced that the effect was genuine.

Here was visual proof that some of the protons streaming down the tube on to the lithium target were not passing through the empty spaces between the lithium nuclei but were actually hitting the nuclei and turning each into two helium nuclei.

Cockcroft and Walton called in Rutherford so that he could see the telltale scintillations for himself. The next day, and the next, they continued the work. Then, on the evening of the 16th, they visited Rutherford's nearby home and there drew up a letter to *Nature* reporting their results. 'It seems not unlikely,' it said cautiously, 'that the lithium isotope of mass 7 occasionally captures a proton and the resulting nucleus of mass 8 breaks into two alpha-particles, each of mass 4 and each with an energy of about eight million electron volts.' In simple words, the nucleus had absorbed a proton; the resulting nucleus had then broken down into two helium nuclei. As Sir Arthur Eddington (he was knighted in 1930) was to say three years later when appealing for funds for the Cavendish: 'The social unsettlement of the age has extended to the world of atoms. An atom which for the last thousand million years has lived peacefully as silicon may tomorrow find itself phosphorus.'

By the time that Cockcroft and Walton were carrying out their experiments in the Cavendish it was known that all nuclei, with the solitary exception of the nucleus of normal hydrogen, contained not only protons but one or more neutrons, particles which had roughly the same mass as a proton but no electric charge. Their existence had been forecast for some years but it was only in 1932 that James Chadwick, also working in the Cavendish, at last identified the particle that was to become the key to nuclear energy. H. G. Wells, writing eighteen years earlier in *The World Set Free*, had placed the vital, but different, discovery as coming in 1933.

While it was Rutherford's intuition which prodded Chadwick on to his discovery of the neutron, its existence had been discussed even before the nuclear atom had fully emerged from the simple certainties of Dalton's billiard-ball atom. In the aftermath of J. J. Thomson's discovery of the electron, Walter Nernst had commented that the existence of a compound of positive and negative electrons, 'an electrically neutral massless molecule', would be of great importance.

'We shall assume,' he wrote, 'that neutrons are everywhere present like the luminiferous ether, and may regard the space filled by these molecules as weightless, non-conducting, but electrically polarisable, that is, as possessing the properties which optics assumes for the luminiferous ether.' There were other, equally hazy suggestions of a neutral particle during the speculations that followed Rutherford's paper of 1911, while shortly after the end of the First World War, Lise Meitner, later the first human being to recognize nuclear fission for what it was, postulated the existence of a neutral particle as a factor in radioactive disintegration.

However, it was Rutherford who steadily worried away at the idea, a dog with a favourite bone he was reluctant to let go, and concerned only by the absence of any experimental evidence to support the idea. In his Bakerian Lecture to the Royal Society in June 1920, he said:

> Under some conditions, however, it may be possible for an electron to combine much more closely with the hydrogen nucleus forming a kind of neutral doublet. Such an atom would have very novel properties. Its external field would be practically zero, except very close to the nucleus, and in consequence it should be able to move freely through matter. Its presence would probably be difficult to detect by the spectroscope and it may be impossible to contain it in a sealed vessel.

Rutherford's ability to explain science in simple terms was well illustrated when, some years later, he was to describe the neutron as 'like an invisible man passing through Piccadilly Circus: his path can be traced only by the people he has pushed aside'.

Shortly after giving the Bakerian Lecture Rutherford invited Chadwick to join him at the Cavendish in the experiments he had begun in Manchester. One reason was that Chadwick had discovered better ways of recording such disintegrations; another, Chadwick has said, was that Rutherford 'wanted someone to talk to, to while away the tedium of working in darkness'. Throughout the next decade Rutherford would from time to time discuss with Chadwick his own view that the existence of an uncharged particle was a theoretical necessity; from time to time Chadwick would attempt to find evidence, some of his efforts being, as he remarked 'so far-fetched as to belong to the days of alchemy'.

The first important clue came in 1930 when Bothe and Becker in Germany and H. C. Webster at the Cavendish found that if the light element beryllium was bombarded with alpha rays from radioactive polonium then a highly penetrating radiation was emitted from the beryllium. Bothe and Becker assumed that this was gamma-ray radiation. To Chadwick, Webster's work suggested that the radiation might be particulate, but experiments failed to produce any evidence of this.

Later in the same year the Joliot-Curies in Paris carried out similar experiments. They found that the 'gamma-ray' radiation was about five times as strong as that from the polonium itself. But while this radiation was hardly affected when it was passed through a thin sheet of aluminium or lead, something very different happened if it was passed through substances such as paraffin which contains hydrogen; then, high-velocity protons appeared to be shot out.

The Joliot-Curies' inexplicable observations were reported in the *Comptes Rendus* early in February 1932, and read by Chadwick one morning in February.

> Not many minutes afterwards, [Norman] Feather came to my room to tell me about this report, as astonished as I was. A little later that morning I told Rutherford. It was a custom of long standing that I should visit him about 11 a.m. to tell him any news of interest and to discuss the work in progress in the laboratory. As I told him about the Curie-Joliot observation and their views on it, I saw his growing amazement; and finally he burst out 'I don't believe it.' Such an impatient remark was utterly out of character, and in all my long association with him I recall no similar occasion. I mention it to emphasize the electrifying effect of the Curie-Joliot report. Of course, Rutherford agreed that one must believe the observations; the explanation was quite another matter.

It was the explanation to which Chadwick now addressed himself. Assuming that the radiation was not electromagnetic but particulate, he studied the effects of the beryllium on atoms of various kinds. Only one conclusion was possible: that the radiation consisted of electrically neutral particles with masses that were equal to the mass of the proton. The importance of this for the world of physics now opening up was that since neutrons had no charge they would not be repelled by the electric field surrounding nuclei and would therefore enter them with comparative ease.

Chadwick's discovery of the neutron helped to explain the existence of isotopes since it soon became clear that while the nucleus of an element always contained the same number of protons and was always circled by the same number of electrons, it would contain different numbers of neutrons, each number indicating a different isotope. But if this helped to build up a more satisfactory picture of the nucleus, the existence of neutrons was important for one very practical reason: it was obvious that a particle with no electric charge could be made to enter a nucleus with its charged protons far more easily than the charged 'bullets' so far used for the job.

Cockcroft and Walton's transmutation of the nucleus with their artificially speeded-up particles and the discovery of the neutron by Chadwick were not the only significant events in the world of physics in 1932. Across the Atlantic Lawrence announced that he had used his cyclotron to accelerate protons to an energy of 1,200,000 electronvolts. This had been done by using a magnetic field to make the particles traverse a spiral path between two hollow semicircular electrodes called 'dees'. At each half-revolution between the dees, the particles received an additional burst of energy from an oscillating voltage, the last addition being made just before the spiral route ended at the target.

If Chadwick's discovery of the neutron, Cockcroft and Walton's artificial transmutation of the atom and Lawrence's success with the cyclotron were the more spectacular events of the *annus mirabilis*, the end of the year witnessed another event which was to have great repercussions less than two decades later. In December Abram Fedorovich Ioffe set up in the Leningrad Physicotechnical Institute a special laboratory for study of the nucleus. Shortly afterwards scientists in the Ukrainian Physicotechnical Institute in Kharkov repeated the Cockcroft–Walton experiment. Within a few months the first Soviet conference on nuclear physics was being held in the Leningrad Institute, where four laboratories were by 1934 working on nuclear physics under Igor Kurchatov, later to become the leading figure in the Russians' nuclear weapons programme. All this suggested that while famine stalked the country and Stalin prepared for the purges which were to cripple it, her scientists were no less interested than others in probing for power within the atom.

The news of these events in the 1930s – particularly of Cockcroft and Walton's transmutation of the nucleus – slowly aroused public interest, partly through an uninformed fear that the scientists might, as Rutherford had jokingly warned, 'make this old world vanish in smoke'. Slightly lesser possibilities were considered by writers and playwrights. Thus Harold Nicolson, whose service with the British delegation to the Peace Commission in 1919 gave him an awareness of political realities, dealt in *Public Faces* with the problems of a future British Foreign Secretary. This was Mr Bullinger who 'knew only that the experts had begun to whisper the words "atomic bomb" and that Professor Narteagle in April of the previous year had explained to the Cabinet (looking mildly at them above rimless pince-nez) that a single Livingstone bomb, no longer than this inkstand (and at that he had indicated the wholly pacifist inkstand of Walter Bullinger), could by the discharge of its electrons destroy New York.' The problem, as the Air Ministry wielded the new weapons as a sign of national virility, culminated in the accidental sinking of a U.S. vessel and a nuclear-produced tidal wave which engulfed part of South Carolina – a fictional mishap with consequences comparable to the change of wind off Bikini atoll in 1954.

Another forecast of things to come arrived a few days before Cockcroft and Walton's experiments in Cambridge. This was *Wings Over Europe*, an ominous play centring on the great deterrent that fails to deter. Desmond MacCarthy was one of those who noted its lesson: 'The destiny of mankind,' he wrote in reviewing the play, 'has slipped (we are all aware of it) from the hands of politicians into the hands of scientists, who know not what they do, but pass responsibility for results on to those whose sense of proportion and knowledge are inadequate to the situations created by science.'

It is by no means clear how science and scientists regarded the matter. Rutherford is the touchstone here. His views on the subject are frequently summed up by the word 'moonshine' with which he appears to have greeted optimistic comments about nuclear energy in 1933. At the meeting of the British Association for the Advancement of Science that year, speaking of the work which had been done since the war, he said:

> These transformations of the atom are of extraordinary interest to scientists but we cannot control atomic energy to an extent which would be of any value commercially, and I believe we are not likely ever to be able to do so. A lot of nonsense has been talked about transmutation. Our interest in the matter is purely scientific, and the experiments which are being carried out will help us to a better understanding of the structure of matter.

The Times quoted him as saying that nuclear transformations provided a very inefficient way of producing energy and anyone who looked on it as a source of power was talking moonshine. But it is not certain whether 'moonshine', a word that in this context went round the world, was Rutherford's or that of *The Times*.

His attitude may in part have been the result of wishful thinking. His devoted pupil Mark Oliphant has said of Rutherford during these years: 'I believe that he was fearful that his beloved nuclear domain was about to be invaded by infidels who wished to blow it to pieces by exploiting it commercially. Also, he disliked speculation about the practical results which could follow from any discovery, unless there were solid facts to support it.' Rutherford, broadcasting in 1933, underlined this pure, almost innocent, approach, saying that the men trying to answer the riddles of the nucleus were not 'searching for a new source of power, or the production of rare and costly elements. The real reason lay deeper and was bound up with the urge and fascination of a search into one of the deepest secrets of nature.'

There is not the slightest doubt that this was true of Rutherford and of many who worked with him. Yet there are at least two indications that as he considered the potentials of the neutron more thoughtfully, he began to see other possibilities; including some, indeed, about which he felt it would be better not to talk too freely. Giving the Watt Anniversary Lecture in January 1936, he noted that 'the recent discovery of the neutron, the proof of its extraordinary effectiveness in producing transformations at very low velocities, opens up new possibilities if only a method could be found of producing slow neutrons in quantity with little expenditure of energy.' At the moment, he went on, radioactive bodies were the only known sources of getting useful energy from atoms, and this was on too small a scale to be of more than scientific interest. However, he now appeared to be thinking of those 'new possibilities'.

Rutherford's belief that significant developments might be just ahead – even if it were not expedient to say so publicly – can be inferred from an incident recalled by Lord Hankey who for years (when he was Sir Maurice Hankey) occupied a key position as Secretary of Britain's Committee of Imperial Defence. After one Royal Society banquet Rutherford drew Hankey aside and said there was something he wished to confide. The experiments on nuclear transformations which he was supervising at the Cavendish might one day turn out to be of great importance to the defence of Britain, even though he could not yet indicate in precisely what way. Perhaps Hankey would keep an eye on the matter.

The date of Rutherford's disclosure to Hankey is not known. But it may have been shortly after a young Hungarian refugee in England, Leo Szilard, had begun propounding what seemed to most physicists to be a most implausible idea, that of a nuclear chain reaction.

Szilard had studied physics in Berlin after being demobilized in 1919. He had emigrated to England when Hitler came to power early in 1933, and had very nearly turned from physics to biology – a step which he was eventually to take a quarter of a century later. Szilard's ebullient character, is illustrated by the occasion when, after the Second World War, he explained to a colleague that he was writing down the facts of nuclear discoveries as they had really taken place – not for publication but for God. It was mildly suggested that God might know the facts, to which Szilard replied, 'not *this* version of the facts'. In the 1930s two contrary views dissuaded him from turning to biology. One was that of Wells's *The World Set Free* and its proposition that atomic energy would eventually be utilized; the other was Rutherford's public avowal to the contrary. Szilard, who had read the Wells novel at the time of Rutherford's 'moonshine' lecture, had his own firm views, writing to the influential Sir Hugo Hirst about *The World Set Free* and stating: 'Of course, all this is moonshine, but I have reason to believe that in so far as the industrial applications of the present discoveries in physics are concerned, the forecast of the writers may prove to be more accurate than the forecast of the scientists.'

Years later he remembered the moment when the great idea had come to him. He was walking in London and as a traffic-light turned green he began to cross the road. Then, he recalled,

> It suddenly occurred to me that if we could find an element which is split by neutrons and which would emit *two* neutrons when it absorbed *one* neutron, such an element, if assembled in sufficiently large mass, could sustain a nuclear chain reaction. I didn't see at the moment just how one would go about finding such an element, or what experiments would be needed, but the idea never left me.

Whatever the element involved, a chain reaction would overcome the problem inherent in the nuclear transformations which had so far been carried out. Whether these transformations were caused by particles from a radioactive element or from artificially speeded-up particles, only a small number of transformations resulted from bombardment by a huge number of particles. This was a justification for Rutherford's rebuke to Oliphant. He had tentatively suggested that it might be possible to get out more energy than was put in: 'Surely I have explained often enough,' Rutherford insisted, 'that the nucleus is a sink, not a source of energy.' But a process in which transmutations multiplied along the lines of 1–2–4–8–16 could be a very different matter. Szilard later said:

I had one candidate for an element which might be unstable in this sense of splitting off neutrons when it disintegrates, and that was beryllium. The reason that I suspected beryllium of being a potential candidate for sustaining a chain reaction was the fact that the mass of beryllium was such that it could have disintegrated into two alpha particles and a neutron. It was not clear why it didn't disintegrate spontaneously, since the mass was large enough to do that; but it was conceivable that it had to be tickled by a neutron which would hit the beryllium nucleus in order to trigger such a disintegration. I told [P. M. S.] Blackett that what one ought to to do would be to get a large mass of beryllium, large enough to be able to notice whether it could sustain a chain reaction. Beryllium was very expensive at the time, almost not obtainable and I remember Blackett's reaction was, 'Look, you will have no luck with such fantastic ideas in England. Yes, perhaps in Russia. If a Russian physicist went to the government and says [sic] "We must make a chain reaction," they would give him all the money and facilities which he would need. But you won't get it in England.'

George Paget Thomson, son of the J. J. Thomson who had run the Cavendish until Rutherford's arrival in 1919, appears to have taken much the same attitude.

Thomson and Blackett were by this time professors of physics in the University of London but their links with the Cavendish remained. Both were close friends of Rutherford and it appears possible that Szilard's 'fantastic ideas', passed on as gossip, were the cause of Rutherford's suggestion to Hankey that he should keep an eye on the subject for the Committee of Imperial Defence. It is known that Szilard eventually secured an interview with Rutherford at which he propounded his theory that a chain reaction could lead to a nuclear weapon. Rutherford's response is not recorded; but it seems unlikely that he would have given any encouragement to such dangerous thoughts, and it is certain that Szilard subsequently crossed the Atlantic to the United States.

However, before leaving Britain he filed a patent dealing with nuclear chain reactions, the first of the numerous patents which throughout the next decade were to prove such an embarrassment to scientists, Governments and national ambitions. Szilard in fact filed three patents. The third was the most significant. The specification began:

This invention has for its object the production of radioactive bodies, the storage of energy through the production of such bodies, and the liberation of nuclear energy for power production and other purposes through nuclear transmutation.

In accordance with the present invention nuclear transmutation leading to the liberation of neutrons and of energy may be brought about by maintaining a chain reaction in which particles which carry no positive charge and the mass of which is approximately equal to the proton mass or a multiple thereof form the links of the chain.

I shall call such particles in this specification 'efficient particles'.

The provisional specification, outlining the process which was to be transformed into a practical possibility by the discovery of nuclear fission in the autumn of 1938, was lodged with the British Patent Office in the summer of 1934. At this stage the details remained secret, since Szilard had post-dated his original application and had then asked for various extensions of time. But two years later the patent would be open to public inspection. Recoiling from this possibility, he assigned the patent to the British Admiralty and in a letter to C. S. Wright, then Director of the Department of Scientific Research and Experiment, explained his reasons.

The object of this Patent has nothing to do with instruments of war, but it contains information which could be used in the construction of explosive bodies based on processes described in the Specification. Such explosive bodies would be very many thousand times more powerful than ordinary bombs, and in view of the disasters which could be caused by their use on the part of certain Powers which might attack this country, it appears very undesirable that such information should be published through the medium of this Patent....

I am fully aware of the fact that if a successful private manufacture is set up on the basis of this Patent, information will leak out sooner or later. It is in the very nature of this invention that it cannot be kept secret for a very long time, and my only concern is that the processes should be developed in this country a few years ahead of certain other countries. This purpose would be served by keeping the Patent secret; and we cannot aim at anything more....

Szilard's letter was accompanied by a note to Wright almost certainly written by Professor Lindemann, then head of the Clarendon Laboratory, Oxford, to which Szilard had for a while become attached.

I am naturally somewhat less optimistic about the prospects than the inventor, but he is a very good physicist and even if the chances were a hundred to one against it seems to me it might be worth keeping the thing secret as it is not going to cost the Government anything.

The recommendation was followed. Szilard's patent for creating a chain reaction was not published until 28 September 1949.

The possibility that information might get into 'wrong hands' – such as those of the Germans – continued to worry Szilard. In June 1935, writing to Lindemann about the possibilities of a chain reaction, he added:

Even if I am grossly exaggerating the chances that these processes will work out as I envisage it at present, there is still enough left to be deeply concerned about what will happen if certain features of the matter become universally known. In the circumstances, I believe an attempt, whatever small chance of success it may have, ought to be made to control this development as long as possible.

Much the same was repeated in letters the following year to Rutherford and to Cockcroft.

Szilard's fear of how nuclear energy might be used if Hitler's Germany learned how to liberate it, and his proposals for secrecy – so much against the grain of a man whom Jacques Monod was to say was 'as generous with his ideas as a Maori chief with his wives' – were both pointers towards the tangled moral problems to be faced by future nuclear physicists. But, in a letter to Enrico Fermi, Szilard also indicated that he, at least, already realized that the exploitation of this new power, when it came, would demand handling in a very special way.

I feel that I must not consider these patents as my private property and that if they are of any importance, they should be controlled with a view of public policy. I see no objection to a commercial exploitation of some such patents, but I believe that the income (if there is any substantial income) should not be used for private purposes, but rather for financing further research or, if the income is very large, for other constructive purposes.

By the time that Szilard had assigned to the British Admiralty the patent for his method of producing a nuclear chain reaction, the release of nuclear energy had been brought nearer by events in Paris and Rome. At the Collège de France, Irène Curie, daughter of the discoverer of radium, had with her husband Frédéric Joliot discovered an unusual result of bombarding certain light elements – notably beryllium, boron and aluminium – with alpha particles. The light elements not only threw out particles when under bombardment but continued to do so when the bombardment had stopped. In other words, the normally stable elements had been transformed into radioactive elements.

At the time the practical implications of induced radioactivity, as it was later called, appeared slight and neither of the Joliot-Curies can have foreseen that within two decades the production of radioactive isotopes would grow into an industry serving medicine, agriculture and industry. Indeed, Frédéric Joliot-Curie's thoughts for the future were a good deal less optimistic, as he made clear when he spoke in Stockholm on receiving the Nobel Prize the following year.

> If surveying the past we look at the progress achieved by science at an ever-increasing pace, we are right to think that researchers, building up or breaking down the elements at will, will know how to bring about transmutations of an explosive character, like chemical chain reactions, one transmutation provoking many others. If such transmutations come to take place in matter, we can expect the release of enormous amounts of useful energy. But, alas, if all the elements on our planet are so infected, we can look forward with apprehension to the consequences of such a cataclysm.

Joliot-Curie's work soon began to throw more light on the internal structure of the nucleus, and no time was lost by physicists in England, Germany, America, and Italy, in repeating and extending the experiments.

Among those who tackled the problem was Enrico Fermi in Rome. Fermi used neutrons as his bullets instead of the French couple's alpha particles and succeeded in creating more than forty different radioactive elements, by this time known to be isotopes of their non-radioactive counterparts. During these experiments he made what he was later to call his most important discovery.

> We were working very hard on the neutron-induced radioactivity and the results we were obtaining made no sense. One day, as I came to the laboratory, it occurred to me that I should examine the effect of placing a piece of lead before the incident neutrons. And instead of my usual custom, I took great pains to have the piece of lead precisely machined. I was clearly dissatisfied with something: I tried every 'excuse' to postpone putting the piece of lead in its place. When finally, with some reluctance, I was going to put it in its place, I said to myself: 'No, I do not want this piece of lead here; what I want is a piece of paraffin.' It was just like that: with no advance warning, no conscious, prior, reasoning. I immediately took some odd piece of paraffin ... and placed it where the piece of lead was to have been.

The result was greatly increased radioactivity. Fermi's hypothesis was that the neutrons tended to bounce into the hydrogen nuclei present in the paraffin wax, were slowed down by the collisions and as a result were more likely to hit the nuclei in the target area. The industrial potentials of the process were realized very quickly, and without delay Fermi and his colleagues took out a patent for their process of producing radioactive elements by slow neutron bombardment. The use of slowed-down neutrons was to become a vital key to the unlocking of nuclear power, but the more immediate importance of the work sprang from scientists who now began to follow it up elsewhere.

During his experiments Fermi had created minute quantities of an unidentified element, and believed this to have a higher atomic number (the number of protons in each atom) than uranium, which, at 92, had the highest atomic number then known. There was very little chance of satisfactorily proving this since the new element had been created in almost unimaginably small amounts. Despite this, the authorities in Rome were anxious to gain a 'first' for Italy and rumours were enthusiastically spread. A Fascist newspaper went so far as to claim that Fermi had 'given a small bottle of element 93 to the Queen of Italy'.

Although there were doubts that Fermi had really achieved the apparently unachievable, there seemed little other explanation of his detailed published results. Only one scientist had an answer to the riddle. She was Ida Noddack, a German chemist. Before assuming that any transuranic element had been created it was necessary, she maintained, to check the new element's identity with all the other elements in the periodic table, not merely with those just below uranium. 'One could assume equally well that when neutrons are used to produce nuclear disintegrations, some distinctly new nuclear reactions take place which have not been observed previously with proton or alpha-particle bombardment of atomic nuclei. ... It is conceivable that the nucleus breaks up into several large fragments.'

It might be conceivable but to most physicists it was so unlikely as not to be worth considering. Research continued in laboratories on both sides of the Atlantic in an effort to discover what it was that Fermi had done.

CHAPTER 3

‘At Variance with All Previous Experience’

The events of the 1930s set the stage for the drama which was to begin with the discovery of nuclear fission in Berlin in the late autumn of 1938, a year after the death of Rutherford. Yet that drama would have unfolded in a different manner had it not been for an event which had taken place in 1933: the appointment of Adolf Hitler as Chancellor of Germany on 31 January. Few political events have so fatefully affected the exploitation of any series of scientific discoveries as the rise to power of a German leader dedicated to the destruction of the Jews. The predominance of Jewish nuclear physicists on the Continent – in strong contrast to the situation in Britain – has long been a subject for discussion; and it is beyond doubt that the anti-Semitic policy of the Third Reich progressively drove out the very men and women who might have produced the ultimate weapon for Germany.

Einstein, who was to play a deeper role in the construction of atomic weapons than is generally realized, found it impossible to return to Germany from the United States in the summer of 1933. Rudolf Peierls, Franz Simon, Nicholas Kurti, Heinrich Kuhn, Max Born and Otto Frisch were only some of the other physicists essential to nuclear development who within the next few years were to settle outside Germany, Hungary, or the Austria which with the *Anschluss* became part of the greater Reich. Leo Szilard and Eugene Wigner were only two of the Hungarians, working in Germany, who soon settled in the United States. Within five years the German Government had effectively driven beyond the frontiers of the Reich most of Europe's most experienced nuclear physicists.

Nor was it only the physicists who were involved, although in the story of nuclear fission it is their experiences which are most frequently quoted. Chemists, and in particular radio-chemists, were to provide many ideas and techniques which were essential to the application of fission for both military and civilian purposes. Many of them, too, were hounded out. Indeed, Lise Meitner, who was first to recognize fission for what it was, had to flee overnight to safety.

In the Kaiser Wilhelm Institute for Chemistry in Berlin, Otto Hahn – also a chemist – had been investigating nuclear reactions with a number of colleagues, notably Fräulein Meitner, an Austrian with whom he had worked for many years. Although Hahn had been deeply involved in nuclear exploration since he had worked first with Sir William Ramsay and then with Rutherford in the early years of the century, it was almost by chance that he was now concentrating on the experiments which were to have such dramatic consequences. In 1917 Hahn and Meitner had discovered the radioactive element protactinium. After Fermi's 1934 experiments, Aristid von Grosse, a German-born chemist working in the United States, suggested that one of Fermi's transformation products might be this very same element. As Lise Meitner's nephew, Otto Frisch, was to say later, 'They felt protactinium was their own baby and they were going to check it.' So Fräulein Meitner who had for a while been working elsewhere, now joined forces with Hahn once again.

Throughout the early spring of 1938 they continued to bombard minute quantities of uranium with neutrons. It was soon clear that protactinium was not the new material. What was not at all clear was the real nature of the element being produced. The amounts were almost infinitely small and quite unsuitable for normal analysis, so mistakes were not only possible but pardonable. Thus although both Hahn and Meitner, each with years of experience, were quite sure that something unusual was taking place, neither had any easy way of discovering what it was.

In mid-March 1938, German troops marched into Austria and the country was quickly incorporated into the Third Reich. Fräulein Meitner automatically became a German citizen and the prospect of dismissal from the Kaiser Wilhelm was the least of the troubles she faced. 'Being Jewish,' Hahn later said, 'she had to wear a yellow badge and, although she was an elderly woman, she was subjected to abuse and even to physical violence. She had received an invitation from Niels Bohr to come to his Institute in Copenhagen. I wrote a letter to our Minister of Culture to ask him that she be permitted to go there but he refused on the ground that if allowed to leave "she would probably not speak well of Germany." '

Her position looked even more dangerous after the physicist Max von Laue reported that Heinrich Himmler had decreed that no university teachers, Jewish or Aryan, could leave Germany. Peter Debye, the Director of the Kaiser Wilhelm Institute for Physics then intervened by writing to a colleague, Dirk Coster of the University of Gröningen. Shortly afterwards, Coster arrived in Berlin. Fräulein Meitner was told that she had just an hour and a half to pack her most precious belongings. Then she was driven to the Dutch frontier where Coster had arranged that a Dutch immigration officer should let her in, even though she had only a now useless Austrian passport and no visa. There were few facilities for nuclear research in Holland and she moved first to Niels Bohr's Institute in Copenhagen and then to Sweden after accepting an invitation to join the Nobel Prize-winner Manne Siegbahn in Stockholm's recently completed Nobel Institute for Physics.

Meanwhile, Otto Hahn had been joined by Fritz Strassmann, and throughout the summer and early autumn the two men continued their efforts to identify the fragments produced from the bombarded uranium. But at the same time Irène Joliot-Curie and Pavle Savitch in Paris were doing similar work and throughout the last months of 1938 the two groups progressed with a series of leap-frogging experiments. The Berlin team found that some of the fragments, still of course minute and extremely difficult to identify, might be either lanthanum or actinium; but although they behaved like lanthanum, they were considered to be actinium, since in these circumstances the production of lanthanum would appear to defy the laws of nature.

In Paris, similar results were obtained. But while, due to problems of impurity, the French were unable positively to identify the fragments as lanthanum, they were able to show one thing: that the fragments were not actinium. The riddle remained.

In Berlin, Hahn and Strassmann repeated their previous work. Earlier they had used barium as a carrier in their experiments but now, on Saturday 17 December, they also used mesothorium, a radioactive isotope of radium, most famous for its humble use on the face of luminous watches. The mesothorium behaved as expected but the unidentified product remained with the barium, and at the end of the day's work Hahn wrote in his notebook: 'Exciting experiment with mesothorium.'

There appeared to be one way of discovering what had happened. If the mysterious substance was an isotope of radium with its eighty-eight protons then the addition of a single proton would turn it into actinium; if it were barium, beta-particle decay which would convert a neutron into a proton would turn it into the 57-proton element lanthanum. As actinium and lanthanum are chemically different a positive identification could then be made.

'It is now practically eleven o'clock at night,' Hahn later wrote to Lise Meitner. 'Strassman will be coming back at a quarter to twelve, so that I can get off home at long last. The thing is: there's something so odd about the "radium isotopes" that for the moment we don't want to tell anyone but you ... Our "Ra" isotopes behave like Ba ... We ourselves realize it [the uranium nucleus] can't really burst into Ba. What we want to check now is whether the Ac isotopes produced out of the Ra behave not like Ac, but like La. All rather tricky! But we *must* clear this thing up.'

Within forty-eight hours the truth was known. The 'actinium' was lanthanum. On the 22nd Hahn and Strassmann reported their work in a paper sent off immediately to *Die Naturwissenschaften*. The editor was so impressed that he held over material already set up, and replaced it with the paper from Berlin.

There was, indeed, sufficient justification for this. At the end of their paper Hahn and Strassmann pointed out that as chemists they should delete from it the symbols for radium and actinium and replace them with those for barium and lanthanum. Yet the puzzlement at the results they had taken so much trouble to confirm was clear from their next sentence. 'However, as "nuclear chemists", working very close to the field of physics, we cannot bring ourselves yet to take such a drastic step which is at variance with all previous experience in nuclear physics.' The puzzlement was only partly resolved by what looked, after all their meticulous work, like the effort of a drowning man clutching at the proverbial straw. 'There could perhaps,' they added, 'be a series of unusual coincidences which has given us false indications.' This was, perhaps, no more than a desperate hope, the reaction of men who, like Wells's Holsten in *The World Set Free* after he had discovered how to create radioactivity, 'felt like an imbecile who has presented a box full of loaded revolvers to a Crèche'.

It is doubtful if either Hahn or Strassmann really believed that unusual coincidences were involved. But the results they reported were so extraordinary that it was natural for them to write a note of caution into their paper. If they were indeed correct, then the impact of a single neutron had split apart the nucleus of the heaviest naturally occurring element. The facts seemed to be there; the only thing missing was an explanation. It was to be provided by Lise Meitner.

Fräulein Meitner's nephew, Otto Frisch, was another refugee from Hitler, who in December 1938 was working in Niels Bohr's Institute in Copenhagen. Frisch believed that his aunt might be lonely during her first Christmas in exile, even though she was staying with friends, and decided to join her. 'In a small hotel in Kungälv near Göteborg I found her at breakfast brooding over a letter from Hahn,' he later wrote. At first Frisch was sceptical of the results from Berlin but his aunt was in no doubt. Hahn, she emphasized, was too good a chemist to have got the facts wrong.

After breakfast the couple went outside, walking up and down in the snow, Frisch on skis and his aunt on foot, discussing what had really happened in the Kaiser Wilhelm.

> It took her a little while to make me listen but eventually we got to arguing about the meaning of Hahn's result, and very gradually we realized that the breaking-up of a uranium nucleus into two almost equal parts was a process so different from the emission of a helium nucleus, that it had to be pictured in quite a different way. The picture is not that of a 'particle' breaking through a potential barrier, but rather the gradual deformation of the original uranium nucleus, its elongation, formation of a waist, and finally separation of the two halves.

At this point, Frisch remembers, they both sat down on a fallen tree. Nuclei, they knew, were positively charged. And if a positively charged barium nucleus with fifty-six protons was suddenly created in intimate contact with another nucleus carrying the remaining thirty-six protons from the original uranium nucleus, then surely since like charges repel each other, they would fly apart with an immense amount of energy? 'The picture,' Frisch was later to explain, 'was that of two fairly large nuclei flying apart with an energy of nearly 200 million electron volts, more than ten times the energy involved in any other nuclear reaction.'

The figure had been obtained from a rough calculation using the known charges on the nucleus. But the 'packing fraction', the factor that came into play when light elements were joined by fusion, was equally relevant if heavier elements were split. The two newly created particles would have a mass smaller than the original nucleus by a mass about one-fifth that of a proton, a loss which when multiplied by the c^2 of Einstein's mass-energy equation would give the energy released. The figure, worked out this way, was almost exactly 200 million electronvolts.

Splitting all the nuclei in a single gram of uranium, it was now realized, would be equivalent to releasing the energy in three tons of coal. However, there was still no way of doing it. It was only the occasional uranium nucleus that was split in two and as long as this was so, the chances of using nuclear energy seemed as remote as ever. But there did appear to be one intriguing possibility. Very little was as yet known about how the particles in the split nucleus disposed themselves; surely, therefore, it was just possible that during such reactions one or more neutrons might be released?

However, even the simple Meitner–Frisch theory was still only a theory. Encouragement came a few days after Christmas when Lise Meitner returned to Stockholm and her nephew continued on to Denmark. He arrived in Copenhagen just as Neils Bohr was boarding ship for the United States where he was to address the Fifth Washington Congress on Theoretical Physics. Hahn's paper had not yet appeared in the *Die Naturwissenschaften* and in the few minutes available Frisch gave him the startling news.

> I had hardly begun to tell him, when he struck his forehead with his hand and exclaimed: 'Oh, what idiots we all have been! But this is wonderful! This is just as it must be!'

He asked if Frisch and Lise Meitner had written a paper about it. Frisch said they had not but would do so at once, and Bohr promised not to talk about it before the paper was out. Then he was off to catch his boat.

Bohr's promise not to talk was the natural reaction of a scientist anxious to ensure that those responsible for a discovery should get their due. But the whole question of secrecy was from now onwards increasingly to be a bone of contention among the scientific community, and for reasons that were obvious as the prospects of war loomed larger.

During the next few days, Frisch and his aunt settled over the telephone the details of a paper for publication in *Nature* which they headed 'Disintegration of Uranium by Neutrons: A New Type of Nuclear Reaction'. In it Frisch called the new process 'fission', having asked William A. Arnold, an American biologist working in Copenhagen, the name that he used for the process of cell division. Frisch recalls that shortly afterwards he explained the fission process to his colleague, George Placzek.

At first, Placzek did not believe the story that these heavy nuclei, already known to suffer from alpha instability, should also be suffering from this extra affliction. 'It sounds a bit,' he said, 'like the man who is run over by a motor car and whose autopsy shows that he had a fatal tumour and would have died within a few days anyway.'

But Placzek did suggest that Frisch should repeat the Berlin experiments. Frisch did so, and 'it was computed that the energy from a single fission could make a visible grain of sand make a visible jump (if we could so direct it).'

Bohr had sailed in the *Drottingholm* for New York carrying in his head the most revolutionary knowledge since Rutherford had conceived the nuclear atom twenty-eight years previously. Lise Meitner in Stockholm and Frisch and Placzek in Copenhagen were also aware of the startling discovery, although Hahn and Strassmann in Berlin were still ignorant of its full significance. Thus only four people as yet knew that mankind had reached the edge of a previously unsuspected precipice. All intended to keep it that way for the time being, basically so that Hahn and Strassmann would reap the reward of undisputable scientific priority when their paper appeared in *Die Naturwissenschaften* later in the month. Meitner and Frisch would do the same with their explanation in *Nature*. The plan was to be disrupted by a succession of mischances, as though Fate was determined that mankind should also be left in no doubt about its position.

Bohr was travelling with Leon Rosenfeld, a young mathematician from his Institute, and as soon as the two men had settled into their cabins Rosenfeld was told the news. Bohr showed him the rough notes remarking: 'We will have to think about it.' What he absentmindedly forgot to mention was that he and Frisch had decided to keep the news secret for the time being.

Bohr had a blackboard set up in his cabin and throughout the voyage worked with Rosenfeld on the various problems that the theory raised. In particular, he asked himself, just why was it that the uranium nucleus, having taken on board the additional neutron, then split into two roughly equal fragments rather than being scattered into a number of smaller ones. Eventually he was satisfied. Lise Meitner's view of the nucleus as a liquid drop, fissioning into two parts much as the biological cell split into two, could be justified on sound mathematical grounds and was established by the time the *Drottingholm* nudged into dock on 16 January.

Waiting on the quay were Enrico Fermi and his wife, who had come to live in the U.S.A. because Fermi was anti-Fascist and his wife was Jewish. With them was John Wheeler, professor of physics at Princeton University, where Bohr was to work for a couple of months. Bohr drew Wheeler aside and told him the momentous news that was, he stressed, to be kept secret: Hahn and Strassmann had split the atom, Meitner and Frisch had explained what happened in the process.

Bohr was to stay the night in New York. Rosenfeld was taken to Princeton by Professor Wheeler, and in Princeton that night attended a dinner of the Journal Club, a small group of physicists. He was asked what was happening in Europe; inevitably, in the circumstances, he told them that Hahn and Strassmann had split the nucleus of the uranium atom.

All those present knew the significance of the news. Many wrote at once to colleagues. Some telephoned that night, George Gamow informing Edward Teller in Washington: 'Bohr has just come in to New York. He has gone crazy. He says [according to Rosenfeld] a neutron can split uranium.' By the time that Bohr arrived in Princeton the following day, physicists in more than one laboratory were considering how they could repeat Hahn's work. Dismayed at what had happened, Bohr immediately cabled Frisch that if he had not yet sent off his paper to *Nature*, he should lose no time in doing so. Then he and Rosenfeld began preparing their own paper which they hoped would put the record straight.

It was not only that the nuclear cat was out of the bag. Bohr now knew that the whole subject of fission would be aired at the conference starting in Washington on the 26th; but Bohr still felt himself bound by his word to Frisch. On the first day of the conference he knew the worst: Fermi had learned of Rosenfeld's statement and was already making his own experiments.

Only the following day was the situation altered. Bohr was shown a copy of *Die Naturwissenschaften* containing Hahn and Strassmann's report. The secret was now officially out and he felt free to announce Meitner and Frisch's explanation of those results which were 'at variance with all previous experiences in nuclear physics'.

Bohr's explanation was hardly finished before some of his audience left the room to telephone their laboratories. Within a few days workers at the Carnegie Institution of Washington, at Johns Hopkins University, and at Columbia University had all confirmed Hahn and Strassmann's results.

It was not only among physicists that the significance of the experiment was appreciated. Robert Potter, one of the two reporters covering the conference, wrote an article that evening which was distributed to the press the following day. 'New hope for releasing the enormous stores of energy within the atom has arisen from German experiments that are now creating a sensation among eminent physicists gathered here for the Conference of Theoretical Physics,' it began. 'It is calculated that only 5 million electronvolts of energy can release from an atom's heart 200 million electronvolts of energy, forty times the amount shot into it by a neutron.'

Bohr's explanation of what Hahn and Strassmann had achieved quickly triggered off experiments not only in the United States but throughout the world. In Warsaw a young Polish physicist, Joseph Rotblat, repeated Hahn's work. So did scientists in the Leningrad Physicotechnical Institute, while the Russian physicist Igor Tamm told his students: 'Do you know what this new discovery means? It means a bomb can be built that will destroy a city out to a radius of maybe ten kilometres.' Joliot-Curie at the Collège de France and scientists in Britain and Germany began to investigate the next question: when a uranium nucleus was split into two were one or more neutrons liberated at the same time? Only when that had been ascertained would it be possible to move on to the next stage: to assess the possibilities of creating the chain reaction to which Szilard had devoted his still-secret patent, now lodged in the archives of the British Admiralty.

However, before any of these investigations got very far Bohr developed the theory of nuclear fission in a way which appeared to stop in its tracks any project for nuclear weapons and to restore to respectability Ernest Rutherford's 'moonshine'. It was known that there were at least three different isotopes of uranium, all inextricably mixed together. Each uranium nucleus contained ninety-two positively charged protons surrounded by ninety-two negatively charged electrons. More than ninety-nine per cent of them also contained 146 neutrons which with the protons thus formed the nucleus of a uranium 238 isotope (92 + 146). (The number of neutrons and protons in an atom is called its 'mass number'.) About seven in every thousand nuclei contained only 143 neutrons and were therefore atoms of the uranium 235 isotope, while an even smaller number contained only 142 neutrons and therefore were uranium 234. But, maintained Bohr, it was only the rare uranium 235 isotope that could easily by split in two when its nucleus was penetrated by a neutron. Bohr's explanation was that when a neutron entered a nucleus with an uneven mass number, such as uranium 235, the nucleus, now made up to an even mass number, would split into two fragments; if the captured neutron brought the mass number up to an uneven figure it would be retained and the nucleus would not split easily.

The implications were devastating for those who believed in the utilization of nuclear energy, since a chain reaction was essential. Yet even if fission did produce 'free' neutrons, there would, following Bohr's theory, be so few fissions that the prospects of a chain reaction would be non-existent.

There was, it is true, one theoretical possibility. The situation would be radically different were it possible to separate the fissionable uranium nuclei from the rest. But even on a laboratory scale the separation of isotopes was a complicated and arduous exercise, and at the end of the day resulted in a product to be measured almost in separate atoms. Whatever the exact amount required to start a chain reaction – and in 1939 it was speculatively thought to be at least some hundreds of pounds – the chance of producing it appeared infinitely remote.

Yet there was always the chance that Bohr might be wrong, unlikely though that seemed to be. Thus the devastating prospect that God had not given man the ability to blow himself to pieces with nuclear energy was taken with reservation by those already moving as fast as possible in the race for nuclear weapons. Prominent among them was the man who had handed the Admiralty the patent for the chain reaction.

Szilard had first learned of nuclear fission while visiting another Hungarian, Eugene Wigner, who had become a U.S. citizen in 1937 and was professor of mathematical physics at Princeton University. Szilard recalls:

> When I heard [of Hahn's discovery] I saw immediately that [the fission] fragments, being heavier than corresponds to their charge, must emit neutrons; and if enough neutrons are emitted in this fission process, then it should be, of course, possible to sustain a chain reaction; all the things which H. G. Wells had predicted appeared suddenly real to me.

Szilard was not a man to waste time and on 25 January, the day before Bohr had even made his startling announcement in Washington, he wrote to the American financier Lewis Strauss, telling him what Rosenfeld had said. 'This is entirely unexpected and exciting news for the average physicist,' he added. 'The Department of Physics at Princeton, where I spent the last few days, was like a stirred-up ant heap.' He thought nuclear power might be feasible although he was not optimistic. 'I see, however, in connection with this new discovery potential possibilities in another direction,' he went on. 'These might lead to a large-scale production of energy and radioactive elements, unfortunately also perhaps to atomic bombs. This new discovery revives all the hopes and fears in this respect which I had in 1934 and 1935 and which I have as good as abandoned in the course of the last two years.'

The immediate task was to find if neutrons were really emitted; and, if so, how many. But this work could not be considered in a political vacuum, particularly by men who had been driven across the Atlantic by the growing strength of German anti-Semitism. And during the next few months self-censorship by the scientists was to be secondary only to their efforts to prove or disprove the potentialities of the discovery made in Berlin.

Back in New York, Szilard conferred with a colleague, Isidor Rabi. Shortly afterwards Rabi approached Fermi with Szilard's proposal that any results pointing to the possibility of a chain reaction should be kept secret. Fermi's classic answer was: 'Nuts!' His explanation when he, Szilard and Rabi met a few days later was simple enough. He believed, presumably as a result of discussion with Bohr, that there was only a remote chance that a chain reaction was possible, a chance that he put at ten per cent when questioned. 'Ten per cent is not a remote possibility if it means we may die of it,' Rabi observed. 'If I have pneumonia and the doctor tells me that there is a remote possibility that I might die, and that it's ten per cent, I get excited about it.'

The question of secrecy was more open to discussion in 1939 than it might be today. To publish without let or hindrance, whatever help it might give to rivals or colleagues, was the professional equivalent of 'open treaties openly arrived at', and to propose anything different tended to create distrust at the least and suspicion at the worst. But now the interests of science, and the staking of the individual's priority in discovery, was counterbalanced by the needs of defence, and particularly of the looming battle against the dictatorships.

In 1939 it was still possible that the problem of nuclear energy was really non-existent. Perhaps there would be no spare neutrons available to release the greatest power on earth; certainly more than one physicist hoped that this might be so – including Lindemann of whom it has been said: 'The idea of such destructive power being available to human hands seemed to repel him so much that he could scarcely believe that the universe was constructed in this way.'

Szilard was taking no chances. On 2 February 1939, he wrote to Joliot-Curie proposing that if the emission of neutrons was confirmed in France, England or America, the fact should be kept secret. At the same time he cabled Lindemann in Oxford, asking him to send to New York a block of beryllium which had been made for him in England and which he had left there on setting out for the United States.

The block, which Lindemann sent at once, was essential for the experiment which Szilard now had in mind. He also required a gram of radium which he was able to hire, and the use of an ionization chamber which had been built at Columbia by Dr Walter Zinn. All was ready by the end of February and on 3 March Szilard and Zinn carried out the crucial experiment. The gram of radium was placed inside the block of beryllium, and its particles, bombarding the beryllium nuclei, caused the nuclear reaction which produced both neutrons as well as more particles. When a uranium nucleus was split by the slow neutron it was clear from the flashes on the cathode-ray oscillograph that fast neutrons were being ejected – an average of about two for each fission. The world, pronounced Szilard, was headed for trouble. Shortly afterwards he reported enthusiastically to Lewis Strauss: 'Performed today proposed experiment with beryllium block with striking result. Very large neutron emission found. Estimate chances for reaction now above fifty per cent.'

Another development during this period was the growing awareness that the chances of fission were very dependent on the speed of the neutrons involved: the slower the speed, the more likely a fission. Yet the neutrons released during fission were of a high speed. Szilard and Fermi therefore devised, and then patented, the idea of a 'lattice-reactor' in which lumps of uranium were embedded in a medium – later known as a moderator – which would slow down neutrons produced by fission so that they became more likely to produce further fissions. This was of course the natural development of Fermi's experiment in Rome with a block of paraffin wax four years previously. Szilard was subsequently offered 'appropriate' compensation, believed to be about $1,000, after the system had been used by the U.S. Government in the huge plants of the Manhattan Project. Disgusted at the sum, he granted the Government the rights for nothing.

Fermi also was by now becoming more optimistic. The reason is not clear since like Szilard he appears to have believed that only the rare uranium 235 isotope was susceptible to fission and no one had yet tackled the problem of separating uranium 235 from 238 and 234. Despite the uncertainty Fermi was successful in securing an interview with Admiral Hooper, technical assistant to the Chief of Naval Operations. Hooper appears to have been visibly unimpressed which suggests that Dean Pegram of Columbia, who had organized the meeting, had chosen the wrong man.

But present at the meeting there had been Ross Gunn, a physicist at the U.S. Naval Research Laboratory. Gunn was not so much concerned with a weapon as with the possible use of fission as a power generator. As he was to point out to the laboratory's director shortly afterwards, the release of energy by fission did not appear to require oxygen. 'This is a tremendous military advantage and would enormously increase the military effectiveness of a submarine,' he said. 'If the method will work it is of outstanding importance ... If it will not work, it is of the utmost importance to determine this fact at the earliest practicable date.' Gunn succeeded in getting $1,500 which he allotted to Merle Tuve at the Carnegie Institution of Washington, and at the same time persuaded Jesse W. Beams, a physicist at the University of Virginia, to investigate centrifugal methods of separating uranium 235 from its other isotopes.

The need for this latter extraordinarily long shot became obvious in March 1939, when Bohr's purely theoretical idea which seemed to bar advance towards either nuclear power or a nuclear weapon was given experimental confirmation. John Dunning of Columbia University, New York, persuaded the University of Minnesota's Alfred O. C. Nier, America's leading expert on mass spectroscopy, to prepare minute quantities of partially separated uranium 235 which he then passed over to Dunning and his co-workers. They quickly made the necessary measurements. These were reported in the 15 March and 15 April issues of *Physical Review* and confirmed beyond reasonable doubt that, as Bohr had prophesied, it was the rare 235 isotope that was largely – at the time it was thought exclusively – responsible for fissioning with neutrons.

Yet in the same month Joliot-Curie's team in Paris showed, as Szilard had already shown, that neutrons were liberated when certain uranium nuclei were split. In the Collège de France there was no hesitation about publication. Once the laboratory figures had been checked, Hans Halban, the physicist in charge of the experiments, wrote a paper on the results, and even rushed it to Le Bourget airport so that it could appear in the first possible issue of *Nature*. But before the end of March the need for secrecy was no longer in doubt. On 15 March Hitler brought the prospect of war nearer by occupying the Czech provinces of Bohemia and Moravia.

Both Szilard and Fermi had already sent papers on the emission of neutrons in the fission process to *Physical Review*, and both had requested that they should not yet be published. This enabled them to keep the work secret while staking claims to the work done. But now the argument about secrecy had become more important. Szilard seemed unlikely to win it and towards the end of the month took matters into his own hands. His friend Victor Weisskopf, an Austrian refugee, had been a close friend of Hans Halban, and the two men now dispatched a cable to Halban. It asked him to tell Joliot-Curie that two papers on neutron emission had been sent to *Physical Review* but publication was to be delayed. 'News from Joliot whether he is willing similarly to delay publication of results until further notice would be welcome', the cable continued. 'It is suggested that papers be sent to periodicals as usual but printing be delayed until it is certain that no harmful consequences to be feared. Results would be communicated in manuscripts to cooperating laboratories in America, England, France and Denmark.'

The cable was delivered the next day. 'It was the morning of 1 April and I was sitting in my bath,' Halban has said, 'when round the door came a hand holding a telegram. I was told that it was 170 words long and that it came from Szilard and Weisskopf. I opened it to find that it consisted of a plea that the Joliot team should not publish any further results.' The belief that the cable was an April fool's joke continued for some hours. 'Then,' says Halban, 'I began to realize that it must have been a very expensive joke; for the first time I wondered if the idea was meant to be taken seriously.'

Szilard might justifiably have expected something more radical. A number of British physicists, including P. M. S. Blackett, John Cockcroft and P. A. M. Dirac had also been approached and it seemed likely that both *Nature* and the Royal Society's *Proceedings* might be willing to restrict publication. In Paris a number of factors, to each of which the various members of Joliot-Curie's team gave their separate emphases, brought a different reaction.

Perhaps the most important consideration was that the French were at this stage concentrating their thoughts not on a bomb but on 'the boiler' as it was to be called: not on an unchecked chain reaction which would produce a cataclysmic explosion but on a controlled reaction which would release a new kind of energy. The idea of a weapon was not ruled out but, as Halban later said, 'From the time of these experiments in March and April we were absolutely bent on creating a nuclear chain reaction which could be used for industrial power.'

In addition, Joliot-Curie had been brought up to regard as unbreakable Marie Curie's rule of always publishing, regardless of the consequences. There was also the constant shortage of funds for research, a shortage which would be at least alleviated if the team could show progress by publication of results. Finally, there had been an unfortunate gap in Szilard's security plans. At the Carnegie Institution of Washington a group headed by Richard Roberts had identified neutrons released from uranium and had reported their results to a news firm called Science Service.

This last point rankled, naturally enough. The Collège de France team was much in advance of any other group in Europe in the search for a chain reaction, and naturally looked on Szilard's cable with suspicion. Szilard's proposition of 31 March was very reasonable they replied, but it came too late. 'Learned last week,' continued their cabled reply, 'that Science Service had informed American press February 24 about Roberts work. Letter follows.' The message was signed by Joliot-Curie, Halban and Halban's unique assistant, Lew Kowarski, and shortly afterwards Joliot-Curie himself closed the argument with nine words: 'Question studied my opinion is to publish now. Regards.'

Szilard and his colleague at Columbia told *Physical Review* to go ahead with publication and their two papers duly appeared on 15 April. A week later a further report from the Collège de France was printed in *Nature*. Headed 'Number of Neutrons Liberated in the Nuclear Fission of Uranium', it reported that the average number was between three and four.

Two views can be taken of Szilard's failure to institute a censorship. It can be argued that once Hahn and Strassmann's paper had appeared in *Die Naturwissenschaften*, sufficient momentum had been given to the search for a chain reaction, and that research aimed at developing nuclear power, or a super-bomb, or both, was now inevitable. Yet the chain reaction was not a theoretical follow-on from the Berlin experiments alone. Even under the spur of war a nuclear weapon was for long considered an impractical figment of scientific imagination.

And when, in the late summer of 1941, the decision was taken in America to build the bomb, it was taken very largely in the mistaken belief that the Germans had already embarked on a similar enterprise. Nevertheless, by 1941 a good deal of the scientific brushwood had been cleared away. The demands of a nuclear weapon in terms of money, man-hours and materials had been investigated in some detail, and its devastating effect had been calculated. A good deal of this sprang from the possibilities launched on the world by the papers from the Collège de France which Szilard and his friends had been unable to suppress.

In one respect – if only one – the story of how nuclear energy came to be unleashed is less satisfactory than any artist would wish: the advance was by a long series of small steps rather than by a single unexpected stride. In Wells's *The World Set Free*, there is a moment when Holsten, the discoverer of the new material that can give limitless power, realizes his unique situation. He finds himself at the door of St Paul's Cathedral and listens, from the door, to the evening service.

> Then he walked back through the evening lights to Westminster. He was oppressed, he was indeed scared, by his sense of the immense consequences of his discovery. He had a vague idea that night that he ought not to publish his results, that they were premature, that some secret association of wise men should take care of his work and hand it on from generation to generation until the world was riper for its practical application. He felt that nobody in all the thousands of people he passed had really awakened to the fact of change, they trusted the world for what it was, not to alter too rapidly, to respect their trusts, their assurances, their habits, their little accustomed traffics and hard-won positions.

In real life, no situation was quite as simple as that. Rutherford and Aston, Szilard and Fermi – as well as Frisch and Rudolf Peierls in the early spring of 1940 – were at least partially shielded from the vision ahead by the high barrier of innumerable scientific and technological difficulties. Nevertheless, the five months that separated Halban's receipt of the request for censorship and the outbreak of the Second World War on 3 September 1939, encompassed a subtle change of approach to what even the most open-minded scientist had tended to regard in his heart of hearts as an idea from the pages of science fiction. Intellectually, it was all very interesting; practically, it had seemed that only the most outside of outside chances would ever allow man to release nuclear energy. But now, as war came nearer, was it not, perhaps, an outside chance that should be investigated.

Superficially, one of the most important advances in theory during the summer of 1939 appeared to lengthen the chances even further. It had early been appreciated that the release of neutrons from the fission of uranium was only one factor in the starting of a chain reaction. Some of the neutrons might hit other nuclei without causing further fissions and some might escape from the lump of uranium without even hitting a target. Thus, argued Francis Perrin (son of Jean Perrin) at the Collège de France, there was a critical mass which had to be reached before a chain reaction could possibly be started. The mass, he estimated, would be a ball of uranium about nine feet in diameter and weighing some forty tons. On the face of it, this ruled out an atomic bomb. On the other hand, the nuclear data which the theory used were extremely imprecise, and more accurate information might provide an entirely different picture.

The Perrin concept of critical size did at least give birth to a university joke much heard throughout the summer. If only the critical mass was much smaller, the world's physicists would have little trouble in disposing of Hitler. Workers in a number of laboratories had only to package up their own parcels of uranium, address them to the Führer, and post them so that they were brought to his desk for inspection throughout the day. When the final package arrived and was placed with the others, the problem of the Führer would be eliminated.

An indication that Perrin's figures were not as pessimistic as they appeared was soon given in Paris and the French team quickly filed five patents.

Joliot-Curie was bitterly against the patenting of anything which he regarded as an application of a law of nature. Things were not handled that way in science. To Joliot-Curie, as to many other scientists, any attempt to limit exploitation of the almost unimaginable power locked within the atom appeared both wrong and stupid. He was eventually won over by two arguments. The first was that Madame Curie had been forced to beg for the radium she needed for research, and that almost half of that used in the Curie Institute had been a gift from America. The second point, almost as strong to any patriotic Frenchman, was that it would be intolerable if in the future France was able to utilize the enormous potential of nuclear power only by using the patents which other countries would no doubt lodge when they caught up with her.

So the French now patented their proposals for making both the bomb and 'the boiler'. The patents described the proportions of uranium and moderator that would have to be assembled, and their homogeneous or heterogeneous distribution in the assembly. Beryllium, with which Szilard had been experimenting at Columbia, graphite, water, and heavy water were proposed as possible moderators which would slow down the neutrons released by fission. Various ways of keeping the chain reaction under control were discussed, as well as methods of extracting the heat which would be generated.

The French were not alone in the early summer of 1939, although they were certainly ahead. In Russia Igor Kurchatov and other physicists persuaded the Soviet Academy of Sciences to increase support for nuclear research. In Germany the Army showed interest. In the United States, Szilard and Fermi began to discuss the respective merits of carbon and heavy hydrogen as moderators, while in Britain the first step along the road to the nuclear age was taken with the purchase of a ton of uranium ore at 6*s*. 4*d*. a pound, an expenditure which caused some shaking of heads in a prudent British Treasury.

The most significant moves during this period when war in Europe appeared inevitable, were made in Britain and the United States, and it is lamentable that on both sides of the Atlantic chauvinism has tended to fog the issue of who did what. In Britain, steps were taken to investigate the practicability of a nuclear weapon, its demands in men, money and materials, the chances of deceiving the Germans about British work, and also of securing world supplies of the essential uranium. As General Groves, head of the Manhattan Project was later to comment: 'It is sobering to realize that but for a chance meeting between a Belgian and an Englishman a few months before the outbreak of the war, the Allies might not have been first with the atomic bomb.'

In Britain the first moves were informal and, in the British way, rather casual. Chain reactions might mean bombs; bombs were normally dropped by aircraft; aircraft of the military variety were run by the Air Ministry, and the various speculations about nuclear fission accordingly found their way to the desk of Sir Henry Tizard, the scientist mainly responsible for the creation of Britain's chain of radar stations which by this time covered the southern and eastern approaches to the island.

The French team's paper reporting that three neutrons were released for each fission of uranium – a figure which later proved to be unduly optimistic – appeared in *Nature* on 22 April. One of the first in Britain to note its importance was G. P. Thomson, 'I began to consider carrying out certain experiments with uranium,' he later said. 'What I had in mind was something rather more than a piece of pure research, for at the back of my thoughts there lay the possibility of a weapon.' However, Thomson needed about a ton of uranium oxide, 'the amount that a pottery firm would use in a year', as he described it. That amount could not easily be obtained so Thomson consulted Sir Will Spens, Master of Corpus Christi College, Cambridge, and a man marked as a Regional Controller for Civil Defence if war broke out. Sir Will consulted Kenneth Pickthorn, the local Member of Parliament, Pickthorn consulted General Ismay, secretary of the Committee of Imperial Defence, and Ismay consulted Sir John Anderson, then Lord Privy Seal. Anderson, a distinguished administrator, had spent a year in Leipzig studying uranium in the early years of the century and was almost the only man of political influence in the country who had the remotest idea of what nuclear energy might mean. An atomic bomb, he told Ismay, was 'a scientific but remote possibility' and Ismay thereupon consulted Sir Henry Tizard, the man who, as chairman of two important scientific committees of the Air Ministry, seemed most likely to help.

Tizard was also the Rector of Imperial College, where Thomson was professor of physics, and had by this time been approached directly by Thomson. 'He arranged for me to meet Air Marshal Sir Wilfred Freeman, then the member of the Air Council concerned with weapons,' Thomson later wrote. 'I shall never forget this meeting. I was so ashamed of putting forward a proposal apparently so absurd. ... However, Freeman and Tizard both heard me out with composure.'

One reason was that Tizard had received yet a third warning of things to come. This had arrived from Professor A. M. Tyndall of Bristol who had written to the Committee of Imperial Defence pointing out that 'as a result of a recent discovery in nuclear physics it may be said that there is a chance that an explosive might be produced setting free an amount of energy ten million times greater per gram than the highest explosive known. Such a weapon in the hands of one nation might be quite decisive.'

With this treble recommendation on his desk Tizard, who was to remain sceptical of nuclear weapons even longer than Professor Lindemann, wrote to David Pye, the Director of Scientific Research in the Air Ministry.

Ismay ... is anxious that some recommendation should be made at once, because strong representations have been made to Ministers that the matter is very urgent. I do not agree with these representations, but now so many people are talking about the subject as a result it is, I think, wise to get ahead. So this letter is to recommend on behalf of the Committee that one ton of uranium oxide should be purchased at once from Messrs Hopkins and Williams Ltd, 16 & 17 Cross Street, Hatton Garden, E.C.1. The price ex-refinery in London is 6/4 a pound.

Eventually, Thomson got his uranium oxide, the cost of £709 6*s*. 8*d*. probably being the first expense which any Government incurred on behalf of nuclear fission. But the difficulty of securing even a single ton was indicative of the trouble ahead if this rare metal, previously used only in very small quantities by industry, should suddenly become a strategic material of war. In 1939 the bulk of the world's known uranium lay in the Belgian Congo where it was extracted by the Union Minière. Smaller supplies were known to exist in Czechoslovakia, by this time entirely under German control. Tizard decided to tackle the problem at source, a process made less difficult by his acquaintance with Lord Stonehaven, a British director of the Union Minière. A meeting was arranged in the company's London offices between Tizard, Stonehaven, Baron Cartier (the Belgian Ambassador in London), and Edgar Sengier (a Belgian director of the company). 'Sir Henry asked me,' M. Sengier later stated, 'to grant the British Government an option on each ton of ore which would be extracted from the Shinkolobwe radium-uranium mine.'

The Belgians would not consent. But Tizard's effort was not entirely fruitless. As he left the meeting he turned to Sengier. 'Be careful,' he said, 'and never forget that you have in your hands something that may mean a catastrophe to your country and mine if this material were to fall into the hands of a possible enemy.' Sengier remembered this a few days later when, back in Brussels, he received a similar warning from Joliot-Curie. He ordered two shiploads of uranium ore, about 1,200 tons in 2,000 steel drums, to be sent to the United States. They arrived in New York shortly afterwards and were stacked in the open at Port Richmond, Staten Island, New York. Here, remakably enough, they remained for nearly two years, marked 'Uranium Ore, Product of the Belgian Congo', but exciting no interest.

Fear that the Germans might already be working on a nuclear project was a main worry and in July Thomson, by now trying to discover if a chain reaction was feasible, made an interesting suggestion to Tizard.

If it is true that the Germans are really trying to make an uranium bomb, they may be anxious as to its possibilities in other hands. Would it be possible to do a bluff, and get them to believe we really had a new and very formidable weapon? If so it might make them hesitate risking a war till they were sure, and so tide us over the dangerous period of the next few months. I imagine our secret service has means which would allow them to plant false information where it would reach high German authorities. If so, it would not be difficult to concoct something with the help of our explosives department which would sound fairly convincing and be difficult to disprove in a short time.

Thomson was soon sending Tizard the alleged draft of a report from Martlesham, an Air Ministry aerodrome and test site in East Anglia. This described the explosion of ten uranium bombs of different sizes, one of them leaving a crater 450 feet across and damaging buildings two miles away. The report was accompanied by a more detailed account which began by stating that the tests had 'been an immense success and there can be no doubt that we have now got a weapon which will revolutionize air warfare'. It went on, prophetically, to emphasize the need for an aircraft to escape from the blast of a nuclear weapon and added that a half-ton bomb 'should give an explosion about equal to the great explosion of Halifax, but in view of the fact that the gross energy available is roughly that of 5 *million* tons of T.N.T. the effect may well be a good deal larger.' There was much detail added and Tizard tried to have the material 'planted' on the Germans. His success is not known, although he later said that the authorities were 'horrified at the thought'.

Meanwhile, Thomson's own work was continuing, soon helped by Professor P. B. Moon, a colleague at Imperial College. Their aim was to discover if a continuing chain reaction could be obtained by using uranium oxide and a moderator of either water or paraffin. The oxide had arrived in wooden boxes; the paraffin wax was provided by a firm of candle-makers. The neutron source was the radioactive gas radon and a small glass ampoule containing beryllium powder prepared in Imperial College in an open mortar in the days before the process was known to be hazardous. A boron trifluoride detector completed the simple set-up. 'Our main work,' Thomson later stated,

consisted of putting the uranium oxide into cast-iron spheres of different sizes. For the paraffin moderator we used scores of ordinary children's night-lights, arranging them in various patterns in the spheres ... By the outbreak of war we had established that an endless chain was *not* possible using uranium oxide and ordinary water, or paraffin, as the second constituent. It seemed likely, though not certain, that it could be done by using heavy water, but this was not available in Britain in large amounts and the military value of [creating a source of power] alone seemed too remote to justify further work in wartime. The second stage [the creation of a bomb] seemed nearly impossible, and if this conclusion now seems disgraceful blindness I can only plead that to the end of the war the most distinguished physicists in Germany thought the same.

This was discouraging, and if anything further were needed to inhibit British research into fission it came in a letter from Winston Churchill to Sir Kingsley Wood, Secretary of State for Air. Churchill had been briefed by Professor Lindemann and after discounting the possibility of using nuclear energy for many years went on to suggest that the Germans might soon be attempting a deception of the very kind that Thomson had proposed to Tizard. 'There are indications,' he said, 'that tales will be deliberately circulated when international tension becomes acute about the adaptation of this process to produce some terrible new secret explosive, capable of wiping out London. Attempts will no doubt be made by the Fifth Column to induce us by means of this threat to accept another surrender. For this reason it is imperative to state the true position.' He went on to point out that only a small percentage of uranium was fissionable and then to suggest what many physicists continued to believe: that even if a large enough mass of fissionable material could be assembled, then 'as soon as the energy develops it will explode with a mild detonation before any really violent effects can be produced'. After suggesting that the British would learn of any large-scale German work and that, in any case, the Germans had access only to the small uranium deposits in Germany, he concluded:

'For all these reasons the fear that this new discovery has provided the Nazis with some sinister, new, secret explosive with which to destroy their enemies is clearly without foundation. Dark hints will no doubt be dropped and terrifying whispers will be assiduously circulated, but it is to be hoped that nobody will be taken in by them.'

This view was of course founded on the Bohr–Wheeler discovery. A chain reaction, or so it appeared, required the separation of the uranium 235 isotope on a scale far beyond the bounds of practicality. Here, then, a point had been reached when search for the greatest power on earth might have been, if not abandoned, then at least put into cold storage until those involved had thought a little longer about the Pandora's box they might be opening. Indeed, with a regretful glance backwards Werner Heisenberg was later to say: 'In the summer of 1939 twelve people might still have been able, by coming to mutual agreement, to prevent the construction of atom bombs.'

With such a possibility in mind it is possible to speculate that such a duodecumvirate might have led on towards the ultimate world government of *The World Set Free* or Leo Szilard's equivalent of it. Just as the world was approaching war once again, Bohr and Wheeler had given it a chance to catch fission in its stride and restrain it until less desperate times. The catch was fumbled.

Certainly while there remained the outermost outside possibility that science fiction might be coming true, the more realistic view was that of C. P. Snow. Summing up the situation in the journal *Discovery* in September 1939 he noted that within a few months science might have produced an explosive a million times more violent than dynamite. 'It must be made, if it really is a physical possibility,' he went on. 'If it is not made in America this year, it may be next year in Germany. There is no ethical problem; if the invention is not prevented by physical laws, it will certainly be carried out somewhere in the world. ... Such an invention will never be kept secret.'

As to the future, Snow was gloomily correct. 'We have seen too much of human selfishness and frailty to pretend that men can be trusted with a new weapon of gigantic power,' he went on. 'Most scientists are by temperament fairly hopeful and simple-minded about political things: but in the last eight years that hope has been drained away. In our time, at least, life has been impoverished, and not enriched, by the invention of flight. We cannot delude ourselves that this new invention will be better used.'

PART TWO

Great Expectations

CHAPTER 4

A Problem for Committees

In the United States, as well as in Britain, the first steps which were to lead to the release of nuclear energy in the University of Chicago, and to the explosions over Hiroshima and Nagasaki, had been taken before September 1939. The most famous, at least in popular legend, was the letter to President Roosevelt signed by Albert Einstein but written by Leo Szilard in August 1939. Yet the significance of Einstein is rather different from that usually taken for granted. As stressed by Vannevar Bush, director of America's wartime Office of Scientific Research and Development, 'The show was going before that letter was even written.' Yet Einstein was to write not one letter but three, of which the third, recommending the creation of what was to be the Manhattan Project, was probably the most important. Moreover the famous first letter now appears to have been written when Einstein was extremely sceptical about the ultimate release of nuclear energy and had even stated to those trying to gain his help that he 'had not been aware of the possibility of a chain-reaction in uranium'.

As on more than one occasion in the nuclear story, events were as due to coincidence as to logical reasoning. In this case the move towards the nuclear age sprang from the fact that Einstein, an amateur violinist, had a few years earlier practised with Queen Elizabeth of the Belgians, by 1939 the Queen Mother following the death of King Albert.

This link was remembered by Leo Szilard when, in July, he began to worry, as Tizard had worried a few months earlier, about the Belgian Congo's stocks of uranium. Szilard had finished his current work at Columbia with Fermi and H. L. Anderson, and the three men had written up their conclusion in a paper on 'Neutron Production and Absorption in Uranium', to be published in *Physical Review* on 1

August. Fermi left for his holidays and Szilard was, as he has described it, left with nothing to do but think.

His new line of thought concerned the use of graphite as a moderator instead of water, an idea on which the French were already working. First calculations looked promising and when Eugene Wigner visited New York from Princeton he and Szilard began to discuss the fresh possibilities. At the same time they thought of what might happen if the Germans gained control of the Congo's uranium. 'So we began to think,' Szilard later wrote, 'through what channels we could approach the Belgian government and warn them against selling any uranium to Germany.'

Only Szilard, operator-extraordinary, would have taken the next step. 'It occurred to me then that Einstein knew the Queen of the Belgians,' he has written,

> and I suggested to Wigner that we visit Einstein, tell him about the situation, and ask him whether he might not write to the Queen. We knew that Einstein was somewhere on Long Island but we didn't know precisely where, so I phoned his Princeton office and I was told he was staying at Dr Moore's cabin at Peconic, Long Island. Wigner had a car and we drove out to Peconic and tried to find Dr Moore's cabin. We drove around for about half an hour. We asked a number of people, but no one knew where Dr Moore's cabin was. We were on the point of giving up and about to return to New York when I saw a boy of about seven or eight years of age standing at the curb. I leaned out of the window and I asked, 'Say, do you by any chance know where Professor Einstein lives?' The boy knew and he offered to take us there, though he had never heard of Dr Moore's cabin.

Szilard and Wigner explained their fears, Szilard going into some detail about the system which he believed might produce a chain reaction. Einstein followed all that was said but then, according to Szilard, said that he had never been aware of the possibility of a chain reaction in uranium. At first glance, the comment seems absurd. However, there is a good deal of circumstantial evidence to suggest that this was no more than the truth. The possibilities of a chain reaction had by this time been widely discussed, but they had been discussed in the context of a weapon or of power production, two subjects in which Einstein had virtually no interest. Physics, pure and unsullied by practical applications, was the matter to which he had devoted his life; he had for many years been more concerned with his search for a viable universal field theory than with nuclear physics; furthermore, he had a great admiration for Bohr, and it was Bohr who as early as February, proposing that only uranium 235 was fissionable, had appeared to rule out the possibility of any chain reaction.

However, there was always the one-in-a-million chance and Einstein was the last man to let the Germans benefit by that. He agreed to help. But during the three-cornered discussion the delicacy of the situation became evident. Here were two Hungarians and a German-born Swiss proposing that they should, in effect on behalf of America, advise a foreign government on the steps it should take. Even Szilard, a man not unduly hampered by the conventions, began to have second thoughts. It was then agreed that a letter should be drafted, that a copy should be sent to the State Department, and that the Department should be given two weeks to object if its Secretary of State did not want the letter sent.

But back in New York Szilard began to have third thoughts. 'I had, however, an uneasy feeling about the approach we had decided upon and I felt that I would need to talk to somebody who knew a little bit better how things are done,' he has written. He first consulted Dr Gustav Stolper, a German refugee who had been a pre-Hitler member of the German Reichstag, and Stolper in turn passed him on to Alexander Sachs, an economist who was well known to be an intimate of President Roosevelt. By a lucky chance Sachs had once heard Rutherford lecture, had as a layman been intrigued by the possibilities of releasing nuclear energy, and was happy to listen to Szilard. Indeed, he would be glad personally to deliver a letter to the President if Einstein would write one.

Once again Szilard visited Einstein at Peconic, this time with Edward Teller, another refugee Hungarian who a decade and more later was to become famous as 'the father of the H-bomb'. Accounts of this second meeting differ, and it is not certain whether or not Szilard had with him when he returned to New York a rough draft, dictated to Teller by Einstein. If so, however, it was merely a set of guidelines for it was Szilard who now wrote not one but two letters, one long and one short. Both were dated 2 August and both were sent to Einstein who signed and returned them with a note. Szilard, said Einstein, should send Sachs whichever he thought would be most effective.

But by this time Szilard appeared to be losing faith in Sachs. He suggested that Colonel Lindbergh might be the best man to approach the President, and Einstein dutifully responded with a letter of introduction. Nothing came of this, however, and Szilard passed on to Sachs, for transmission to the President as soon as possible, the longer of the two letters that Einstein had signed, together with a covering memorandum. The letter read as follows:

Sir,

Some recent work by E. Fermi and L. Szilard, which has been communicated to me in manuscript, leads me to expect that the element uranium may be turned into a new and important source of energy in the immediate future. Certain aspects of the situation seem to call for watchfulness and, if necessary, quick action on the part of the administration. I believe, therefore, that it is my duty to bring to your attention the following facts and recommendations.

In the course of the last four months it has been made probable – through the work of Joliot in France as well as Fermi and Szilard in America – that it may become possible to set up nuclear energy chain reactions in a large mass of uranium, by which vast amounts of power and large quantities of new radium-like elements would be generated. Now it appears almost certain that this could be achieved in the immediate future.

This new phenomenon would also lead to the construction of bombs, and it is conceivable – though much less certain – that extremely powerful bombs of a new type may thus be constructed. A single bomb of this type, carried by boat or exploded in a port, might very well destroy the whole port together with some of the surrounding territory. However, such bombs might very well prove to be too heavy for transportation by air.

The United States has only very poor ores of uranium in moderate quantities. There is some good ore in Canada and the former Czechoslovakia, while the most important source of uranium is the Belgian Congo.

In view of this situation you may think it desirable to have some permanent contact, maintained between the administration and the group of physicists working on chain reaction in America. One possible way of achieving this might be for you to entrust with this task a person who has your confidence and who could perhaps serve in an unofficial capacity. His task might comprise the following.

(a) To approach Government departments, keep them informed of further developments, and put forward recommendations for Government action, giving particular attention to the problem of securing a supply of uranium ore for the United States.

(b) To speed up the experimental work which is at present being carried on within the limits of the budgets of the university laboratories, by providing funds, if such funds be required, through his contacts with private persons who are willing to make contributions for this cause, and perhaps also by obtaining the co-operation of industrial laboratories which have the necessary equipment.

I understand that Germany has actually stopped the sale of uranium from the Czechoslovakian mines which she has taken over. That she should have taken such early action might perhaps be understood on the ground that the son of the German Under Secretary of State, von Weizsäcker, is attached to the Kaiser Wilhelm Institute of Berlin, where some of the American work on uranium is now being repeated.

Yours very truly,

A. Einstein

Some of the views in this letter appear to have been the opposite of those of the man who had signed it. Far from expecting that uranium might provide 'a new and important source of energy in the immediate future', Einstein himself later maintained: 'I did not, in fact, foresee that it would be released in my time. I believed only that release was theoretically possible.' In August 1939, this may well have been true. But during the next few months, as little progress was made with the Szilard–Einstein initiative, Einstein became increasingly worried that something should be done. It is only reasonable to assume that what he had initially regarded as an outside chance, steadily began to look less unlikely.

It was only on 11 October that Sachs saw Roosevelt. He took with him a letter outlining the dangers as he saw them and referring to the Szilard–Einstein letter which he planned to leave with the President. He had waited for the right moment and he stage-managed the occasion well. With him he took a volume which included a lecture by Aston on 'Forty Years of Atomic Theory', delivered three years previously. He now showed this to Roosevelt, 'with a view to highlighting that, as with other fruits of the tree of knowledge, there is an ambivalence to atomic power with poles of good and evil'. Aston's concluding sentences were these:

> There are those about us who say that such research should be stopped by law, alleging that man's destructive powers are already large enough. So, no doubt, the more elderly and ape-like of our prehistoric ancestors objected to the innovation of cooked food and pointed out the grave dangers attending the use of the newly discovered agency, fire. Personally, I think there is no doubt that subatomic energy is available all around us, and that one day man will release and control its almost infinite power. We cannot prevent him from doing so and can only hope that he will not use it exclusively in blowing up his next-door neighbour.

Roosevelt turned to Sachs and said: 'Alex, what you are after is to see that the Nazis don't blow us up.' Sachs said 'Precisely,' and

Roosevelt thereupon called in his secretary, General Edwin M. Watson, and told him: 'This requires action.' Sachs then went out with Watson, leaving his own letter, and the Szilard–Einstein letter. Interestingly enough, Sachs's opening testimony to the Senate's Special Committee on Atomic Energy in November 1945 made no mention of the President having read either letter before his vital decision was taken. Aston, speaking to an audience in Cambridge three years earlier, may have started more than he was to suspect.

Watson worked quickly and within a few hours a three-man committee had been created to decide what should be done next. Colonel Adamson represented the Army, Commander Hoover the Navy, while as the central figure there was Dr Lynam J. Briggs, Director of the Bureau of Standards.

At this point the search for nuclear energy might be thought to have started seriously. This was not so, and for a variety of reasons. Even the most optimistic of the physicists involved were quick to produce reservations, qualifications and doubts. Most of them, moreover, admitted that very considerable sums would have to be spent before the problem of ultimate success or failure could be solved. More than five years later, in March 1945, Lord Cherwell (as Professor Lindemann had now become) commented: 'No one can be sure that it will go off – there's many a slip 'twixt cup and lip. Think what fools the Americans will look if the bomb does not work, now that they have spent 1,600 million dollars.' And even if success could be guaranteed, how difficult was it for laymen to understand exactly what the effects would be. Churchill was to regard it merely as an 'improvement' on conventional explosives while Admiral Leahy, Roosevelt's Chief of Staff, remarked even after the successful test at Alamogordo: 'It sounds like a professor's dream to me,' adding later, 'I knew of no explosive that would develop the power claimed for the new bomb.' And Leahy, it should be remembered, was an explosives expert. For these reasons, and possibly for a residual feeling that refugees from Europe might be more worried than was genuinely necessary, it at first seemed likely that the release of nuclear energy was not being brought very much closer.

On 21 October, the Briggs committee, as it was soon called, held a meeting in Washington attended by Szilard, Wigner and Teller. Einstein was absent, and although the official history of the U.S. bomb implies that he had been invited, Szilard's own papers suggest this was not the case. This was possibly as well. Sachs's account of the meeting reflects a caution which was felt by quite a lot of physicists not only in the United States but also in Britain and France.

Many scientists were there who were not as concerned as [the] refugee scientists, for ... the latter, in addition to their interest in the advancement of science, were interested in the imperilled position of the United States and civilization. They were infused with a concern, in the Quaker sense of the word, of devoted interest and responsibility. Many of the other scientists said: 'This is very remote; we have got to wait and see; there are other lines of progress rather than the chain reaction that may be more attractive.' The discussion wandered all over attractive side issues.

Colonel Adamson commented at one point that it usually took two wars to develop a new weapon and, in any case, it was morale rather than arms that brought victory. It appeared that progress, if any, would be slow.

The Briggs committee reported on 1 November. It recommended that the position should be thoroughly investigated and that the work in the different universities should be co-ordinated. Furthermore it recommended that four tons of pure graphite should be obtained at once and that if early results with this looked promising then fifty tons of uranium oxide should be bought. But the recommendations were made with little enthusiasm and when Briggs was told, two weeks later, that the President wished to keep the report on file for reference, he probably saw which way the wind was blowing.

There was to be no further word from the White House for two and a half months and then it came not from Roosevelt but from General Watson who told Briggs that he intended to bring the report to the President's notice again. Briggs tentatively wondered whether Sachs had any ideas for giving further impetus to the work. Sachs thought once again of Einstein, arranged to visit Princeton and here found that Einstein, like himself, was 'dissatisfied with the scope and the pace of the work and its progress'. There were further meetings and it was then decided that Einstein should write a letter to Sachs which could be taken to Roosevelt – the first of two in which the lines of the Manhattan Project can be dimly discerned. In this Einstein recalled his letter of 2 August and then continued:

Since the outbreak of the war, interest in uranium has intensified in Germany. I have now learned that research there is being carried out in great secrecy and that it has been extended to another of the Kaiser Wilhelm Institutes, the

Institute of Physics. The latter has been taken over by the Government and a group of physicists, under the leadership of C. F. von Weizsäcker, who is now working there on uranium in collaboration with the Institute of Chemistry. The former director was sent away on a leave of absence, apparently for the duration of the war.

Should you think it advisable to relay this information to the President, please consider yourself free to do so. Will you be kind enough to let me know if you are taking any action in this direction?

Dr Szilard has shown me the manuscript which he is sending to the *Physic Review* [sic] in which he describes in detail a method of setting up a chain reaction in uranium. The papers will appear in print unless they are held up, and the question arises whether something ought to be done to withold publication.

I have discussed with Professor Wigner of Princeton University the situation in the light of the information available. Dr Szilard will let you have a memorandum informing you of the progress made since October last year so that you will be able to take such action as you think in the circumstances advisable. You will see that the line he has pursued is different and apparently more promising than the line pursued by M. Joliot in France, about whose work you may have seen reports in the papers.

This letter – which has all the signs of having been actually written by Einstein in contrast to that of 2 August – politely pointed a pistol at the President's head: either 'the uranium question' was important, in which case the work on it should be speeded up; or it was not important, in which case Szilard would be publishing information which might be of use to the Germans.

However, no one in the White House was taking the uranium question very seriously and the Briggs committee recommended that the whole matter should be held in abeyance until its members had received a report from Columbia of the work that Szilard and others were doing there. Only under further pressure from Sachs did Roosevelt decide to call a further meeting to consider the work once again. This time he noted in his letter to Sachs that General Watson would fix 'a time convenient to you and Dr Einstein', taking it for granted that Einstein himself would be coming along. Einstein had other ideas. Szilard saved the situation by telling him that 'we shall prepare a polite letter of regret in English which you can use if you think it advisable'.

Einstein did think it advisable. And after explaining that he was unable to attend, and referring to the work of Wigner and Szilard, he went on:

I am convinced as to the wisdom and urgency of creating the conditions under which that and related work can be carried out with greater speed and on a larger scale than hitherto.

I was interested in a suggestion made by Dr Sachs that the Special Advisory Committee supply names of persons to serve as a board of trustees for a non-profit organization which, with the approval of the governmental committee, could secure from governmental or private sources, or both, the necessary funds for carrying out the work.

Given such a framework and the necessary funds, it (the large-scale experiments and exploration of practical applications) could be carried out much faster than through a loose co-operation of university laboratories and Government departments.

This letter was dated 25 April 1940, a fortnight after the 'phoney war' had ended with the German invasion of Denmark and Norway, and while bitter fighting was continuing in Norway itself. It was to lead to a tightening of the rather casual voluntary censorship of nuclear information that many scientists had begun to operate in September 1939 and it opened the way to a reorganization of the Briggs committee two months later when it was brought under the newly created National Defense Research Committee.

The eight months since the outbreak of the European war had therefore witnessed an increase in American interest in nuclear weapons. However, this interest was limited by the fact that the country was not yet at war and that even to enthusiasts like Szilard and Fermi the use of nuclear energy in any form was still something that should be investigated rather than considered as a likely prospect of the not-too-distant future. Much the same view was held elsewhere, although contrasting military, strategic, economic and scientific circumstances gave a different emphasis to development elsewhere in the world. In Italy, where considerable theoretical advances had been made under the leadership of Fermi, the prospects for a nuclear weapon never emerged. In Japan, where the standard of nuclear research was high, the Army was sponsoring investigations into a fission weapon before the end of 1940. Factors which held it back included, after Pearl Harbor, belief that the war would end before a nuclear weapon could be made and, possibly most dominant factor of all, lack of the industrial muscle which it was seen from the start would be an essential of success.

In Russia, purely theoretical research was pushed on under a surprisingly bright searchlight, newspapers and magazines publishing information that would have been censored in Britain, France or Germany, and frowned upon in the United States. This was very different from what had happened between the wars. Before August 1939, when the apostle of the anti-Christ in the shape of Stalin linked arms with the bloody murderer of the proletariat in the shape of Hitler, information from Mother Russia had tended to be uncertain and spotty. Combined with the Russian state's coralling of scientific advance to political ends – highlighted by the rise of Lysenko – it was to breed a distrust of Soviet scientific ability. This was understandable if misplaced: the damage suffered by Russia during the First World War had been compounded by that of the Civil War and of the purges that followed. Yet as early as 1922 V. I. Vernadskii, founder of the Radium Institute in Leningrad, had said: 'We are approaching a tremendous revolution in human life with which nothing hitherto experienced can be compared. It will not be long before man will receive atomic energy for his disposal, a source of energy which will make it possible for him to build his life as he pleases.' This may, of course, have been a forecast more optimistic than Vernadskii genuinely felt was justified. Nevertheless, the Russians had followed closely the nuclear research in the Cavendish, in Niels Bohr's Copenhagen laboratory, and in the United States.

The country was still poor and every effort was being concentrated on the first Five-Year Plan. Yet even before the impetus given by the *annus mirabilis* of 1932, A. F. Ioffe had felt it essential to begin work on the atom nucleus. 'We were worried,' he later wrote of the situation in May 1930, 'because it was the middle of the year, when appropriations for our work had already been made, and the new researches we outlined required an additional expenditure of several hundred thousand rubles. I went to Sergei Ordzhonikidze, who was chairman of the Supreme Council of National Economy, put the matter before him, and in literally ten minutes left his office with an order signed by him to assign the sum I had requested to the Institute. Once started, we have continued work on the atomic nucleus for fifteen years as an essential part of our plan.'

This was written with the hindsight of 1945 – although before the Anglo-U.S. success with nuclear weapons had been revealed – yet there can be no doubt of Russian interest in the release of nuclear energy in the 1930s. Interest continued to grow, Russian pride in her own achievements more than counterbalancing any need for secrecy. Thus the separation of the uranium 235 isotope and the production of heavy water were both openly discussed in a number of papers when the Academy of Sciences held the First All-Union Conference on Isotopes in the spring of 1940. In March 1940, Ioffe publicly stated that the atomic nucleus was one of three major subjects of contemporary research. Two months later the Leningrad cyclotron was described in detail, public lectures on fission continued to be reported, and in a New Year's Eve review under the title 'Uranium 235', *Izvestia* stated:

> Physics stands at the threshold of discoveries of boundless significance. We are confronted not only with the fact that mankind will acquire a new source of energy surpassing by a million times everything that has been hitherto known. Nor is our perspective merely that we shall have a 'fuel' which will substitute for our depleting supplies of coal and oil and thus rescue industry from a fuel famine. The central fact is that human might is entering a new era ...
>
> Man will be able to acquire any quantity of energy he pleases and apply it to any end he chooses.

Yet it is clear that whatever the scientists themselves might think, the Soviet Government – like the U.S. and British Governments – tended to regard this as essentially a long-term affair. It is true that a Uranium Commission was set up in June 1940 by the Academy of Sciences, but its proposals were modest and an appeal by the physicists Kurchatov and Khariton for a major attack on the problem of creating a chain reaction aroused little interest.

When the Fifth Nuclear Conference was held in Moscow five months later, Kurchatov's renewed appeal for concentrated effort produced from one of the senior Russian physicists a reaction which almost startlingly echoes much of the previous year's reaction in Washington and London. He was reported as saying that

> it was too early to speak of the industrial generation of atomic energy. Some young physicists, in particular the students of the lecturer, were so captivated by uranium projects that they forgot about the needs of the present day. The lecturer had shown that if the isotopes of uranium were separated (and that was still unrealizable in practice) or if tens of tons of heavy water were accumulated instead of the existing tens of kilograms, then a chain reaction would become possible. But 'possible' and 'practicable' were not concepts of equal value. The practicability of a nuclear chain reaction had not yet become established even in the laboratory. Unfortunately, it had to be acknowledged that uranium energy was a distant prospect, still a beautiful dream.

It is possible that the Soviet Government, and privileged Establishment scientists, were already sceptical about their German allies' intentions, and thus anxious that any national resources should be husbanded for use in an uncertain future. It is more likely that once the uncertainties of nuclear research were seriously considered the odds of success appeared so small that research was almost abandoned for the time being. Whatever the factors in 1940, all prospects of serious nuclear research were to be temporarily halted the next year as in the quiet of a warm June night thousands of German tanks with muffled treads moved through the darkness across the Russian frontier.

What, meanwhile, had been happening in the Third Reich, Hitler's Germany where Hahn and Strassmann had kicked the pebble that was to start the nuclear avalanche? The answer is that more had been going on than is generally realized even today. The reasons for this comparative ignorance are complex. The most obvious explanation is that the Germans failed. Despite the head start given them by Hahn, and the quick initial response which made them, by the outbreak of war in September 1939, the only country with an office devoted exclusively to the military applications of nuclear fission, they never reached even the first practical stages of making a nuclear weapon.

It was therefore easy for the Allies to overlook or to undervalue the very real theoretical progress made in Germany, and made in the face of handicaps considerably greater than those which hampered Britain, France and the United States. As for the Germans themselves, many were only too eager after the war to emphasize that they had failed. After all, decent German physicists had from the first been unwilling to unleash on the world indiscriminate weapons such as atomic bombs. If they could not entirely hold up progress, then they could at least play their parts in an honourable go-slow, prevaricating where they could not openly oppose, and ensuring that, time after time, unexpected stumbling blocks barred the way ahead.

At least, that was the story. It does appear that among the large number of physicists involved there were some who, while able to stomach the Nazi programme of aggression and genocide, did feel that Hitler should not have the benefit of nuclear weapons. Professor Wirtz, judged by the secretly recorded reactions of German physicists during internment in 1945, exclaimed: 'I'm glad we didn't have it.' Weizsäcker was more positive: 'I believe the reason we didn't do it was because all the physicists didn't want to do it on principle. If we had all wanted Germany to win the war we could have succeeded.' But Hahn was not so sure. 'I don't believe that,' he declared, 'but I am thankful we didn't succeed.' And quite as trenchant was Dr Bagge. 'I think it absurd for von Weizsäcker to say he did not want the thing to succeed,' he said; 'that may be so in his case, but not for all of us.'

Certainly there was from mid-1939 onwards a great debate in Germany about the development of nuclear weapons, but how much ethics, politics or military expediency was involved is even now not completely clear. However, it seems likely that many if not most German physicists entered the race for reasons very similar to those of scientists elsewhere: in the hope that their country would be able to command a unique weapon. However, this general feeling had none of the desperation which powered Britain's nuclear investigations during the first years of the war. The success of the German campaigns in Poland and France suggested there was no need for a new and untried weapon which could, in any case, be developed only at immense cost in men and materials. Later in the war the main German need was for an arm which could be developed quickly. Heisenberg said later, 'We would not have had the moral courage to recommend to the government in the spring of 1942 that they should employ 120,000 people on such a project.' After 1943, as the great Allied air onslaught gathered weight, any chance of successfully operating the huge plants needed for making a nuclear weapon was effectively ruled out.

Yet in spite of these handicaps, which prevented the German effort from achieving practical results, the effort did attract recruits, and for one unusual reason. 'We were delighted [with the nuclear project],' Professor Josef Mattauch has said, 'as it gave us a chance of protecting our young men from call-up, in order to continue scientific research in the manner to which they were accustomed.' And von Weizsäcker has been quoted as admitting that he accepted a nuclear research contract from the German War Office because his other commitments would not have kept him out of the forces.

The first German moves had come only a few days after the publication in *Nature* on 22 April 1939 of the French team's paper reporting that fission of uranium created a surplus of neutrons. In Göttingen, Professor Wilhelm Hanle read a paper on the employment of uranium fission in a reactor which would produce energy, a paper followed by a letter to the Ministry of Education which controlled the universities. The Ministry acted without delay and a secret meeting under the chairmanship of Professor Abraham Esau, head of the Reich Bureau of Standards, was called for 29 April. Here, only a week after the French paper, it was decided to secure all available stocks of uranium, to co-opt the necessary physicists, and to start work on the 'uranium project'.

By this time, however, an independent move had been made by Professor Paul Harteck of Hamburg and his colleague Dr Wilhelm Groth. They wrote not to the Ministry of Education but to the War Office. 'We take the liberty of calling to your attention,' they said, 'the newest development in nuclear physics, which, in our opinion, will probably make it possible to produce an explosive many orders of magnitude more powerful than the conventional ones.' The German War Office reacted as swiftly as the Ministry of Education. Within a few weeks a uranium laboratory had been built at Gottow, the Army Ordnance Department's proving ground outside Berlin, and a second research team was at work.

Like the Russians, the Germans at first took a relatively casual attitude to security, and in the summer of 1939, *Die Naturwissenschaften* published an article by Dr Siegfried Flugge headed 'Can the Energy Contained in the Atomic Nucleus be Exploited on a Technical Scale?' and specifically drew attention to the prospects for a bomb. Flugge gave an impressive series of figures and concluded by deciding 'that one cubic metre of uranium oxide would suffice to lift a cubic kilometre of water (total weight: 1 million million kilograms) twenty-seven kilometres into the air!'

While the Germans were to make considerable theoretical advances during the following few years, they failed to realize that the key to the bomb lay in the use of fast rather than slow neutrons. Thus in the German view the first task was to build a slow-neutron reactor; once it had been done, they believed, it would be necessary to build a smaller and more active reactor that would in effect be a bomb. Combined first with the belief that Germany really had no need of nuclear weapons in order to win the war, then with the difficulties imposed by the Anglo-U.S. air offensive, the Germans failed to exploit the theoretical lead which had been provided by Hahn and Strassmann.

Little of this was known to the Allies who consistently overrated German ability in nuclear physics. This very largely accounted for the temporary worry which followed Hitler's speech in Danzig two weeks after the outbreak of war. In this he harangued the Allies on the dangers of continuing the war and according to the official translation stated that 'the moment may come when we use a weapon which is not yet known and with which we could not ourselves be attacked. Let us hope that we will not be forced to use this means. It is to be hoped that no one will then complain in the name of humanity.' More than one worried man took this as a sign that even if the enemy did not already have a nuclear weapon to hand he hoped to have one in the not-too-distant future. Even the suggestion that Hitler had not said 'a weapon with which we could not ourselves be attacked' but 'a weapon against which there is no defence' failed to bring reassurance.

It was only when R. V. Jones, a physicist attached to Air Ministry Intelligence, was seconded to discover what the weapon really was, that daylight began to appear. His first move was to get a record of the speech from the B.B.C.'s listening post outside Reading and to have it played back to a German scholar. 'He complained of Hitler's bad grammar,' says Jones, 'but came to the conclusion that the Führer had not in fact spoken of a "secret weapon" at all.' This was found to be so. The alarm had been caused by an error in translation. What Hitler had meant by 'Waffe' – 'Eine Waffe gegen die es keine Verteidigung gibt' – had merely been the Luftwaffe, already regrouping after its successful campaign in Poland.

The Germans were by now deciding that heavy water would be the best material for a moderator, and were preparing to secure the necessary supplies. But the French were also turning in the same direction. At first the Collège de France team had concentrated on the prospects for nuclear power. It is true that when M. Sengier, fresh from his interview with Tizard in London, had met the French team in Paris they had discussed nuclear weapons. 'They proposed,' he has said, 'to effect the fission of a uranium bomb in the Sahara. I accepted in principle, and agreed to furnish the raw material and part of the financing.' Yet Halban remembered the meeting as mainly concerned with a reactor which would produce power, and it is clear that production of nuclear power was the first priority.

It was some while before the French experiments showed a glimmer of hope. Blocks of uranium oxide three inches wide and fifteen inches long were built into a sphere which was then drenched with water. The result, it was discovered with mounting excitement, was that fissions in the middle of the sphere did begin to spread through it. But there was no need for Joliot-Curie and his colleagues to use any of the methods for controlling the reaction that they had ready to hand. The nuclear fire died down of its own accord.

Despite this, the French had shown what had not yet been shown anywhere else in the world: that a nuclear fire could at least be lit, even though it was not self-sustaining. Common sense now suggested that if only natural uranium and the right sort of moderator could be arranged together in the right way, then it would be possible to keep the fire alight. To ensure that the French team's priority would not be overlooked even if scientific publication became impossible, Joliot-Curie deposited a sealed document at the Académie des Sciences on 30 October 1939, signed by himself and by Halban and Kowarski. It dealt with the possibility 'of producing, in a medium containing uranium, unlimited chain reaction'.

At this crucial point the French authorities tried to side-track the team on to what was considered more vital work. Joliot-Curie was not easily side-tracked. He quickly sought an interview with the Minister for Armaments, thus bringing on to the scene a personality without whom the nuclear story would have been very different.

The Minister, Raoul Dautry, was a man of conservative beliefs. Even in the first months of 1940 he had little faith in the aeroplane, and at first glance would have been the last man to support Joliot-Curie. Yet years earlier – possibly during the period of wild speculation which followed Cockcroft and Walton's experiment – M. Dautry had read in a popular paper that if it were possible to disintegrate an ordinary kitchen table, the energy locked inside it would be enough to blow up the world. He was, unexpectedly, fascinated by Joliot-Curie's explanation of the work which had been going on. Joliot-Curie was too honest a man to claim that the prospects for the bomb were as bright as those for 'the boiler'. But he pointed out that nuclear power did not have only civil uses: it could, quite conceivably, be utilized in a submarine.

Dautry promised all possible help. Graphite was the material in short supply at this point, but the Minister quickly discovered an unexpected source in Grenoble and before long Halban and other members of the Collège de France team were working on a stack of pure graphite five feet square and ten feet high. Their results were inconclusive and they now turned to the use of heavy water – that is, water which has been treated so that more of its hydrogen than usual is heavy hydrogen, or deuterium.

Early in 1940 heavy water was being produced on an industrial scale only at Rjukan in central Norway. Dautry was easily persuaded that the Norwegian stock should be cornered, in much the same way that Tizard had tried to secure the Belgian Congo's uranium ore. While preparations were being made, the French received what appeared to be alarming news. Until the late summer of 1939 the Germans had shown little interest in the Rjukan plant. Yet now, French intelligence learned, they were not only trying to buy the existing stocks in Norway but were offering the Norwegians a contract for large and regular supplies. Were the Germans, like the French, mainly looking for nuclear power? Or had they, perhaps, plans for a super-bomb? Whatever the answer, it became even more important to keep the heavy water out of their hands.

This was to bring a cloak-and-dagger touch into the story of nuclear fission. So far, it had been largely a record of one group of physicists leap-frogging another in laboratory research or in back-of-an-envelope calculations. Everywhere blind chance appeared to be an important determining factor, and this was still to be the case with the entrance on to the scene of a young Frenchman, Lt Jacques Allier.

Allier was a successful businessman who had been called up as an officer in the French Deuxième Bureau, one of the country's intelligence services. He was able to explain to Dautry what the French knew of German nuclear research. But he also had another more important qualification; he was a member of the Banque de Paris et des Pays-Bas. The significance of this was that the bank held a large financial interest in the Norsk Hydro Company which owned the Rjukan heavy-water plant.

Early in March, Allier left Paris for Oslo, travelling via Stockholm where he gained the assistance of three other members of the Deuxième Bureau, Captain Muller, Lieutenant Mosse and a M. Knall-Demars. Arrived in Oslo, he quickly arranged a meeting with Dr Axel Aubert, General Manager of the Norsk Hydro who had, he learned, been suspicious of German replies when asked why they wanted the heavy water.

Allier took a different line, as he later reported to Dautry. 'Dr Aubert was very glad that I dealt with the subject quite openly,' he wrote,

> and he showed the greatest sympathy with every point which I revealed to him. Himself passionately interested in scientific research (he had given his name to a fund put at the disposal of Oslo University), he was eager to know to what use France would put the heavy water produced by his company. And at the end of our meeting he declared himself quite ready to negotiate with France in view of the German 'hedging'. This first meeting was extremely cordial and assured. I well knew the sense of responsibility, the great integrity, of this man, as well as his quite legitimate wish to ensure that his dignity and freedom in the matter were respected. Thus the only attitude I was able to adopt, and the one best likely to succeed, was to appeal to his better feelings;

and, by dealing with the matter on this level, and by appealing to his friendship for France, I gained his assent to more things than we had hoped.

In fact the agreement which the two men signed a few days later ensured not only that the French were to have free use for the duration of the war of the 185 kg of heavy water then at Rjukan but, possibly more important, a priority claim on the Rjukan plant's entire output.

Later in the war, after the German occupation of Norway, it was realized that the Rjukan plant would have to be put out of operation, since Intelligence reports revealed that production of heavy water was increasing. A disastrous failure in an attack on the plant in November 1942 was followed in February 1943 by a successful raid which disrupted the works for many months.

In the spring of 1940 Allier and his colleagues faced two problems despite the Norwegian cooperation. The first was to have the heavy water sealed up in the right kind of containers. The second was to get them to France without arousing German suspicions.

Before leaving Paris Allier had been warned that precautions had to be taken about the transport of the heavy water. Special cans might have to be made, and if so it was essential that neither cadmium nor boron should be used in the welding process since even minute quantities of either would so contaminate the heavy water as to make it unusable. The difficulty here was that any large firm in Oslo, ordered to weld a number of containers to such very strict specifications, would be given a good clue as to why they were needed. The job was therefore farmed out to a one-man welding yard on the outskirts of the city which produced, it was later discovered, a result that could not have been bettered. The containers were sent to Rjukan, filled, then returned to Oslo where they were held for a day or two in a house owned by the French Embassy.

At first it was planned to take the cans to France by submarine. This appeared impracticable and early in March the four Frenchmen and half of the heavy water, contained in thirteen canisters, arrived at Fornebu airfield on the outskirts of Oslo. Allier and Mosse then prepared to leave on the daily plane for Amsterdam. Shortly before take-off the canisters were loaded on to a second plane on which the two men had booked seats under assumed names. They boarded this and soon were in the air, heading for Edinburgh.

Allier explained to the pilot that he and his companion were French officers in civilian clothes, and when an unidentified plane was reported hovering on their tail they decided to cross the North Sea in cloud cover. There was so much of this that the plane was diverted north from Edinburgh to Montrose. Here they calmly loaded their thirteen canisters on to a local train for Edinburgh, spent the night in the Scottish capital and were joined the following morning by their two colleagues with the other thirteen canisters.

From here the party and their twenty-six cans travelled almost the length of the British Isles, going first to the French Military Mission in London, then continuing the following day on to the Channel coast. By 16 March, Allier and his colleagues were back in Paris, and the world's entire supply of heavy water was safely lodged in the cellars of the Collège de France until a special air raid shelter, designed to withstand a direct hit from a 1,000 lb bomb, had been completed to protect it.

By this time there had been decisive developments in Britain, following a period when disillusion with the prospect of nuclear weapons, important in August 1939, had been quietly deepened. First there had come a paper from Rudolf Peierls, Professor of Mathematical Physics in the University of Birmingham. Peierls, one of the very few men who both made a crucial contribution to the first nuclear weapons and retained a balanced view of their possibilities, good and bad, was a Berliner who had come to Cambridge in 1933 on a Rockefeller Scholarship, occupied a series of increasingly prestigious posts, and become a naturalized Englishman shortly before the outbreak of war.

Approaching the problem of a possible chain reaction in a rather different manner from that followed by Perrin in Paris, Peierls came to much the same conclusion. The size of a potential bomb, he later stated, 'was of the order of tons. It therefore appeared to me that my paper ["Critical Conditions in Neutron Multiplication"] had no relevance to a nuclear weapon. ... There was of course no chance of getting such a thing into an aeroplane, and the paper appeared to have no practical significance.'

To this discouragement there was soon added another. Otto Frisch, who had been visiting Britain when war broke out, had decided to stay in Birmingham. Early in 1940 he emphasized in a note printed in the Chemical Society's *Annual Reports on the Progress of Chemistry* that even if uranium could be enriched with ten times its natural percentage of uranium 235, a chain reaction would not be fast enough to cause an explosion. 'Fortunately,' he wrote, 'our progressing knowledge of the fission process has tended to dissipate these fears [of nuclear weapons] and there are now a number of strong arguments to the effect that the construction of such a super-bomb would be, if not impossible, then at least prohibitively expensive and that furthermore the bomb would not be so effective as was thought at first.'

These views were to be changed in a chance and roundabout way that started when the Cabinet ruminated on the awful possibilities beneath Hitler's apparent threat of secret weapons. Lord Hankey, forewarned by Rutherford a few years previously and now Minister without Portfolio, was asked to look into the matter. Hankey asked Dr Appleton, head of the Department of Scientific and Industrial Research, to do the same. Appleton, forgetting that Tizard had earlier supplied George Thomson with uranium oxide, turned to Professor Chadwick, the discoverer of the neutron and now the head of the Physics Department at Liverpool University.

On the face of it Chadwick, who had just put into operation Britain's first cyclotron, was as cautious as his contemporaries. 'I think one can say,' he reported, 'that this explosion is almost certain to occur if one had enough uranium. The estimates of the amount vary from about one ton to thirty or forty according to the data adopted in the calculations. I am sorry,' he went on, 'that I can give no definite answer to this question for it is a very interesting one indeed. The amount of energy which might be released is one of the order of the energy of the well-known Siberian meteor.' This verdict, indecisive as it might seem, was enough to bring from Hankey the comment: 'I gather that we may sleep fairly comfortably in our beds.'

However, Chadwick's cyclotron could possibly provide information on one aspect of uranium about which very little was known. This was the cross-section of the nucleus, a concept which physicists used to describe the chances of a neutron splitting a nucleus when it hit the target, or of merely bouncing off. 'For example,' Peierls has said, 'if I throw a ball at a glass window one square foot in area, there may be one chance in ten that the window will break, and nine chances in ten that the ball will just bounce. In the physicists' language, this particular window, for a ball thrown in this particular way, has a "disintegration cross-section" of 1/10 square foot and an "elastic cross-section" of 9/10 square foot.' It was the great difference in estimates of nuclear cross-section which largely accounted for the wide gap between Chadwick's one ton and thirty to forty tons. He and his assistants therefore began to concentrate, throughout the early weeks of 1940, on getting more accurate cross-sectional figures for the various isotopes of uranium when these were subjected to either slow or to fast neutrons.

While Chadwick was trying to obtain these basic data on fission, a lone attempt to produce a chain reaction – very comparable to Thomson's – was being independently made in Ottawa, a forerunner of Canada's later significant place in the story of nuclear energy.

During the furore which had swept through the physicists' world in 1939 Dr George C. Laurence of the Canadian National Research Council had speculated on the fact that Canada was one of the two world sources of uranium. The country lacked the industrial or financial capacity for embarking on the Herculean task of isotope separation. Why not, then, investigate the possibilities of natural uranium?

Early in 1940 Laurence therefore began experimenting with what has been somewhat optimistically called 'the building of Canada's first experimental pile'. This was a bin lined with paraffin wax and containing a lattice arrangement of powdered uranium oxide and carbon. A mixture of metallic beryllium and radium was inserted to initiate fissioning. Although a good deal of interesting data was collected, the experiment produced no sign of a chain reaction. Nevertheless it was continued throughout the year and before the end of 1940 there came help from an unexpected quarter. Lord Melchett, deputy chairman of Imperial Chemical Industries, already intensely interested in the possibilities of post-war nuclear power, provided a gift of $5,000, made through Canadian Industries Ltd, a company with which I.C.I. had close connections.

While Laurence's experiments were in their early stages, the entire prospect of nuclear weapons was transformed. The change came, almost literally overnight, in Birmingham, where Frisch and Peierls were together mulling over the figures which each had produced a few months previously. What would happen, they speculated, if one were to consider not natural uranium enriched by a greater-than-natural percentage of uranium 235 but a lump of pure uranium 235?

'Any competent nuclear physicist,' Peierls has said, 'would have come out with very similar answers to ours if he had been asked: "What is the likely fission cross-section of pure uranium 235? What critical size for separated uranium 235 follows from this? What will be the explosive power of such a mass? How much industrial effort would be needed to do the separation? And would the military value be worth while?" The only unusual thing that Frisch and I did at this point was to ask questions.'

Here was of course a classic example of the need to ask the right question. But no one had yet isolated more than minute quantities of any isotope. It was taken for granted that a uranium bomb would be based on material weighed in tons, and until now no one had thought of finding out if this was correct.

'We sat down to work out the answer to these questions on the back of the proverbial envelope,' Peierls has said. 'There were two things that had to be done. First we had to guess the nuclear cross-section of this particular isotope of uranium. Then, when we had done this, we had, as it were, to fit the dimensions of this cross-section into the formula which I had already published in the Cambridge *Proceedings*.'

Both Frisch and Peierls arrived at a figure for the cross-section which was too large, a miscalculation which made their problem simpler than it would otherwise have been. Yet this was a small matter compared with the staggering fact which emerged from the calculations. For it became clear that the critical mass of uranium 235 was not weighed in tons, as had been taken for granted by all those who had even considered the matter. It was not even weighed in hundredweights. The answer was measured in pounds. 'In fact,' Peierls has said, 'our first calculation gave a critical mass of less than one pound.'

One other obvious theoretical problem remained about a nuclear weapon made of uranium 235. 'It was the problem,' Peierls has said, 'of whether the bomb would explode or whether it would swell up and go off with a pop instead of a bang. In fact, it was really a question of which of two factors would win a race – the chain reaction which would have been started when the two pieces of uranium came together; or the swelling-up of the critical mass itself.' For if an appreciable fraction of the atoms in the critical mass underwent fission within a very short time, two things would happen; the temperature of the uranium would rise to many millions of degrees and the pressure within it would rise to many million times that of the atmosphere. The mass of metal would expand with great rapidity; and, as it did so and as its density decreased, so would the chances of the newly released neutrons creating fresh fissions decrease in like manner. But if the chain reaction did win the race, if the vast multiplication of neutrons and the resultant liberation of energy did take place so quickly that the critical mass was not, as it were, expanded into explosive uselessness, what quality of explosion would be created?

This was a purely mathematical problem and it was Peierls who solved it. 'I worked out the results of what such a nuclear explosion would be,' he has said. 'Both Frisch and I were staggered by them.'

The technical problem of separating a pound or so of an isotope which had previously been separated in little more than sub-microscopic quantities was still of course immense. But it was now a problem of a different order of magnitude. Before, it could have been likened to filling the Empire State Building or the Royal Albert Hall with individual grains of sand, one by one. Now the task was that of filling a barrel.

A vast amount of work, theoretical as well as practical, still had to be carried out. Immense problems still obscured the path ahead. And not for another five years, until the test bomb erupted in the New Mexico desert, could the scientists be sure that their figures had been correct. Yet despite this the calculations made in Birmingham on a February day in 1940 marked the real watershed in man's progress towards the release of nuclear energy. Before the Frisch–Peierls calculations the possibilities of success had been so slim that no rational man believed it sensible to start assessing the problem in terms of the men, money and materials required to make a nuclear weapon; after the calculations, no rational and informed man involved in a life-and-death struggle with a ruthless enemy could afford not to make such an assessment.

The main problem that Frisch and Peierls now faced was that of knowing what to do next. Neither was British by birth and Peierls had only recently been naturalized. Neither had been involved, even peripherally, in the mass of defence work, largely concerned with radar, which was by this time occupying most physicists in Britain. They wanted to help but were unaware of how best to do it.

Luckily Mark Oliphant, Rutherford's colleague of the Cavendish days, was Professor of Physics in the University of Birmingham. He was developing the cavity magnetron which was to form the heart of airborne radar, and although neither Frisch nor Peierls knew of the closely guarded secret, they knew Oliphant's reputation as a physicist. They approached him, showed him their figures, and asked for his advice. 'Write to Tizard,' he advised.

The upshot was that there shortly arrived on Sir Henry Tizard's desk what might be called if not the birth certificate of nuclear fussion then at least the marriage certificate that encouraged the conception.

There were two parts to the document, one headed 'On the Construction of a "Super-bomb": Based on a Nuclear Chain Reaction in Uranium', the other a 'Memorandum on the Properties of a Radioactive "Super-bomb" '. The first gave the figures and arguments supporting the authors' view that a bomb in which two small hemispheres of uranium 235 were suddenly brought together would liberate as much energy as several thousand tons of dynamite. In view of later suggestions that the dangers of radiation were not considered until after Hiroshima, it is significant that the Frisch–Peierls memorandum outlined them in some detail. 'Most of it will probably be blown into the air and carried away by the wind,' this said.

> This cloud of radioactive material will kill everybody within a strip estimated to be several miles long. If it rained the danger would be even worse because active material would be carried down to the ground and stick to it, and persons entering the contaminated area would be subjected to dangerous radiations even after days. If one per cent of the active material sticks to the debris in the vicinity of the explosion and if the debris is spread over an area of, say, a square mile, any person entering this area would be in serious danger, even several days after the explosion.

The question of lethal radiations was also emphasized in the second half of the document, and in a manner which throws significant light on the morals of mass-destruction then held. 'Owing to the spreading of radioactive substances with the wind,' this said, 'the bomb could probably not be used without killing large numbers of civilians, and this may make it unsuitable as a weapon for use by this country. (Use as a depth charge near a naval base suggests itself, but even there it is likely that it would cause great loss of civilian life by flooding and by the radioactive radiations.)'

The authors went on to admit that there was some uncertainty about the critical size, which if appreciably larger than the estimates would certainly increase difficulties. But that size could be found as soon as a relatively small amount of uranium 235 had been separated, and it was suggested that this should be done without delay.

Tizard was still sceptical. But this was not a subject which could be ignored, and he replied encouragingly to Oliphant. 'What I should like,' he said, 'would be to have quite a small committee to sit soon to advise what ought to be done, who should do it, and where it should be done, and I suggest that you, Thomson, and say Blackett, would form a sufficient nucleus for such a committee, and if you like to bring someone else in please make a suggestion.'

CHAPTER 5

Progress with a Plan

It is ironic that the memorandum which was to lead from Birmingham to preparatory work throughout Britain and then by a long chain of events to the destruction of Hiroshima and Nagasaki should have arrived on the desk of Sir Henry Tizard. The scientist who had done more than any one man to ensure that Britain's radar chain was operating in time, Tizard was among the most sceptical of those who in 1940 considered the possibility of nuclear weapons. 'You must realize,' he said in the spring of the year to an Assistant Chief of Air Staff, Sir Philip Joubert de la Ferté, 'that any project which is going to have some influence on the course of the war must have been examined, tested and put into execution already. There are about five of these, and the effort needed to apply them against the enemy will absorb all the available men and manufacturing power.' The judgement was sound enough. 'The war' in 1940 was the war against Germany alone, and it was to end two months before the world's first nuclear weapon could even be tested.

Tizard had since 1935 chaired his own scientific committee, the Committee for the Scientific Survey of Air Defence which in the first months of the war had been merged with a similar group, considering air offence, to form the Committee for the Scientific Survey of Air Warfare. It was this group – meeting regularly in Balliol College, Oxford, since the outbreak of war and christened the Balliol Beagles on the grounds that it was always letting loose fresh hares before it had caught the first lot – which was now asked by Tizard to consider what should be done about this new, and rather outlandish possibility of using the power locked within the atom. The Balliol Beagles duly responded and set up a small subcommittee under Thomson with a brief 'to examine the whole problem, to co-ordinate work in progress and to report, as soon as possible, whether the possibilities of producing atomic bombs during this war, and their military effect, were sufficient to justify the necessary diversion of effort for the purpose.'

Thus there were by the spring of 1940 four groups at work: the Germans, whose civilian and Army members had by this time been merged under Army control; the semi-informal university teams working under the auspices of the Briggs Committee in the United States; Joliot-Curie's team in France; and the British group whose members appear to have felt that their task was more to rule out the chances of a German bomb than to suggest that a British weapon could be made. Thomson wrote to Cockcroft on 5 April 1940 that Oliphant had 'raised the question of the possibility of separating the isotopes of uranium and so producing an atomic bomb. He seems to think this is feasible though I am doubtful.' And when his committee finally reported in the autumn of 1941, Thomson wrote that they had 'entered the project with more scepticism than belief, though we felt it was a matter which had to be investigated'.

Thomson's committee – the Subcommittee on the U-Bomb of the Committee for the Scientific Study of Air Warfare as it officially was – met for the first time at 2.30 p.m. on 10 April 1940 in the rooms of the Royal Society off Piccadilly. At least, this is today the received version. Certain papers record the birth of the subcommittee a week or so later and Thomson's own notes suggest that the meeting on 10 April was at the time regarded as a completely informal occasion. In addition to Thomson those present included his assistant from Imperial College, P. B. Moon; Oliphant, who had forwarded the Frisch–Peierls memorandum to Tizard; and John Cockcroft, the specialist in atomic disintegrations who had been called away from the radar work on which he had been engaged since the start of the war. There was also, as guest, Jacques Allier, who with the precious heavy water now safely stowed away in the cellars of the Collège de France, had crossed the Channel with a special letter of introduction from M. Paul Montel, the director of the Franco-British mission attached to the French Ministère de l'Armament.

The importance of M. Allier is clear from the brief minutes of the meeting, an almost casual introduction to the nuclear age written in George Thomson's neat handwriting.

M. Allier made a statement as to the efforts which the Germans were believed to be making to obtain information about work done in France on the U-bomb, and to obtain heavy water. He asked that all details should be kept strictly secret and not mentioned outside the meeting.

The possibility of separating isotopes was then discussed and it was agreed that the prospects were sufficiently good to justify small-scale experiments on uranium hexafluoride. It was decided to ask Dr Whitelaw Gray to provide a specimen for test. Professor Oliphant was asked to find a chemist with suitable experience to do the preliminary work at Birmingham.

It was agreed that Dr Frisch should be informed of the importance of avoiding any possible leakage of news in view of the interest shown by the Germans. It was agreed that he should be informed that his proposal was being considered but should not be told details. There was however no objection to his continuing scientific work on his own lines.

Allier had three main aims which he saw – rightly as it turned out – as vital to the race for nuclear energy now being run by the Germans and the French. He wanted to emphasize to the British the value and extent of the French work. He wanted to stress how German reaction to the heavy water operation suggested that the Germans were hard at work on the nuclear problem. And he hoped to prepare the way for permanent nuclear collaboration between the two countries aimed at making a weapon before the Germans succeeded in doing so.

In one way Allier was unlucky in his timing. Only a few hours before the meeting in the Royal Society, German troops had invaded Denmark and Norway. Heavy fighting was in progress, the phoney war in the West had been transformed overnight, and a hypothetical weapon which at the best could not be made for many years now seemed even less important than it had a few days earlier.

Tizard, to whom Allier was taken after the meeting, reflected its less than urgent feeling in a note sent to the Offices of the War Cabinet later in the day.

M. Allier seems very excited about the possible outcome of uranium research, but I still remain a sceptic and think that probability of anything of real military significance is very low. On the other hand it is of interest to hear that Germany has been trying to buy a considerable quantity of heavy water in Norway. I think you ought to make enquiries from the Belgian company whether Germany has also been trying to get uranium. You will remember that the head of the Belgian company promised to inform us whether there were any unusual demands from Germany. I have received no such information and I suppose you have not, otherwise you would have passed it on. It would be a good thing to find out the facts now.

If Germany is trying to buy heavy water and not trying to buy uranium, I cannot think why they want the heavy water. It may be, however, that they are terrified at intelligence of our own work in England and think they had better corner the heavy water so that we should not have it. This is always quite a possibility.

If there were any doubt about the British attitude, it was removed by a note which Tizard wrote a few days later to Brigadier Charles Lindemann in Paris, Professor Lindemann's brother and a scientific attaché at the British Embassy. He had seen Allier, he wrote, before adding: 'We have taken all necessary action, but for your own information I may say that the French are unnecessarily excited.'

The attitude was rather casual – but no more so than the polite suggestion that Frisch 'should avoid any possible leakage of news'. Indeed, despite the formidable security restrictions automatically imposed on all aliens, including those working on the uranium bomb, it was only eight months after formation of the subcommittee investigating the bomb that Thomson commented that it seemed to him 'desirable that those engaged on the investigation should sign an undertaking not to divulge information outside members of the Technical Subcommittee and those actually engaged on the work'. However, once ordered, the ruling was rigidly enforced. 'The lab boy will be paid £1 a week,' it was reported from Cambridge soon afterwards. 'He has already signed the "secrecy" form before being paid.'

Scepticism towards nuclear power was not limited to the British. It was evident a few weeks later when Professor A. V. Hill, one of the secretaries of the Royal Society recently attached to the British Embassy in Washington, technically as special air attaché but in fact as scientific adviser, reported back on the current American attitude to nuclear energy.

It is not inconceivable that practical engineering applications and war uses may emerge in the end, but I am assured by American colleagues that there is no sign of them at present and that it would be a sheer waste of time for people busy with urgent matters in England to turn to uranium as a war investigation. If anything likely to be of war value emerges they will certainly give us a hint of it in good time. A large number of American physicists are working on or interested in the subject; they have excellent facilities and equipment; they are extremely well disposed towards us; and they feel that it is much better that they should be pressing on with this than that our people should be wasting their time on what is scientifically very interesting, but for present practical needs probably a wild-goose chase.

However, the Thomson committee was not to be deterred and at a meeting on 24 April decided that further investigations should be made at either Birmingham or Liverpool. The primitive stage of the work is emphasized in the minutes of the next meeting when it was recorded, almost with surprise, that uranium hexafluoride was not only violently corrosive but attacked glass when wet. 'It was decided,' said the minutes, 'to make up a glass apparatus and also a copper one and a gold-plated copper one with gold tubes'; but not, it would appear, until an estimate of the cost had first been made. The meeting was not without its significant moments; in a single sentence Thomson recorded that 'Dr Frisch produced some notes to show that the uranium bomb was feasible'.

The almost conspicuously off-hand attitude of those involved in the first practical attempts to discover whether a nuclear weapon might be made was not due to any scientific miscalculation of the energy involved. Nor was it due to the imponderables of engineering which were for long thought to make construction of a bomb impossible. Far more important in cutting the project down to size was the small percentage of the available energy that it was thought could be utilized. John Cockcroft, among the most far-seeing and level-headed of those involved, was writing in mid-May of a uranium bomb which might be as destructive as 100 tons of T.N.T., which was of course impressive when compared with the two- and three-tonners which were then the largest bombs available. Nevertheless, it was very different from, for instance, Professor Tyndall's concept of a weapon which 'might be quite decisive'.

The German invasion of France and the Low Countries on 10 May was crucially to affect the British nuclear effort, subsequent Anglo-U.S. collaboration and, eventually, the slender chances of international co-operation. Before this chain reaction of events was to be started by the transfer to Britain of the French Collège de France team there was an incident which, viewed with the hindsight of history, retains an air of light comedy.

On 16 May, Cockcroft received a letter from his physicist friend, O. W. R. Richardson, saying that Richardson had received a cable from Lise Meitner in Sweden which read: 'Met Niels and Margarethe recently both well but unhappy about events please inform Cockcroft and Maud Ray Kent.' 'Who,' Richardson asked Cockcroft, 'is Maud Ray Kent? Perhaps you can deal with that part of it.'

The first half of the message clearly showed that Niels Bohr and his wife wished to tell British physicists through their friend Cockcroft, that they were safe following the German invasion of Denmark the previous month. The second part of the message appeared inexplicable – at least, until Cockcroft devoted his mind to it. 'You will see,' he wrote to Chadwick in a letter repeating the cable, 'that the last three words are an anagram for "radium taken". This agrees with other information that the Germans are getting hold of all the radium they can.' An even more ingenious solution appears to have struck Cockcroft soon afterwards, since a military intelligence note in the official files comments: 'Since last three words do not appear to make sense Prof. Cockcroft suggested they might contain a hidden reference to some sort of ray', an idea somewhat languidly taken up by Dr Gough, the scientific adviser to the War Office. Another expert saw the three last words as an anagram of 'Make Ur Day Nt', an explanation which as Tizard was told was 'all very wild but just sufficiently reasonable to make one worry'. Among the physicists, only Oliphant appears to have been mildly sceptical. 'The cryptic message from Bohr, through Miss Meitner, seems almost certainly to refer to the uranium problem, but the number of solutions is unfortunately large,' he submitted before asking what to many minds must have seemed a rather obvious question: why not ask the British Embassy in Stockholm to find out from Miss Meitner what it was she wished to convey? The answer was provided only three years later when Niels Bohr arrived in Britain and asked whether his message had ever reached his old governess, Miss Maud Ray, who then lived somewhere in Kent.

By this time, however, Thomson's committee had gone down in the records as the M.A.U.D. Committee, a name given it on 20 June 1940. Thomson explained the name to Pye by saying: 'In order to avoid having to make all letters about the subcommittee secret, we have decided to call it by the initials M.A.U.D.' The story that the name had sprung from Bohr's cable was not unduly discouraged in later years. This is understandable. Edward Appleton, the Secretary of the Department which was to become formally responsible for Britain's nuclear research, has written: 'M.A.U.D. means *M*inistry of *A*ircraft (Production) *U*ranium *D*evelopment', thus inferring that the committee must have had one of the most fragile cover-names in history. But uncertainty appears to have remained. Five years later, after 'The M.A.U.D. Committee' was quoted in a memo to Churchill, Lord Cherwell was asked: 'What meaning?' 'M.A.U.D. was a code name,' he replied. 'We have not been able to find out what the letters represented.'

On the day that the committee was named M.A.U.D. its members discussed with some anxiety what had happened to the world's entire stock of heavy water, last heard of in Britain as being safe in the cellars of the Collège de France. On the same day Chadwick wrote to Professor Lindemann:

> I believe that a sufficient quantity of uranium metal and heavy water *will* satisfy the conditions for a chain reaction by the thermal fission process, but I am not convinced that the thermal fission process has such very dangerous potentialities. But there is a very definite risk in it and I think it would be folly to allow the Germans to get the heavy water, especially as they were extremely keen about it. Can you do anything to see that this does not happen? The best thing is of course to bring the heavy water here; if that is not possible, then to put it down the drain ... I hope you don't mind my worrying you on this matter but there is a very real danger that the uranium process can be made to work.

But the Germans were advancing rapidly through France, the French Government was on the point of collapse, and the best that could be hoped for in Britain was that the French had been able to dump their precious material in the Seine or the sea. Its eventual fate was instead to have repercussions, scientific, political and international, that not even the most ambitious of Frenchmen would have dared suggest in the summer of 1940 as the Germans moved west and then south through France against resistance that was at first diminishing, then non-existent.

As France faced defeat, M. Dautry, the French Armaments Minister, had much on his mind. Nevertheless he did not forget the precious heavy water, the material that in a not entirely indirect way might be used to restore French prestige when the tide had turned. Joliot-Curie was ordered to ensure that the cans did not fall into enemy hands, and Halban was in turn ordered to take them to Mont-Dore, the spa in central France, on the first leg of a journey to Clermont-Ferrand, where they could be immured in the vaults of the Banque de France.

'My instructions,' he has said, 'were to take the heavy water, radium, and certain documents. I put my wife and one-year-old daughter in the front of the car, the one gram of radium [from the laboratory] at the back, and, in order to minimize any possible danger from radiation, the cans of heavy water in between.'

Halban's German accent added to his difficulties when he arrived in the pleasant little spa of the Massif Central, and it was only after some argument that his story was believed and that he was allowed to lodge the cans of heavy water in the safety of the town's women's prison.

The following morning, after they had been moved for greater safety to the condemned cell of the prison in nearby Riom, Halban began to set up shop in a small villa. 'Luckily,' he says, 'all that was really vital to us was running water and some elementary equipment. The water was available in the kitchen and the bathroom and the equipment was erected by army engineers detailed for the task.' In these humble surroundings preparations continued for experiments which it was hoped would lead France along the path towards limitless power. For no doubt the enemy would eventually be halted, just as they were in 1914.

Shortly afterwards all appeared to be ready. The rest of the team had arrived and a small celebratory luncheon was organized at which the coming work could be discussed. As it was about to begin a small Simca drew up outside the building. Out stepped Lt Allier. There would, he made clear, be no further experiments, at least in Clermont-Ferrand. The enemy was still advancing and the French Government had given orders for the heavy water to be evacuated. 'I was,' Halban later said to British officials, 'told to make all possible efforts to do this work in Canada or the United States so as to make sure it would not be interrupted a third time, also to obtain safety for the heavy water.'

In Riom the prison governor at first refused to hand over the cans without a military order. Allier drew his revolver and said that that was his authority. A number of prisoners serving life sentences were thereupon ordered to move the world's total stock of heavy water from the condemned cell into the waiting vehicle.

The evacuation was to be through the port of Bordeaux. Here Halban and Kowarski found the *Broompark*, an inconspicuous British coaler. And here they were met by the 20th Earl of Suffolk, a remarkable character of great daring and courage. Destined to be blown to pieces the following year with his secretary and chauffeur while defusing a German mine, Suffolk was liaison officer with the French and was now rescuing a number of rare machine-tools, two and a half million pounds worth of industrial diamonds, some fifty French scientists thought too valuable to be left in German hands, and the cans of heavy water. 'Suffolk had got the crew too drunk to sail until the machinery and ourselves were aboard,' Kowarski was later to say. 'The ship was a coal ship, so none of us was in dinner jackets. Suffolk was unshaven for several days. There was sea-sickness; there were twenty-five women aboard. Suffolk was pouring them champagne. "This is the perfect remedy," he said. The ship stayed in harbour for a day, where we were bombed, then moved out to the mouth of the estuary where we waited again.'

Joliot-Curie, who reached Bordeaux after the *Broompark* had been moved from the quay, decided to remain in France, where he was to become a formidable leader of the Resistance, another decision that was to affect post-war nuclear history. It was, therefore, only two members of the Collège de France team who guarded the heavy water. Suffolk at once supervised the construction of a wooden raft to which there were lashed the packets of industrial diamonds and the cans of heavy water. If the *Broompark* were sunk, a not unlikely possibility, the raft with its invaluable cargo was to be cut free and every effort made to save it. As the ship edged down the Gironde estuary amid the chaos the vessel next to her was sunk by a mine. But she escaped damage and four days later anchored in the comparative safety of Falmouth Roads off the south coast of Cornwall.

Eventually the heavy water reached London. So did the physicists where their traditional feelings about the strangeness of the British were confirmed. They found it difficult to believe that the 'B.E.F.' scrawled on troop-trains did not stand for British Expeditionary Force but, as many soldiers maintained, 'Back Every Friday'. And, although allies, they were told that since they had no identity cards they were legally unable to move anywhere. Even when allowed to drive to the Cavendish Laboratory at Cambridge they were forbidden to use maps, although signposts had been removed in view of the invasion scare. Kowarski solved the problem with admirable ingenuity by learning and memorizing the names of the public houses in all the villages on their route. He or Halban had merely to ask for 'the Blue Anchor' or whatever unusually named inn lay next on their route.

Meanwhile the 180 kg of heavy water, contained in fifteen aluminium and eleven tin cans, was moved to the safety of London's Wormwood Scrubs prison, already taken over by the War Office. For reasons that are not quite clear the bulk was then moved to the cellars of Windsor Castle where the Librarian, Owen Morshead, signed a receipt for it and was told of its unique significance for a purpose that was not described to him.

The heavy water was during the next few months to present

considerable problems to the British civil service which steadily built up an inches-thick file about its fate. The name itself aroused some surprise in the members of a not-too-scientific body, and for some while it was merely called, with obvious surprise, 'a substance known as "Heavy Water" '. There were different accounts of who owned it. According to one file it appeared to be the property of the Norwegian Hydro-Electric Nitrogen Society; according to another it was owned by M. Allier's Banque de Paris. Its value was estimated at various figures from £22,000 to £100,000 and a main civil service problem concerned how it could be insured. Cover under the emergency wartime scheme was impossible since, it was pointed out, 'Dr Halban is not a person carrying on business as a seller of goods, nor does he own the Heavy Water in the course of his business.' The Private Chattels Scheme was ruled out for other reasons while the plan that Halban should himself insure this unique potential for nuclear power was ruled out by one simple fact: Halban, who had arrived with little more than he stood up in, had no available money. Eventually a suitable solution was discovered: ownership of the world's entire stock of heavy water was vested in the Custodian of Enemy Property.

Long before this ingenious solution had been found, plans were in hand for the cans to be moved to the Cavendish Laboratory where the French team had been installed. Its members had not been greeted with quite the enthusiasm they expected. This was partly because their efforts appeared to have been concentrated more exclusively on 'the boiler' than on 'the bomb' although in these early days no one could be quite sure that 'the boiler' might not have unexpected wartime possibilities. 'I remember,' Cockcroft said later, 'one member of the committee suggesting that perhaps the heat developed in the boiler might be used to produce a powerful searchlight.'

In addition to the feeling that if the boiler was to be considered it should only be considered after the war, there was some *frisson* of regret among the British when it was known that the French had not only patented their work but had entered negotiations with a commercial organization, a process which in the eyes of at least some British scientists put them if not beyond the pale, at least in a category rather different from those who sought scientific knowledge for its own sake.

The first details had been revealed, as early as 10 July, only a few days after the French team had arrived in Britain. The M.A.U.D. Committee had first agreed that it was worth while 'to carry out

Halban's experiments with heavy water as a source of power and for the information they may give as regards fission processes'. Then the man himself was called in to address the committee; and only now, it appears, did its members learn of the events which were eventually to bedevil negotiations between the British and the Americans, and to cause a good deal of international unpleasantness in the world of post-war nuclear diplomacy. The minutes of the meeting record that 'Dr Halban described the events that had led up to his forming a society with Union Minière for the exploitation of Atomic energy from uranium.'

A more detailed account was given by Halban some months later to the Ministry of Aircraft Production when the enormous potentials of post-war nuclear power had already introduced a Byzantine element into negotiations the Ministry was conducting with industry for the manufacture of nuclear weapons.

The French, says Halban's account to the Ministry, had opened talks with the Union in May 1939.

Joliot had already some personal contact with this society, which had always generally supported (and on the other hand largely profited from) Madame Curie. We had a first conference with the General Director of this Society, Sengier. We came very quickly to the following arrangement: we were to found together a Syndicate whose task it should be to prepare the scientific and half-technical experiments; our contribution to the Syndicate would be the patents; the contribution of the Union would be up to 50 tons of product and 4 grams of radium, both loaned for the duration of the experiments, with full risk taken by the Union. The Union would take immediate responsibility for one million francs [then £6,000] for the cost of the experiments. In the event of a success, the Union would charge itself with all commercial and administrative actions, and fifty per cent of all gains would come to us (this right was to be ceded to the Caisse de Recherche).

There were of course two ways of regarding such dealings: one was to consider them in no way different from the exploitation of coal and electricity which had throughout the world been handled for private gain; the other was to regard them as commercial exploitation of the laws of nature, a practice which earned little sympathy from many scientists – especially Lew Kowarski.

However, the French team was to get much assistance in Britain: any limitations sprang from the fact that its members had arrived in the country at a time when the nuclear enterprise still aroused little

Government interest, despite the change which had swept Chamberlain from office and replaced him with Winston Churchill. One of Churchill's first deeds was the creation of the Ministry of Aircraft Production under whose wing the M.A.U.D. Committee now found itself. But this was a mixed blessing: compared with the desperate shortage of aircraft, the work of a scientific committee trying to discover if some newfangled sort of weapon might be built, was not of much importance.

The comparatively casual way in which the bomb was still considered is illustrated by Professor Lindemann's admonition to one of his assistants, James Tuck, who was to represent him at an air defence committee where Tizard – whose scepticism still tended to dampen the whole enterprise – was to be present. 'Tizard is sure to try and pump you as to what I am up to here,' Lindemann warned Tuck. 'You say that our big line is a uranium bomb. This is just the sort of thing that Tizard would expect me to be thinking about. You know enough about it to make a plausible story.' The strange thing, Tuck later said, was that Tizard did buttonhole him after the meeting and ask him what Lindemann was up to. 'I duly whispered the astonishing secret,' says Tuck. 'He duly snorted and said: "Ha! Just what I would expect! It will never be used in this war, if it works at all!" '

Another fact, which in the summer of 1940 made the integration of the French into the M.A.U.D. effort something of a problem, was that they were, if not enemy aliens, at least non-British. After the use that the Germans were alleged to have made of Fifth Columnists on the Continent, now known to have been grossly overrated, it was natural that the British should be cautious of all those born beyond the Channel, even though this led to some piquant situations. At the top level there was the situation of Frisch and Peierls, without whom the M.A.U.D. Committee would never have been set up. At first they had been excluded from discussion of their own paper of February 1940, a droll situation which led Peierls to tell Tizard that while he fully understood the security problems, he did not 'think that any useful purpose can be served by trying to keep our own ideas secret from us'.

Much the same problem arose with the French team. Eventually, the ludicrous situation was solved by creation of a Technical Committee at which only scientific matters were discussed, but were discussed without restriction, while questions of high policy, and those which might conceivably affect military operations, were limited to the smaller M.A.U.D. Committee.

At a different level, restrictions remained. Thus Rotblat and Frisch working in Liverpool to discover the war-winning weapon, had to be indoors by 10.30 every night, and when critical measurements had to be taken over long periods, Frisch had to obtain a special 'late pass' from the police. When either man had to take top-secret documents to meetings of the M.A.U.D Committee in London, special travel permits had first to be obtained. Even the problem of getting to work was difficult at times since no alien in Liverpool was allowed to own a bicycle. In less sensitive areas, where aliens were allowed to use them, various regulations had to be observed, and those working towards the ultimate weapon had regularly to set aside time for keeping up to date the Bicycle Registration list – number and name of make – which had to be submitted to Chief Constables.

Similar oddities affected such men as Egon Bretscher, a Swiss working in the Cavendish. Having been invited to a meeting of the M.A.U.D. Committee by telephone, he had to plead for a written invitation: it would have to be shown to the Chief Constable in Cambridge before he would be allowed to travel to London.

Halban, the man staking out claims for post-war nuclear power, was not allowed to enter any of the specially protected areas which covered large parts of Britain, nor to have a large-scale map. He outlined the outcome in a letter to the Ministry of Aircraft Production in which he said that it was essential that he went from Cambridge to Birmingham. 'By train,' he said, 'this takes two days. So I will go by car. This means that I must make a roundabout way round Nottingham, which is a protected area. In order to find the roundabout way I need a good map. In order to take a map with me, I need an Englishman in the car, which means a day's work lost for Fenning (who will of course enjoy the ride in the car).'

The need to compile bicycle registers, and to get an Englishman to get a map to get round Nottingham, were not the only activities which unexpectedly occupied members of the M.A.U.D. Committee on occasion. It was, for instance, discovered that a good deal could be found out from studying the lecture lists still being published in the *Physicalische Zeitschrift* and it was suggested that the lists might be duplicated and circulated. However, when application was made to the Government organization still receiving issues, the librarian 'seemed to think that because of some international copyright agreement, it was not permitted to make photostatic copies of any matter from periodicals'.

Despite such irritating problems, the members of the M.A.U.D. Committee steadily came to grips with the enormous problems of finding out just how practicable the building of a nuclear weapon might be. To this extent, they were rather different from those involved elsewhere in nuclear research. In America, to which the news of the Frisch–Peierls memorandum had not yet percolated, the emphasis was entirely different, despite the continuing warnings from refugees. 'In the summer of 1940,' says the official historian of the American project, 'American scientists saw it [a chain reaction] first as a source of power. All of them, certainly, had thought of the possibility of a bomb. Some believed that in achieving a chain reaction they might gain understanding of what it took to make a bomb. But scientists in America did not direct their thinking primarily toward a weapon. When [George] Pegram and Fermi outlined the research plans for the Columbia team in August, they listed their objectives only as power and large amounts of neutrons for making artificial radioactive substances and for biological and therapeutic applications.' In France and Germany, equally unaware that the critical mass of uranium 235 was to be weighed in pounds rather than tons, nuclear power produced by slow neutrons was the prospect, a harbinger of industrial supremacy once the war had been won by weapons already to hand or at least under development. Only in Britain, her armies now pushed from the Continent, was nuclear energy now being considered in terms of a bomb which could be made and dropped on an enemy in the current war. In practical terms the British were the first to contemplate marching across the nuclear threshold, and it is pertinent to ask if they had any qualms about the ethics of the matter. Did they, during this period, think about this aspect at all?

With one exception, the short answer appears to be 'No'. The exception was Dr Charles Darwin, Director of the National Physical Laboratory. Darwin was a grandson of the author of *The Origin of Species* and had worked with Rutherford in Manchester after graduating in 1911. In the autumn of 1941 when the British had reported that an atomic bomb was feasible and should be built as soon as possible, he was head of the British Central Scientific Office in Washington. He had talked to Dr Vannevar Bush and Dr James B. Conant, the Americans on whose shoulders the burden of U.S. nuclear research now rested, and it was quite clear to him that the British effort was now to be augmented by America's massive industrial potential. But this, he wrote to Lord Hankey, by this time supporting the British project, posed a difficult question. If the bomb could be made, should it be used? 'For example are our Prime Minister and the American President and the respective general staffs willing to sanction the total destruction of Berlin and the country round, when, if ever, they are told it could be accomplished at a single blow?'

No reply seems to be on record and from the silence it appears that the answer to the question was yes. Yet one point should be made in an age which considers megadeaths with barely a twitch of conscience. It is that views on bombing from the air – the obvious means of delivering a nuclear weapon – were in a state of continuous evolution throughout the first decades of the twentieth century. When Marshal of the Royal Air Force Sir Hugh Trenchard proposed in 1928 – in a memorandum on 'The War Object of an Air Force' – that the aim of such a force was 'to break down the enemy's means of resistance by attacks on objectives selected as most likely to achieve this end', his proposal was strongly attacked by the other Chiefs of Staff. Field Marshal Sir George Milne, the C.I.G.S., protested that 'the impression produced by the acceptance and publication of such a doctrine will indubitably be that we are advocating what might be termed the indiscriminate bombing of undefended towns and of their unarmed inhabitants'. Admiral of the Fleet Sir Charles Madden, First Sea Lord and Chief of the Naval Staff, was even more condemnatory, maintaining that the policy was an unjustified departure from the accepted principles of war and that 'civilian life [would] be endangered to an extent which had hitherto not been contemplated under International Law'.

Even on the outbreak of the Second World War the bombing policy was, on both sides, still limited to military, or militarily important, targets. The first British attacks on Germany were deliberately planned to avoid the danger of civilian casualties. In 1940, the destruction of Rotterdam by the Germans, if not their earlier destruction of Warsaw, about which argument continues, altered all that. And in 1941 the head of Bomber Command was instructed that with the exception of diversionary attacks he was to 'direct the main effort of the bomber force, until further instructions, towards dislocating the German transportation system and to destroying the morale of the civil population as a whole and of the industrial workers in particular'. When, in the autumn of 1943, the British Government was trying to claim that the bomber offensive was not aimed at civilians, the head of Bomber Command, Air Chief Marshal Sir Arthur Harris, maintained that the truth should be told and that the 'aim of the Combined Bomber Offensive ... should be unambiguously stated [as] the destruction of German cities, the killing of German workers and the disruption of civilized community life throughout Germany'.

Thus within the psychological context of the 1940s the production of more effective bombs was even to most scientists only a logical step, and its questioning as unreasonable as condoning bombardment by light artillery but condemning the use of heavier guns.

Moreover there was always the chance that nuclear weapons could not be built at all. Darwin appears to have been one of the few willing, in the desperate days of 1941, to ask what would happen if the reverse was the case.

A few men did believe that they should have nothing to do with such an apocalyptic idea, whatever the chances of success. One was Max Born, a pioneer of quantum mechanics. He had left Germany after Hitler's rise to power, become a naturalized British subject in 1939 and was in 1940 Tait Professor of Natural Philosophy at Edinburgh. 'I cannot say that at that time I was a pacifist, since I wished the downfall of Hitler with all my heart,' Born has said. 'But I could never reconcile myself to the methods used, e.g. the pattern bombing of cities which means the killing of defenceless people, women and children, in order to weaken the military power of Hitler, although he had started this (Rotterdam, Coventry, Warsaw).' Born was a man whose views were so well known that he was never directly asked to help the M.A.U.D Committee. But he had two young German assistants, one of whom received a letter from Professor Peierls asking him to join a team working on a very secret project. 'We knew what it was,' says Born. 'He was inclined to accept. I told him of my attitude to such kind of work and tried to warn him not to involve himself in these things. But he was filled with a tremendous hatred of the Nazis, and accepted.' So the young German, Klaus Fuchs, left Edinburgh for Birmingham and the road that was to lead through Los Alamos to a British prison and then to a post-war East German research centre in Leipzig.

The work to which the M.A.U.D. teams devoted their efforts from the spring of 1940 onwards was a shadowy pre-play of what was to take place in the United States two years later. But the differences of scale were so great as to make them qualitative rather than quantitative. The money authorized for the on-the-ground, practical research was comparable to the early tentative funding of the Briggs Committee, but the parsimonious way in which it was doled out was somewhat bizarre; universities whose Fellows were investigating what could well be the most powerful weapon of all time were forced to submit monthly lists of their expenses which included items such as 2*s*. 6*d*. for a tin box and £1 0*s*. 2*d*. for three set squares. Flexibility within universities helped; so did Imperial Chemical Industries which supplied a good deal of the research material and debited the cost to their own general research account.

Shortage of hard cash for essentials was equalled by shortage of staff for essential, if mundane, jobs. 'I had to explain that I really could not go on typing all the documents myself,' Peierls has said. 'I asked whether I might engage a part-time secretary, and this was allowed.' First one, then a second desk calculator was yielded up by His Majesty's Stationery Office to help the enterprise, together with the operators to work them. Qualified specialist staff were far more difficult to find since during the first nine months of the war most scientists had been drawn into radar or some other comparable enterprise. The Ministry of Labour maintained a Central Register of scientists but by this time most of those who remained on it were for one reason or another unable to fit into the official machine. But the Register produced unexpected surprises as Peierls discovered when he interviewed one potential recruit, a mathematician born in Russia and only recently naturalized. Anxious that the young man should understand the difficulty of the work involved, Peierls wrote down one of the equations whose solution had escaped him for some while. 'He left me to travel back home,' Peierls has said. 'The following morning I received a note from him in the post – together with the answer to the equation. He had worked out the solution in the train.'

Yet if the limitations imposed by a country already stretched by war made the British enterprise very different from the American one which followed, an even more important factor was the difference in knowledge between 1940 and 1942. Frisch and Peierls in Birmingham had been forced to estimate the cross-section of the uranium nucleus; but on this depended the all-important critical mass, while on the detailed characteristics of the nucleus – still comparable to Churchill's description of Russia as 'a riddle wrapped in a mystery inside an enigma' – depended the force of a nuclear explosion. Secondly, the vastly complicated theoretical aspects of what exactly happened when nuclear fission was repeated multi-millionfold to release immense energy within a minute fraction of a second were still not completely understood. And if nuclear physicists were by this time difficult to find, those who were also mathematical physicists, the kind needed for this work, were even more scarce.

There was also the totally different practical problem to be solved. How did one separate from 1,000 atoms of natural uranium the seven that contained 143 neutrons wrapped up in their nuclei, and not 146 or 142, when those seven atoms are chemically, and in every way except that of nuclear structure, identical with the other 993?

While Frisch and Peierls in Birmingham and Chadwick and his team in Liverpool tackled the theoretical work, a start was made on the all-important problem of isotope separation. In practical terms, this was for the moment the heart of the bomb problem as Frisch emphasized early on. Although various laboratory methods had been used, the amounts involved were minute. Producing pounds was entirely different. 'It was like getting a doctor who had after great labour made a minute quantity of a new drug, and then saying to him: "Now we want enough to pave the streets," ' as Frisch put it.

Some twenty years previously Aston and Lindemann had discussed isotope separation in a famous paper and it was a curious coincidence that Lindemann, now Churchill's scientific adviser and soon to become Lord Cherwell, had already played a significant part in the careers of three of the four men who were now to tackle the separation problem with considerable success.

The most important of these was Franz Simon, who had left Germany after the rise of Hitler since he no longer considered it, as he said, 'the place in which to bring up children decently'. Simon had after his arrival in England joined the Clarendon Laboratory at Oxford supported by an I.C.I. research grant arranged with the help of Lindemann, who was head of the laboratory as well as Professor of Experimental Philosophy in the university. He was soon joined by two other refugees from the Continent whose I.C.I. grants had also been engineered by Lindemann: Dr H. G. Kuhn from Germany and Dr Nicholas Kurti from Hungary. As aliens all three had been debarred from taking up radar work, and the same was true of a young American, Dr H. S. Arms. During the last few months of 1939 and the first of 1940, when no nuclear project officially existed and the prospects for the world's most powerful explosive were not, technically, of security value, Simon was visited more than once by Peierls. He was then slowly brought into the work, this being possible since although enemy aliens could not be employed by the Government on secret work, the ban did not apply to those hired by universities. The M.A.U.D. work was subcontracted to the universities and the greatest of all Britain's wartime secrets could thus be entrusted to men who were excluded for security reasons from other war work.

Simon's first service to the M.A.U.D. Committee involved not the problem of isotope separation but the winning over of Lindemann, still distinctly sceptical that God had built the world so as to make nuclear weapons possible. Simon persuaded his friend Peierls to try converting Lindemann, and on 2 June, Peierls wrote him as follows: 'Simon tells me he has seen you about the question of the uranium bomb, and that you would like a brief statement on the reasons why we believe that it should now be taken seriously.' Although admitting that success was not certain, 'the probability of its working is sufficiently high to make it important to investigate the matter as rapidly as possible,' he went on. '... Another important aspect of the matter is that, while there is no evidence that the Germans realize the potentialities of a uranium 235 bomb with fast neutrons, it is quite possible that they do, and for all we know they may have almost completed its production. It seems, for this reason, imperative to study the effects that such a bomb would have, in particular the physiological effects of its radioactive radiations, and to prepare such protective measures as are possible against it.'

The two men met, but after the interview Peierls was by no means certain that he had succeeded. 'I do not know him sufficiently well to translate his grunts correctly,' he reported. However, 'I feel sure,' he added, 'I have convinced him that the whole thing ought to be taken seriously.' As was often the case, Lindemann's grunts meant more than appeared obvious. When the time came for crucial decisions, it was Lindemann who persuaded Churchill that despite the uncertain prospects of success the bomb project must continue.

Peierls's quiet reminder of the need to investigate radioactive hazards was one of the few to be made, either in Britain or the United States, while work on the bomb was being continued. As early as May 1940, Cockcroft had visited Dr Mayneord of the Institute of Cancer Research to gain his views. They were ominous. 'Dr Mayneord,' it was reported, 'said that a few micrograms of dust in the lungs may prove fatal. In addition, 10^{-10} curie/cc of Radon inhaled over a period of 200 days would prove fatal, as far as can be ascertained from existing evidence.' Little more appears to have been done and although there was a proposal during the blitz of 1940 that the Germans might use bombs including radioactive material, the next serious suggestion of radioactive dangers came in 1944. Then it was suggested that the

Normandy invasion beaches might be protected by the Germans with radioactive devices, and special units equipped with Geiger counters went ashore with the leading troops. There was no need for them.

While Simon was invoking the aid of Rudolf Peierls to influence Lindemann he was also carrying on with his pioneer work on the gaseous-diffusion method of isotope separation. His genius for simple explanation was illustrated when he arrived in the Clarendon one morning with a simple kitchen sieve. 'What we want,' he said, holding the metal mesh up to the light which came through the myriad of small holes, 'is something like this, only with very much finer holes.' What he visualized was the industrial application of the method used on a laboratory scale by Professor Herz in Berlin a decade previously for separating the isotopes of neon. If uranium in some gaseous form were passed through a sieve of holes, the lighter 235 isotope would pass through more quickly than the heavier isotope and the gas on the far side of the sieve, or barrier, would contain slightly more than the natural 0·7 per cent of the 235 isotope. The difference would be extremely small and it was obvious that the number of barriers, or membranes as they were later called, would be very large indeed.

The Clarendon workers had not yet been officially brought into the project and the first experiments, forerunners of the acres of diffusion barriers at America's Oak Ridge separation plant, were of a satisfying elementary crudity. 'The first thing we used,' Dr Kurti has said, 'was "Dutch cloth" as I think it is called – a very fine copper gauze which has many hundreds of holes to every inch. These holes were of course very much larger than those which we were to utilize later, but we felt that the material would show us if various ideas worked.' Choice of a test material was not easy. It was correctly thought that the only practicable material would be uranium hexafluoride; but this was only available in minute quantities and was, in any case, known to present major handling difficulties which would have to be solved before it could be used. For the experiments needed to solve the problems involved, Kurti has said, 'it was necessary to use something that had components of two fairly different molecular weights, and we decided to use a mixture of water vapour and carbon dioxide – in other words something much like ordinary soda-water.'

Only when this work had been done was it possible to consider the formidable implications of any success. The most complex calculations would be necessary to discover how many tens of thousands of membranes would be required. Special precautions would have to be taken to deal with the activity of uranium hexafluoride which not only affected metal but also the rubber and the lubricating oils which would form parts of a gaseous-diffusion plant. In addition there were the immense engineering problems involved in pumping a uniquely corrosive gas through the immense number of membranes.

Throughout the late summer and early autumn of 1940, as the issue of victory or defeat was fought out over the fields of southern Britain, plans for making the world's first nuclear weapons began to crystallize. At the same time a mission under Sir Henry Tizard visited Washington to lay the foundations for interchange of scientific information between America and Britain. With him went Cockcroft who attended a meeting of Briggs's Committee, visited a number of universities where work on nuclear fission was in progress, and used the machinery of the Tizard Mission to ensure that the Americans would be kept informed of any progress made by the M.A.U.D. Committee.

Despite the Ministry of Aircraft Production's preoccupation with more immediate and urgent matters, its M.A.U.D. Committee steadily assumed a more official shape; the help of old friends who willingly took on work in their laboratories and financed it as best they could was replaced by formal contracts. Thus on 12 September, the Ministry signed a £2,000 contract with Liverpool University for: 'An experimental investigation at the Liverpool University under the direction of Professor Chadwick into the possibilities of developing a new explosive. The work, which is being supervised by the M.A.U.D. Committee on behalf of the Ministry of Aircraft Production is to be to the satisfaction of the Director of Scientific Research.' This was David Pye: an engineer who did more than most men to make the Whittle jet engine a reality, Pye's faith in nuclear fission as a practical source of energy was distinctly minimal. Despite his scepticism, he loyally supported the M.A.U.D. enterprise and contracts were signed which put the work at Oxford, Cambridge and Birmingham on an official basis. Nevertheless, as late as October Thomson felt justified in telling Pye that although a thoroughgoing investigation of materials would need £80,000 to £100,000, he felt it was 'still premature to think of anything of this kind'.

CHAPTER 6

An Awkward Decision of State

Thomson's rather gloomy prognostications in the autumn of 1940 matched the spirit of the times. Invasion of Britain certainly had been called off and the Battle of Britain had been won. But the night blitz by the Luftwaffe was gathering force, Germany controlled the western seaboard of Europe from the North Cape to the Pyrenees, and the prospects of an Allied return across the English Channel appeared remote. Survival might have been ensured, at least for the time being and if the nightly hammering from the air could be endured, but it was difficult to see how the war could actually be won.

Yet before the end of 1940 two developments had taken place which radically altered the chances of the Allies producing nuclear weapons in time to win the war; or, perhaps more significantly, to break the stalemate which many saw extending into the years ahead.

The first was described in a thick report from Simon which outlined in fairly specific terms the design for an isotope separation plant. This was, for its time, a quite remarkable document. It not only described the methods by which many of the chemical and engineering problems could be overcome but forecast the amount of electricity and the number of men needed to run the plant, as well as the costs of building and operating it. The report dealt with materials and processes which had been used only rarely, and then on a laboratory scale, but would now have to be dealt with on an industrial scale. When a comparable plant came to be built in Britain after the war, it was found that Simon's estimate of cost was, when the revised value of the pound sterling was taken into account, wrong by only a factor of two or three.

But it was not only the Simon report that arrived on George Thomson's desk during the second half of 1940. There also came one from Cambridge; in its own way it was quite as remarkable.

Like all other M.A.U.D. teams, the French and their colleagues had been short of money, equipment and staff. For some while their assistants numbered three: a recently qualified B.A., a mechanic and a laboratory assistant. Even the making of new aluminium containers for the heavy water had been as difficult as in Norway.

It was in this string-and-sealing-wax operation that one of the nuclear project's most influential discoveries was to be confirmed. It was a discovery that not only settled the fate of Nagasaki four years later but was to bind the destructive and peacetime potentialities of nuclear energy with an indissoluble link which was to make international control such a suspect subject. It also tempted into the nuclear arena the forces of private enterprise whose efforts to gain a monopoly of post-war power was such an important factor in souring Anglo-U.S. co-operation.

All this sprang from a possibility which physicists in both Britain and the United States had been considering since the middle of 1940. No more than a small cloud on the horizon, it retained for some while the trappings of science fiction, yet was suddenly to be transformed into a second method of unleashing the huge power within the atom. The priorities are confused, but it appears that the first hint of things to come was given in a paper which Louis Turner of Princeton's Palmer Physical Laboratory sent to Szilard on 27 May 1940. In this Turner suggested that if Szilard was successful in creating a chain reaction with uranium, the process would produce another element and that this itself could be fissionable. The first part of this suggestion was shown to be true shortly afterwards, in a paper published in the 15 June issue of *Physical Review*. Written by the American physicists E. M. McMillan and P. H. Abelson, it reported that experiments in the Berkeley cyclotron showed that the result of bombarding uranium was that some atoms of uranium 238 – the isotope forming more than ninety-nine per cent of natural uranium – were transformed into an element, of mass number 239, which had never been observed before. This new element, as yet unchristened, appeared to be stable.

At roughly the same time, the French team in Cambridge was proposing something very similar to the M.A.U.D. Committee. But the French, notably Halban, suggested that this new element might itself be fissile if subjected to neutron bombardment, a possibility which had occurred to Chadwick and Rotblat in Liverpool as soon as they read McMillan and Abelson's paper. The same view, it was later learned, was taken by the German physicist, von Weizsäcker, who before the end of the year had prepared a paper for the German War Office suggesting that a nuclear explosive might thus be obtainable for uranium 238.

The next step appears to have been taken almost simultaneously in Britain and the United States. Three members of the Cambridge team decided that it would be convenient if the new element were given a name. Uranium itself, discovered by Klaproth in 1789, had been named after the planet Uranus, recently seen for the first time. It was by now known that the new, and potentially fissile, element was not formed directly from uranium 238. The bombarded uranium atoms first transformed into atoms of yet another previously unknown element. This transitory element was called neptunium after the planet whose orbit lay next to Uranus and the new stable element was named after Pluto whose orbit lay next again in the solar scheme. The Americans, it was later learned, acting completely on their own, had also christened the new element plutonium.

So far, the potentialities of plutonium were no more than a crock of destruction at the end of the road since without a self-sustaining chain reaction its production could be no more than an interesting but, for practical purposes, theoretical operation. But the French team's experiments in the late autumn of 1940 radically altered the situation. They had begun to experiment with uranium oxide powder suspended in a revolving sphere of aluminium containing heavy water. The apparatus consisted basically of two metal hemispheres, each about two feet in diameter, and made watertight with an annular rubber seal that was held between the two halves as the spheres were locked together. 'In one experiment in which there were 360 heavy hydrogen atoms to one of uranium,' as Halban reported it to a later meeting of the M.A.U.D. Committee, 'thirty per cent of the primary neutrons were absorbed by the system and there was a 6 ± 2 per cent increase in the neutrons emitted over those emitted by the primary source. In the other experiment there were 160 heavy hydrogen atoms to one of uranium. Sixty to seventy per cent of the primary neutrons were absorbed in the system and there was a $5 \pm 1{\cdot}5$ per cent increase in the neutrons emitted over those emitted by the primary source. Both cases were positive proof of a chain reaction.'

Perhaps 'positive proof' was a phrase with a tinge of undue optimism. Nevertheless, it was something vastly more encouraging than anything that had previously come to light.

By mid-December there could no longer be any reasonable doubt:

uranium oxide, used with heavy water as a moderator, would produce a self-sustaining chain reaction if the amounts were large enough. A few tons of uranium oxide and between three and six tons of heavy water would be needed to confirm the experiments, and as neither were available, the results had to be taken on scientific trust. But, as the introduction to the Cambridge team's report spelled out, if plutonium were fissile then 'the boiler', the power-producer of the future, could provide an easier route to the bomb. Chadwick's prophecy to Cockcroft a few months earlier – 'I do not believe that the problem can be separated completely into two parts – fast neutron and slow neutron ...' – had come true in an unexpected way.

'Two distinct lines of research have been followed,' wrote Halban of the Cambridge work,

> (i) by Halban's and Kowarski's team on the conditions necessary for the realisation of a divergent chain of fissions produced by slow neutrons, and (ii) by Bretscher and his collaborators, in the Cavendish H.T. Laboratory, on problems bearing upon the realisation of a divergent chain with fast (or medium fast) neutrons. It was assumed at the outset that only (ii) had immediate relevance to the problem of producing super-explosives, but, as the following Reports show, it now appears likely that the success which has already been achieved in (i) ... will enable us to approach (ii) with the greatest chance of success, also. Briefly, this comes from the following considerations. Early work was based on the expectation that separation of the rare isotope U235 was the necessary preliminary to the practical realisation of the divergent-chain super-explosive. It was considered likely that the fission properties of U235 would be satisfactory in this respect – but a difficult practical problem was encountered in showing conclusively that this was the case. ... On the other hand, general considerations indicate that the body [plutonium] would probably be even more satisfactory as regards fission properties than U235 and the realisation of the slow neutron chain process makes possible the production of this body even more readily than U235 may be produced by isotope separation. For preliminary work on the fission properties of [plutonium], it seems that sufficient material may be obtained by irradiation of large quantities of uranium on the Cavendish H.T. set. [Plutonium] is left after two successive disintegrations of U239, formed from the *abundant* uranium isotope U238 by slow-neutron capture. The advantage of depending upon a reaction of the abundant isotope (99·3 per cent) does not require elaboration.

'This was the first time,' said Thomson after reading the report, 'that anyone had specifically said that a nuclear reactor would work – in other words that it would really be possible to utilize atomic energy. The thing was not completely certain – the amount of heavy water available was not large enough to justify that claim – but it was ninety-five per cent certain. If Halban's paper had been published, it would have been enough to convince almost any nuclear physicist that the thing would work.'

No chain reaction had as yet taken place. Maybe none every would take place. But if it did, then the production of plutonium would do far more than create a second nuclear explosive. As a chemically different element from the uranium out of which it was being made, it could be separated with comparative ease, thus removing, in a single step, the enormous problems of isotope separation which would otherwise have to be solved. The future problem was thereby spelt out: any country which generated nuclear power would contain within its national apparatus the potential for producing nuclear weapons. Despite later public relations exercises, this is still the case.

News of Halban's paper immediately aroused the interest of the Americans, who since the early autumn of 1940 had continued to receive reports of the M.A.U.D. Committee's progress. Thomson had passed on news of Halban's report to Ralph Fowler, British scientific representative in Washington. Fowler in turn informed the National Defense Research Committee (N.D.R.C.), and was asked to discover, as quickly as possible, whether Chadwick was prepared to confirm the Cambridge team's results. If so, it appeared that the United States might be prepared to spend up to a million dollars on a plant for the production of as much heavy water as could be made. Since Halban's figures referred to the chain reaction rather than to the fissile properties of plutonium, which were still only conjecture, it appears that the Americans were at that time more interested in power than in a nuclear explosive.

In March 1941, failing sufficiently strong confirmation from the sensibly cautious Chadwick, Briggs encouraged Fowler to send a hastener to Britain. 'So much is [the Halban report] worrying Briggs,' Fowler wrote to Cockcroft, 'that he asked me to go and see him yesterday and asked me whether, in view of the fact that Halban's experiments with heavy water were perhaps finished for the time being, his heavy water could be sent over here, so that confirmatory experiments could be done to check the figure and enable the "go ahead" to be given for the heavy hydrogen project.' Fowler did not contemplate with pleasure the risk of losing the heavy water in the Atlantic, and Thomson therefore sent, instead, a copy of Halban's actual report.

Even this failed to assuage the Americans' thirst for information and in April Thomson received a somewhat anguished cable from the British office in Washington saying: 'critical study by Chadwick might carry necessary weight for decision. Please expedite.'

However, there now fortuitously appeared an even better way round the problem. Professor K. T. Bainbridge, who had worked with Aston in Cambridge in the early 1920s and had investigated isotope separation in Harvard early in 1940, was visiting Britain on behalf of the N.D.R.C. Permission could surely be obtained for him to attend a special meeting of M.A.U.D. There was to be no problem. 'This being purely academic scientific work,' pronounced a Ministry of Aircraft Production minute, 'it scarcely seems necessary to obtain further authority.'

In the words of the M.A.U.D. Committee's minutes, since Professor Bainbridge was in Britain, 'the present meeting had been called to discuss the question of heavy-water production in his presence'. Chadwick, Frisch, Peierls, Cockcroft, Thomson, Halban and Kowarski were among those present, and virtually every aspect of the British effort was discussed in some detail. But the main object of the exercise was clear. 'The Chairman,' say the minutes, 'stated that Dr Halban's work showed that a plant using a large quantity of heavy water would give a chain reaction but it would be useless as a bomb. It might, however, be a satisfactory source of power and it would, of course, be an advantage in obtaining the necessary appropriations of money for the bomb project if we could show some fission process actually working.'

Two important points stood out from the meeting and were obvious to Bainbridge. One was that all those present believed it would be possible for Britain, perhaps with the help of Canada, to make a nuclear weapon before the end of the war, estimated to continue for at least another three years. Quite as significant, it was felt that Britain now had the technical knowledge needed to produce nuclear power, even though wartime conditions ruled out diversion of the electricity needed to provide the necessary graphite or heavy water for use as a moderator. But when the war was over everything would be very different. This was at once obvious to Vannevar Bush, soon to get an account of the meeting from Bainbridge together with a copy of the minutes.

The importance of plutonium was to be confirmed within a few weeks. In the United States, Lawrence at Berkeley had been urged to prepare samples and to make the necessary measurements. Before his work was finished Glenn T. Seaborg and Emilio Segrè at the University of California had bombarded uranium with neutrons, identified plutonium, and shown that it was subject to fission. Within a fortnight Seaborg was able to report to Briggs that plutonium was about 1·7 times as likely as uranium 235 to be fissionable with slow neutrons. Only a few weeks later Lawrence spelt out the new situation.

> An extremely important new possibility has been opened for the exploitation of the chain reaction with unseparated isotopes of uranium. ... It appears that if a chain reaction ... is achieved, it may be allowed to proceed ... for a period of time for the express purpose of manufacturing element 94 [plutonium]. ... If large amounts ... were available it is likely that a chain reaction with fast neutrons could be produced. In such a reaction the energy would be released at an explosive rate which might be described as [a] 'super-bomb'.

However, it was in Britain that these new prospects for post-war nuclear power were to produce the most interesting reaction. This was nothing less than an attempt to draw into the hands of a single company control of what its able leaders realized would be a key to post-war power. Looking back with the hindsight of forty years, the idea may seem audaciously enterprising even for the most enterprising of men. In the climate of 1940, when even Churchill could regard the nuclear weapon as only a rather more powerful explosive, the proposal was no more than that of a courageous merchant venture. Indeed, it came within a whisker of success.

From the first it had been unquestioned that Imperial Chemical Industries was the only company in Britain with the technical equipment, the financial resources and the experienced staff required if the country was to implement plans for either the bomb or the boiler. The company had been brought in at least as far back as November 1940 when a small amount of uranium hexafluoride was needed to test early theories on the separation plant. There were great difficulties in getting it, and as a result Simon had written to Lindemann to ask if he would use his influence with Lord Melchett to get him to take a personal interest in the production of the first three kilograms. Lindemann had obliged. Then, in March 1941, an Energy Co-ordination Committee was set up inside I.C.I. 'with the purpose of co-ordinating research and development work in I.C.I. with work going on in the universities on the project for utilizing nuclear energy'. Meetings were to be 'arranged as required with workers in I.C.I.' and 'research workers from the universities [would be] invited to attend meetings in which they are interested'.

The advantage of such close and mutually helpful co-operation was seen in an early stage of work on the isotope-separation plant planned by Simon when Dr Arms, the American in his team, approached I.C.I. with details of the membranes required – metal barriers in which there were 400 holes to the linear inch, or 160,000 holes to the square inch. The holes had to be 0·0003 inch (3/10,000 inch) in diameter and there would have to be a tolerance of no more than ten per cent in the mean hole size. In itself, this was a formidable requirement. But the gas to be passed through the membranes was uranium hexafluoride, about which only one fact appeared to be certain: that it was dangerously corrosive. 'Gumming up' of a very small percentage of the 160,000 holes in any one square inch would cause trouble, while the membranes would have to be produced not in thousands, not in tens of thousands, but in millions of square feet.

The problem was passed on to S. S. Smith, the research manager of I.C.I. Metals Division at Witton outside Birmingham who in the early spring of 1941 had a stroke of inspiration which proved to be one of the first steps in solving the intractable problem of separating isotopes on an industrial basis.

'I was coming to work one day,' he has said, 'and I can even remember the exact place in Birmingham where I looked up and saw an advertisement poster on a hoarding'. The poster included a photograph reproduced by the half-tone process, enormously enlarged; so enlarged, in fact, that the minute dots which comprise the photograph, and of which there are perhaps 150 to the linear inch, were blown up on the poster to the size of moth-balls, golf-balls and tennis-balls. This was the genesis of the idea, and it was printers and engravers who after more than a year's development work, made membranes for the prototype separation plant eventually built beside a government poison-gas factory in North Wales.

The links between Thomson's committee and the company were therefore similar to, and no more unusual than, those between the large aircraft firms and the officials of the Ministry of Aircraft Production and it is against this background that Halban, Simon, Dr Tuck who represented Lindemann, and a number of senior I.C.I. staff met in the company's Nobel House on 24 April to hear certain proposals. These proposals were, in brief, that I.C.I should take over the development of nuclear power and, in the process, Dr Halban. The company's vision of the future is suggested in a euphoric note on the 'Halban Scheme' later lodged with the M.A.U.D. Committee. 'Halban's work on nuclear energy has been principally concerned with the development of a source of power which might be used as an engine to drive a ship or a large aeroplane without any fuel consumption, and might be used as an industrial power source in peace or war,' it began. 'It would make the reorientation of world industry possible because the coal-mine would no longer be necessary.'

Since the French efforts in Britain might affect the bomb, they had been paid for by the Ministry of Aircraft Production, the note went on, but now that Simon's scheme for isotope separation appeared hopeful, it was being suggested that the French team's work might continue in the U.S.A. I.C.I. should repay the Ministry, then send Halban across the Atlantic where they would arrange for him to work with Du Ponts, probably in collaboration with an engineering firm. I.C.I., concluded the memorandum, 'would agree to let the Government have free user rights on any patents connected with the Halban scheme for military, naval or air-force purposes and similar arrangements would be made between Du Ponts and the U.S. Government.'

Nearly a month passed before I.C.I's. suggestions for a nuclear power take-over produced any official response. Meanwhile, however, the company was told that contracts were to be drawn up for a pilot isotope-separation plant to be built by I.C.I. and Metropolitan-Vickers. Only in June did the Ministry ask the company to produce a paper dealing with the broad prospects for civil nuclear power. No second request was required and there now began a period of very persuasive lobbying.

News of the fresh development did not spread very quickly outside an extremely limited circle. When it did, reaction among many scientists was critical and was epitomized by a letter which Oliphant wrote to B. G. Dickins, secretary of the M.A.U.D. Committee. 'I am very unhappy about the question of patent rights in connection with uranium work,' he said.

> I am told that there are arrangements specific to this particular work whereby many who are working specifically for the committee, and paid by the committee, are able to take out secret patents and retain a large share of the rights in these for themselves. This is entirely alien to the spirit of the times, and I feel strongly that arrangements should be made whereby the results of all work paid for by the Government remain the property of the Government whether patented or otherwise. This is of course the arrangement existing in

all other Ministries. Any other scheme of things will undoubtedly lead to untold difficulties and troubles if there should be peace-time applications which are profitable. At the same time M.A.P. will be laying itself open to large claims for compensation after the war.

Two days later the M.A.U.D. Technical Committee gathered for a meeting that was to be more important than even Thomson can have imagined. It was, in effect, to be a 'dummy run' for the final M.A.U.D. report and those attending included not only Lindemann but Dr Slade of I.C.I. and, the guest of honour as it were, Professor C. C. Lauritsen of the N.D.R.C., an ordnance specialist taking part in the exchange of information that had followed the Tizard mission.

The report's recommendations were passed unanimously: work should be pressed ahead with all speed for building plant in Britain to separate the uranium 235 isotope by the gaseous-diffusion method: at the same time every effort should be made to construct a uranium bomb when sufficient uranium 235 had been separated. This in itself was important, but more significant to the release of nuclear energy was the fact that within a few days Lauritsen was back in Washington. On 10 July he was explaining to a somewhat surprised Vannevar Bush that the British were confident not only that a nuclear weapon of immense power was a theoretical possibility but that it could very probably be made.

However, uncertainties still remained, and not only about the basic question of whether a self-sustaining chain reaction could be started. Even the effects of a nuclear explosion were in doubt. This was brought out in a paper from the United States which might have stopped the whole nuclear enterprise in its tracks. Had it done so, the incentive to restart it in post-war years would have lacked the spur of war, and the release of nuclear energy might have been reserved for the twenty-first century. The paper was by George Kistiakovski, a specialist in explosives and later an American President's scientific adviser. Kistiakovski's conclusion, put briefly, was that even if a uranium bomb did explode, its effects would be staggeringly less than anticipated. The reason was not connected with the scientists' recurrent fears that the bomb would swell up rather than explode, but rather with the way that energy would be produced and dissipated.

The only comparable blast which came to mind was that of the 7,000 tons of explosives which blew up in Halifax harbour during the First World War. But the explosives were loaded in a number of ships, and different results could be expected from an explosion started in an entirely different manner. The chairman of the M.A.U.D. Committee dealt with the matter in a typically informal way. 'I was working on various committees for the Ministry of Supply which dealt with the effects of ordinary chemical explosives,' Professor Sir Geoffrey Taylor has said, 'when Sir George Thomson told me there was the possibility of releasing a very large amount of energy in an explosion, and I guessed he was thinking of nuclear fission. But there was the problem of what exactly would happen. Sir George asked what mechanical effects such an explosion might produce.'

The results were discussed in a paper circulated to the Civil Defence Research Committee of the Ministry of Home Security; if not all that could be hoped for, they were reassuring – an atomic bomb would be only half as efficient, as a blast-producer, as a high explosive releasing the same amount of energy. But it would be enough. The search for a nuclear weapon was at least still worth while.

As Thomson prepared to finish his report in the summer of 1941, Lord Melchett threw his personal influence into the scales. No decision had yet been taken on the I.C.I. plan but Melchett may have sensed a feeling, at least among some scientists, that nuclear power was hardly a field for private enterprise. Yet hope remained as long as the whole nuclear operation was kept afloat, and it was essential that the proper decision should be taken on the M.A.U.D. report. Melchett's endeavours to see that the decision went the right way were justified by his earnest belief that the defeat of Nazi Germany came before any other consideration; and if he failed to appreciate how unacceptable private control of nuclear power would be in the post-war world his failure was no greater than that of most ordinary men in appreciating the implications of that power. As the official historian of Imperial Chemical Industries has said: 'Between June and October 1941 the small group in I.C.I. who knew of the atomic project in its entirety worked hard to get the development of nuclear energy for industrial purposes in the United Kingdom handed over to I.C.I. They very nearly succeeded.'

On 2 July, Lord Melchett sent to Thomson the rough draft of a suggested summary of the committee's deliberations which he hoped would be of help. After stating that the proposal was to make nuclear bombs, it continued: 'The practicability of the preparation of such bombs, and scientific methods of obtaining the sources of such nuclear energy, on an industrial scale, has been agreed to and approved by a

Scientific Committee of the highest standing in their report of 2 July, with the support (on the industrial side) of Messrs Imperial Chemical Industries Ltd.' The summary went on to compare the costs of nuclear bombs with those of conventional weapons and suggested that, tonnage for tonnage, nuclear weapons would cost two-thirds of conventional bombs. More significant, and less open to doubt, was the assertion that while delivery of 54,000 tons of conventional one-ton bombs would require thousands of sorties, an equivalent nuclear attack would need only thirty-six sorties.

Melchett, who had his ears very close to the ground, knew that whatever the M.A.U.D. Committee recommended it would be considered by the Scientific Advisory Committee of the Cabinet. He knew that whatever the S.A.C. decided, the final decision would rest with the Prime Minister. And he knew who would have a considerable responsibility for guiding Winston Churchill on such matters. 'Dear Professor Lindemann,' he therefore wrote on 2 July in a letter which enclosed the summary he had already sent to Thomson, 'I would particularly like to draw your attention to the last two paragraphs which I believe constitute a most important element in arriving at a decision.'

These two paragraphs went as follows:

8. This explosive effect has never yet been produced, and doubt might be cast upon the possibility of the explosion occurring in spite of the recommendations of the eminent scientists, practical chemists and engineers mentioned above, but the position must also be viewed in the light of the fact that it is known that the enemy are working on similar lines. As there is no defensive answer to this form of attack the first side to perfect this scheme will gain a decisive and crushing victory.

9. It therefore seems that the question of proceeding with the scheme is determined by this issue rather than the calculation of the chances of failure or success.

The M.A.U.D. report was finally signed by Thomson in mid-July. It consisted of a discussion on the use of uranium for a bomb, together with technical evidence; a note by I.C.I. on nuclear energy as an explosive; various other appendices including the long paper by Simon on the proposed isotope-separation plant; Professor Taylor's paper on the blast to be expected from a uranium explosion; and a report on the use of uranium as a source of power together with two appendices on the subject from I.C.I. Although the committee did not consider that the power project was 'worth serious consideration from the point of view of the present war', Halban and Kowarski might continue their work in the United States since 'some of their conclusions may have a bearing on our problem'.

I.C.I., who appear to have been aware of the M.A.U.D. Committee's findings well in advance of the Ministry of Aircraft Production, took the paragraphs on the power project as an encouraging sign. 'On 23 July 1941,' says the company's official history, 'the Secret War Committee of the I.C.I. Board agreed to start negotiations with the Ministry of Aircraft Production. Somewhere about this time McGowan [chairman of I.C.I.] discussed the M.A.U.D. Report, characteristically over lunch, with Lord Hankey, Chairman of the Government's Scientific Advisory Committee. It is impossible to believe that McGowan would have missed that opportunity to take soundings.'

However, it was with their recommendations about the bomb that the M.A.U.D. Committee took the decisive step. 'The committee,' it stated, 'considers that the scheme for a uranium bomb is practicable and likely to lead to decisive results in the war. It recommends that this work be continued on the highest priority and on the increasing scale necessary to obtain the weapon in the shortest time and that the present collaboration with America should be continued and extended especially in the region of experimental work.'

Here was something very different from the tentative German efforts, from the American initiative, and from the work which the French team had been able to carry out before the Germans had occupied their country. Here was a recommendation not in nebulous terms but in terms of usable hardware for something far more revolutionary in warfare than had ever before been considered.

However, the proposals were so far only recommendations. Not everyone took the view of Thomson and his colleagues, while the report itself had been produced, it should be remembered, by a subcommittee of a subcommittee about whose activities those in power knew very little indeed. Dr Pye, the Director of Scientific Research in the Ministry of Aircraft Production, which had inherited the M.A.U.D. Committee from the Air Ministry, officially received the report on 29 July. He sent it within a few days first to Tizard and then to the Minister, Colonel Moore-Brabazon, a brave pioneer aviator but comparatively innocent of scientific knowledge.

Dr Pye had many misgivings, believing that a fully developed bomb might require the work of a decade. Tizard – and Blackett, who produced a minority report – thought much the same, and as the

M.A.U.D. report began its journey through the official channels, the prospects of acceptance looked rather bleak.

Moore-Brabazon, delighted that he did not have to adjudicate on such an unfamiliar subject, formally passed the report over to Lord Hankey, Secretary of the Scientific Advisory Committee to the Cabinet, late in August. However, Hankey had been forewarned, and not only by Dr Pye who had informally sent him a copy of the report as soon as he had received it. Lord McGowan had discussed the subject with Hankey. So, apparently, had Lord Melchett and Professor Lindemann, by now raised to the peerage as 1st Baron Cherwell, a circumstance which caused Tizard, putting his own private views to Hankey, to comment: 'I hear that you have been discussing "Maud" in the highly optimistic atmosphere created by the enthusiasm of three well-known peers!'

Hankey had thus been well briefed when, late in August, he was formally asked that the S.A.C. consider the matter. A special panel consisting of Sir Edward Appleton, now Secretary of the Department of Scientific and Industrial Research (D.S.I.R.), Sir Edward Mellanby, head of the Medical Research Council, Sir Henry Dale, President of the Royal Society, and the Society's two Secretaries, A. C. Egerton and A. V. Hill, was set up to consider the M.A.U.D. findings. But while this august body was still deciding what should be done about nuclear fission, the decision to press ahead with the bomb was taken quite independently by Churchill on the advice of Cherwell.

On the day that Moore-Brabazon had officially sent the M.A.U.D. report to Lord Hankey, Cherwell effectively by-passed the entire official machinery by sending his own briefing to Churchill. A masterpiece of analysis and summary, two and a half pages long instead of the 'half-page' which Churchill normally demanded, the minute, written on 27 August, succeeded in describing nuclear possibilities, the dangers as well as the opportunities, in words which any educated man could understand.

The Government was really faced with two questions, Cherwell pointed out. The first was whether nuclear research should continue. The answer to that was yes; at least for six months, after which the situation should be reviewed. Secondly, should a search be made now for a suitable site for the isotope-separation plant, or should the search only start in six months' time, if the decision was then made to continue with the work. Start now, was the recommendation, after which Cherwell went on to suggest where that site might be.

We have worked in close collaboration with American scientists. Nevertheless, I am strongly of opinion that we should erect the plant in England or at worst in Canada. The reasons against this course are obvious, i.e. shortage of manpower, danger of being bombed, etc. The reasons in favour are the better chance of maintaining secrecy (a vital point) but above all the fact that whoever possesses such a plant should be able to dictate terms to the rest of the world. However much I may trust my neighbour and depend on him, I am very much averse to putting myself completely at his mercy. I would, therefore, not press the Americans to undertake this work; I would just continue exchanging information and get into production over here without raising the question of whether they should do it or not.

These proposals will go before Lord Hankey's Committee which will no doubt endorse them. He is in favour, as is also Sir John Anderson. The latter, I gather, started life as a physical chemist and by a strange coincidence researched on uranium in early life. I understand he would be glad to take charge of the whole thing on the ministerial level.

People who are working on these problems consider the odds are ten to one on success within two years. I would not bet more than two to one against or even money. But I am quite clear that we must go forward. It would be unforgivable if we let the Germans defeat us in war or reverse the verdict after they had been defeated.

Three days later Churchill sent Cherwell's brief to the Chiefs of Staff. With it went the following note: 'General Ismay for Chief of Staffs Committee: Although personally I am quite content with the existing explosives, I feel we must not stand in the path of improvement, and I therefore think that action should be taken in the sense proposed by Lord Cherwell, and that the Cabinet Minister responsible should be Sir John Anderson.'

On 3 September Churchill met the Chiefs of Staff. Always in favour of improved explosives, they recommended that neither time nor labour, neither materials nor money, should be spared in building a uranium bomb as soon as possible. But, like Cherwell, they were not anxious to see Britain dependent on another power. It was therefore decided that she should press on alone, without U.S. help, and should tackle, at the height of the war, with all her resources dangerously stretched, a task which was to test the country severely after the war.

The fact that the Scientific Advisory Committee to the Cabinet had thus been pre-empted by Churchill and Cherwell was not as surprising as it sounds; when the committee had been set up slightly less than a year previously, Churchill had agreed to its formation only on condition that 'we are to have additional support from the outside, rather than an incursion into our interior'.

But the different, although linked, question of nuclear power had not yet been decided, even by Churchill and Cherwell, and throughout September I.C.I. still believed that it would be granted permission to take over Halban – and possibly other members of the Cambridge team. Plans were therefore drawn up by the company to co-ordinate all nuclear power under the I.C.I. War Committee.

I.C.I. had every reason to feel confident. Thomson and Cherwell were not the only influential men who believed that the prospects for post-war nuclear power should be pressed ahead with all the force of private enterprise. On 12 September the Minister for Aircraft Production wrote a 'Highly Secret' note to Hankey. After dealing with the portion of the M.A.U.D. report which dealt with the bomb, he continued:

I am a little worried lest what I will call the 'Commercial' aspect, for want of a better term, is side-tracked, a contingency which I think would be a national disaster, especially having regard to the great interest shown in the study of the subject in America and in other countries. My own view is that my Ministry has no direct responsibility for this commercial development. Nor has any Government Department the means at its command for the rapid development of a project of this kind. My suggestion, therefore, is that the whole question of development should be placed in the hands of Imperial Chemical Industries under terms to be agreed with the Treasury Solicitor acting on a general directive issued by the War Cabinet.

I do not regard this Ministry as being the proper Department to handle the matter. In order, however, to prevent any avoidable delay, I am having examined, entirely without prejudice, a draft Agreement, submitted by Imperial Chemical Industries, under which my Ministry releases in favour of I.C.I. such rights as it may have in the inventions already filed, or to be filed in future, in return for proper reservations in favour of H.M. Government.

I am impressed by the extreme need for speed in dealing with the matter, and should, therefore, be grateful if you could let me have your views as quickly as possible.

Yours, ever,

(Sgd.) Brab.

The S.A.C.'s special panel was meanwhile concentrating its energies on the needs of the war, and taking evidence from a number of experts – Cherwell, Tizard, Fowler, Blackett, Pye and Dr Guy of Metropolitan-Vickers who had been supervising work on the separation plant. Their report was finished before the end of the month and delivered to

Anderson on the 25th. They had come to two main conclusions: the first was that 'the development of the uranium bomb should be regarded as a project of the very highest importance'. The second was that nuclear power was a long-term matter over whose research and development the British Government should maintain close control. 'The matter,' it stressed, 'should not be allowed to fall into the hands of private interests but should be pursued in close collaboration with the Canadian and United States Governments.'

This was of course the opposite of Moore-Brabazon's opinion. In between there lay the view of the Minister's Director of Scientific Research, sophisticated enough to appreciate the problems that might arise. There was no doubt in his mind, he said in a minute dated 29 September,

> that I.C.I. Ltd is peculiarly fitted to undertake the development of a scheme of this kind; and the decision to be taken on the Company's proposals appears to depend chiefly on the desirability of putting into the hands of any industrial organization the monopoly of what might turn out to be a revolutionary source of power, especially when it is realized that during the development work which will be involved discoveries of a military value might conceivably be made. The question to be decided, therefore in my opinion is not whether I.C.I. Ltd shall be entrusted with the development of the scheme, but whether they shall proceed with or without Government control. The agreement proposed by I.C.I. Ltd would take control entirely out of the Government's hands.

Yet a decision still had to be taken on the report and throughout the first part of October, while its recommendations were still technically secret, I.C.I. began hiring staff for the reactor project, and McGowan wrote to 'Dear Brab.' what was evidently intended to be a clinching letter.

'In view of your discussion with the Lord President, I think that I ought to put in writing the substance of our conversation yesterday,' he began.

> My Company has made a careful study of this new potential development of the use of Uranium for the production of energy, and we feel that it is one of extreme importance. ...
>
> In Dr Halban this country is fortunate in having one of the leaders in this new development, but he can achieve nothing without adequate scientific and, in the near future, substantial technical assistance.
>
> We do not think that the development of this process can be undertaken at a

sufficiently rapid rate by the Department of Scientific and Industrial Research or any other Government organization because ... the whole of their scientific resources are being used, obviously rightly, in the investigation of more direct military interest. Likewise our Research Laboratories are almost entirely occupied in the investigation of problems connected with National Defence but I am confident that, with the help of our organization in the United States, and our very close relationship with the Du Ponts and other large American companies, we can arrange for Dr Halban to carry on his work there, where facilities for work of this kind are still available.

Apart from this it is our belief and experience that the normal procedure of Government research organizations is not best suited for a development of this kind because decisions, involving the expenditure of large sums of money, have to be taken at very short notice. To a considerable extent such expenditure is, in the early stages at any rate, in the nature of a gamble and it is not easy for an official body to spend money in this way whereas it is part of the normal research routine for an industrial company like mine.

However, despite this cosy arrangement between I.C.I. and the Ministry, the company was told that the Government's mind was not yet made up. They would have to wait another week. The seven days passed. Then it was pointed out that the decision rested with Anderson, the Lord President of the Council to whom the S.A.C. formally reported. Any work undertaken by I.C.I. would meanwhile be carried out at the company's own financial risk.

The final decision did not come for a further week and from the few details even now available it appears that a good deal of in-fighting was involved. But after meetings on 15, 16 and 17 October between Anderson, Appleton, whose D.S.I.R. had been given responsibility for work on the uranium bomb, and I.C.I. officials, the decision was given: there would be developed through the mechanism of the D.S.I.R. all aspects of uranium work. The operative word was 'all'.

'So,' notes the official I.C.I. history, 'ended I.C.I.'s bid to become the United Kingdom Atomic Energy Authority. It was made under a total misapprehension. McGowan had no idea – nor had anyone else – how big a thing he was offering to take on. It is hardly conceivable that I.C.I. could have contained both nuclear energy and the post-war chemical industry, and a sustained attempt to do so might have been disastrous.' There also appears to have been another misapprehension, hinted at in the I.C.I. history which prints McGowan's letter with certain cuts. Some 'rather nervous phrases about the Lord President in McGowan's letter to Moore-Brabazon', it comments, 'give good reason for thinking that Sir John Anderson, though in the past and again in the future a director of I.C.I., was himself against the idea.' The 'though' is a nice gloss on the loss of great expectations.

Lord McGowan was now informed that while I.C.I. would of course continue to be employed on the bomb project, the boiler project would remain under Government control. Halban would remain a civil servant. And while the boiler project might eventually be pursued – as it later was at Chalk River, Canada, where the first reactor outside the United States provided the experimental tool around which Britain's post-war nuclear work was to grow – it would be pursued partly in the hope of producing a second string to the nuclear bow in the form of plutonium.

The Scientific Advisory Committee's recommendation that nuclear power should remain under Government control sprang from an instinctive feeling that if it were viable at all it would represent far too immense a power to be left in private hands. There was, however, another factor which would have made any other decision at least unfortunate and very probably disastrous. This was the American suspicion that while Britain was ostensibly concentrating all nuclear research on a weapon to win the war, she was in fact simultaneously trying to gain a lead in the race for post-war nuclear power. The papers now available do not remove this suspicion. Had the Anglo-U.S. controversy that was to grow during the last three years of the war been merely a circumscribed argument between allies, the matter would have been of comparatively little enduring importance; as it was, the mounting distrust made almost inevitable the clutch of post-war failures to deal with the perils of nuclear fission in either war or peace.

The S.A.C.'s advisory panel had been heavily weighted with medical experts and their conclusions contained one paragraph which to some extent remedied the M.A.U.D. Committee's failure to give much consideration to the effects of radiation, and at the same time cast a long shadow ahead. 'The Medical Research Council,' this said, 'should be invited to nominate experts to co-operate with the M.A.U.D. Committee and with the Experimental Station at Porton in a fuller study of the range and extent of the radioactive effects of the bomb explosion and the feasibility of obtaining such effects by a more gradual release of the energy.' The danger of radioactive fall-out was to become one of the great arguing points in later debates on nuclear energy, both in war and in peace; and the suggestion of 'gradual release', which did not then appear feasible, was to resurface three decades later in the proposals for a neutron bomb.

As a result of the S.A.C. recommendations there was, in October 1941, a swift reorganization of Britain's nuclear effort; so swift that many of the scientists learned that they had been transformed overnight from academics to civil servants only when the deed had been done. The new body set up to handle all aspects of nuclear energy was 'Tube Alloys', the code-name for a division of the D.S.I.R. which was to operate under a consultative council chaired by Anderson. To run the new organization there was Wallace Akers, the Research Director of I.C.I., and his assistant Michael Perrin. 'This,' declared a senior civil servant to Moore-Brabazon, 'verges on the Gilbertian! The Government decide to turn down I.C.I.'s request to be allowed to run the developments, and then ask I.C.I. to let them have the chief protagonist of the I.C.I. request, to manage the matter for the Government.

'There is nothing we can do about it.'

The two men were, in fact, by far the best qualified for the jobs and the autumn of 1941 thus saw Britain efficiently preparing for the tasks ahead. In the United States, also, it was now being realized that the release of nuclear energy appeared to be a practicable proposition and, despite Britain's reluctance to place herself, in Cherwell's words, at the mercy of her neighbours, some merging of the two efforts might eventually take place. Russia, now unexpectedly still holding the main thrust of the German forces, would surely have to be brought into the enterprise? How lucky, it must have seemed, that the necessities of war would push men into a planned and concerted effort to release and control this key to a bright new future. How fortunate that chance was opening the way to exploitation of this arbiter of the post-war world by a company of united nations.

But it was not to work out exactly like that.

CHAPTER 7

All Honourable Men

During the first months of 1942 few men realized the size of the problems which the release of nuclear energy would raise. The scientists were not entirely certain to what extent theory could be translated into practice. Frisch and Peierls might be satisfied on intellectual grounds that two sub-critical pieces of uranium 235 would produce an enormous explosion when brought together. Halban might be equally convinced that a slow-burning nuclear chain reaction could be kept alight in the right sort of assembly to produce almost limitless power. Yet neither they, nor any other nuclear physicist, would have been prepared to guarantee that unexpected barriers would not block the path ahead. As for the engineers, they were still faced with difficulties they would not have been prepared to tackle in peace-time and which they were prepared to tackle now only because fear of a nuclear-armed Germany presented such awful possibilities. Businessmen on both sides of the Atlantic considered nuclear energy, when they considered it at all, as a long shot on which it would be justifiable to gamble quite an amount; within the last century men had made fortunes in exploiting electricity, the petrol engine, the wonders of the synthetic dyestuffs industry and the marvels of wireless. Who could deny that this latest discovery might not provide an even more important item on the balance sheet?

Few of the sailors, soldiers and airmen at the top of the Service ladder had been briefed with even the vaguest hint of nuclear possibilities: naturally enough, they wanted the most efficient weapons for winning the war into which the politicians had allowed them to be thrust with so little time for proper preparations. As for the politicians themselves, it was not entirely that they failed to glimpse the future. Had they done so they would have been in good company – even

Lincoln could admit: 'If we could first know where we are and whither we are tending, we could better judge what to do and how to do it.' But Roosevelt's admonition to take care 'that the Nazis don't blow us up' and Churchill's urbane assurance that he did not wish to 'stand in the path of improvement' when it came to better explosives, both suggested something less than complete understanding of what nuclear fission meant. And, in any case, there were more pressing matters for attention. The war had to be won, decisions of life and death had to be taken, and the daunting path between triumph and disaster had to be trodden with all the care that could be concentrated.

Even so, the failure which marks the nuclear story from the start of Anglo-U.S. collaboration is not entirely rooted in ignorance and overwork. Were this the case, the element of tragedy would be missing. It appears, rather, as the inevitable result of the caution, if not the suspicion, that underlies the best of alliances; of industrial rivalry; of differing estimates of Russian intentions; and of a failure to realize that in the nuclear age to which all were being driven by mutual fear, one of Benjamin Franklin's sayings was particularly apposite: 'If we do not hang together we shall all hang separately.' Thus neither Roosevelt nor Churchill, neither Stalin nor Mr Mackenzie King, the Canadian Prime Minister, nor in later years Truman or de Gaulle, were able to halt a growth in distrust and suspicion that brought in the Cold War, the nuclear arms race, the economically crippling policy of peace through terror, and the threat of the 'plutonium society'.

Yet this fumbling of the greatest catch in human history arose from the actions of honourable men. During the last third of a century a diverse collection of attempts have been made to allocate blame. Yet it is easy to see the logic, almost the inevitability, of even the most controversial acts and attitudes. General Groves, the American leader of the immense Manhattan Project which built the world's first nuclear weapons, was later to maintain that within two weeks of taking charge, he had no illusion 'but that Russia was our enemy and that the project was conducted on that basis'. Considering the already long record of Communist tyranny, and the disastrous consequences of the Russo-German Pact of August 1939, the attitude was not unnatural, even if it did not encourage tripartite collaboration. Lords McGowan and Melchett of I.C.I., staking a claim for the company's monopoly of post-war nuclear power, were planning no more than the best thing for the company's shareholders. The Americans who were to force Halban from control of the Canadian project on which Britain's post-war plans were to be built, were wrong in doubting the security of those they termed 'foreign *émigrés*'; but the Englishman Allan Nunn May in Montreal and the German-born Klaus Fuchs who had with British security clearance entered Los Alamos, the nuclear holy of holies, showed that their caution was not entirely unwarranted. American distress on learning that the French team had not only taken out patents on various nuclear processes but had come to a private agreement with the British Government, was reasonable if upsetting. For their part, the Americans secretly bought Canada's entire output of heavy water without informing the Canadian Government of the fact, an operation carried out for the most patriotic of motives. As to suspicions that even at the height of the war Britain was prepared to divert effort to post-war nuclear power, these were not entirely unjustified.

The interlocking fears and suspicions – that under cover of working on the bomb the British were planning for the post-war nuclear power race; that I.C.I. was intent on stealing an industrial march on Du Pont; that an inefficient British security system might allow American secrets to reach German hands and that the politicians might even present material to the Russians – were soon to be affected by two other factors. Before the end of 1941 it became evident that fissile plutonium could be produced in a nuclear reactor, and with this the wartime bomb project and the peace-time boiler became inextricably linked. Secondly, the power balance radically changed: the United States, previously anxious to gain all possible information from the British, was by the end of 1942 already devoting so many men, so much money and material to the project that Britain had ceased to count. All this might not have been disastrous if merely the wartime project had been involved, and if the rift between the Western Allies and Russia had not steadily grown. As it was, the decline in nuclear trust which ran parallel to the Americans' nuclear and engineering success prepared the way for the post-war future.

When Tube Alloys was set up in the autumn of 1941, the first tentative moves towards Anglo-U.S. collaboration in unleashing nuclear energy were already more than a year old. The Tizard Mission of September 1940 had paved the way not only for the appearance of Bainbridge and Lauritsen at meetings of the M.A.U.D. Committee, but for the quasi-official exchange of documents that increased as the Americans became aware of the progress being made in Britain. Their interest was such that at the end of July 1941 Bush and Conant had proposed to Dr Charles Darwin, in Washington, that nuclear research should be undertaken as a joint project by the United States and Britain. Darwin passed on the proposal to Lord Hankey, adding his own suggestion that a British mission should be sent to Washington to discuss the matter. The British, anxious not to rely too heavily on even their best friends and equally anxious about security in a country not yet at war, did not reply at once. Indeed, they did not reply until more than seven months later, when on 23 March 1942 Anderson sent a non-committal answer. 'The British,' the official British historian has noted, 'haughtily replied that they mistrusted American security.'

Long before that, the prospects of collaboration had been dramatically changed. October 1941 was the vital month: while the British were at the highest level debating whether I.C.I. should be allowed to control the post-war nuclear energy industry, the Americans were being made increasingly aware of Britain's encouraging nuclear prospects. On the 3rd, Thomson in Washington personally handed copies of the M.A.U.D. report to Bush and Conant.

The impact of the report, and the reason for its effect on a country not yet at war, have been succinctly outlined by the official historians of the American enterprise.

> This report gave Bush and Conant what they had been looking for: a promise that there was a reasonable chance for something militarily useful during the war in progress. The British did more than promise; they outlined a concrete program. None of the recommendations Briggs had made and neither of the two National Academy reports had done as much. The scientists at work in the United Kingdom were no more able or advanced than the Americans. Fundamentally, the trouble was that the United States was not yet at war. Too many scientists, like Americans in other walks of life, found it unpleasant to turn their thoughts to weapons of mass destruction. They were aware of the possibilities, surely, but they had not placed them in sharp focus. The senior scientists and engineers who prepared the reports that served as the basis for policy decisions either did not learn the essential facts or did not grasp their significance. The American program came to grief on two reefs – failure of the physicists interested in uranium to point their research toward war and a failure of communication.

Less than a week after Bush had read the M.A.U.D. report he told Roosevelt of the British belief that a nuclear weapon could be made from uranium 235 produced by gaseous diffusion. Roosevelt was convinced. He endorsed complete interchange of information with the British and told Bush to consider policy on the matter within a group that consisted of Roosevelt himself; the Vice President [Henry Wallace]; Stimson, the Secretary of War; General Marshall [Chief of Staff]; Bush and Conant. Finally, as he was about to leave, Bush was told to draft a letter for the President's signature which would open discussion 'at the top'. Two days later Roosevelt signed the letter to Churchill which Bush had drafted.

Meanwhile, the Americans had been moving with their customary speed and efficiency. Less than a fortnight after Bush and Conant had received the M.A.U.D. report, plans had been completed for George Pegram and Harold Urey to visit Britain and assess for themselves what the state of the British work really was. They arrived in London on 27 October, met Appleton, Akers and Perrin, and were given complete details of the entire British project. In Liverpool Chadwick explained the work that had been going on for the last two years. In Bradford they were shown the first membranes for isotope separation and in Oxford they heard from Simon and Peierls how the problems of incorporating millions of membranes in a separation plant were being overcome. In Cambridge they were convinced by Halban and Kowarski that only lack of material had prevented the team from creating the world's first self-sustaining chain reaction. The Americans were handed by Halban a technological memorandum outlining the entire ideas of his group on the power project and on plutonium. And in Cambridge Harold Urey, the discoverer of heavy water, was handed a can containing a gallon of the substance which previously he had only seen in microscopic quantities. Pegram and Urey reported enthusiastically to Washington. While this was happening, F. L. Hovde, the U.S. scientific liaison officer in London, was handing over the full text of Roosevelt's suggestion of co-operation. 'It appears desirable that we should soon correspond or converse concerning the subject which is under study by your M.A.U.D. Committee and by Dr Bush's organization in this country, in order that any extended efforts may be co-ordinated or even jointly conducted,' it read. 'I suggest for identification that we refer to the subject as MAYSON.'

Hovde now elaborated on the President's intentions to Anderson and Cherwell. Roosevelt, he explained, 'thought it desirable that this work should now be pressed forward with all possible speed, and he fully endorsed the view taken by Dr Bush. A review of the scientific possibilities was now being prepared for the President's information: there would be no difficulty in making available all the funds required for the expansion of research work now in progress: and Dr Bush would be prepared to assume personal control and direction of the whole work.'

Churchill appears to have relied on Hovde to give Roosevelt his reactions. Certainly the President received no hurried reply. There is even doubt as to whether he received any reply at all. In the British files there exists, it is true, the carbon of a letter dated 'December 1941' and running as follows:

My dear Mr President
Thank you so much for your letter of October 11 on the subject of MAYSON.
I need not assure you of our readiness to collaborate with the United States Administration in this matter. I arranged for Mr Hovde to have a full discussion with the Lord President (Sir John Anderson) and Lord Cherwell, and I hope it will be possible for them shortly to hand him a detailed statement for transmission to America.

But two years later Cherwell, reminding Churchill of Anglo-U.S. exchanges, told him: 'There is, however, some doubt whether your letter was ever sent.' An archivist of the Franklin D. Roosevelt Library at Hyde Park, New York, states, moreover: 'We have checked our file of Roosevelt–Churchill correspondence and have not found a copy of the December 1941 letter ... or any indication that a reply to the 11 October letter was received.'

While Roosevelt was waiting for 'official' British reaction, Norman Brook, Anderson's personal assistant at the Privy Council Office was reassuring Akers: 'We can proceed at once to draw up the statement of our concrete suggestions both as regards security and otherwise.'

This was the situation at the end of November 1941. It was the British who had most to give and who from the higher ground could suggest to the Americans how they might best manage their security and their 'otherwise'. The situation was to be transformed, not on 8 December as the Japanese catapulted the Americans into the war at Pearl Harbor, but during the months that followed. Even before December 1941 was out it had been decided greatly to expand the American nuclear effort. By the late spring the building of the bomb was being put on a military basis and in September General Groves was appointed to run what became the Manhattan Project. Isotope separation was to be tackled by a gaseous-diffusion plant which would cost two million dollars in the first year and an unlimited amount afterwards, by a centrifuge plant costing 38 million, and an electromagnetic separation plant costing 12 million. In addition one or more plutonium-producing piles were to be built at a cost of 25 million each, as well as a heavy-water plant costing 2·8 million dollars.

Thus within less than a year after Roosevelt had tentatively inferred that the Americans might be grateful for Britain's technical expertise, their own industrial investment in nuclear weapons had far outstripped anything that Britain could even contemplate. She might, as Churchill, Cherwell and Anderson all wished, carry on with her own endeavours as planned. But as the full scale of the effort needed became more obvious, so did the prospects of this shrink. From the start of 1942 it became apparent that the British would be leaning on the Americans rather than the other way round. As it turned out, 1942 was to be a year of increasing breakdown in Anglo-U.S. collaboration in nuclear affairs; so much so that Churchill was eventually considering the barren prospect of Britain continuing on her own.

It has often been assumed that the main trouble lay in control of the U.S. enterprise by the Army in general and General Groves in particular. This is not so. It is true that Groves, a prickly and self-opinionated Army engineer, had a built-in mistrust of scientists and no outstanding liking for the British. Yet his position in the demonology of Anglo-U.S. relations is considerably overrated and it is notable that James Chadwick, a model of the reserved, aloof Englishman, had a respect for Groves that verged on admiration. Even on the controversial subject of compartmentalization, Akers noted in a letter to Halban, 'I must say that I like [Groves] and feel that he is honest, even if very misguided in our view.' The problems which arose, and which by the end of the year had brought Anglo-U.S. co-operation to the point of complete breakdown, had a more complex origin than that of simple soldier versus intellectual, as a chronological account makes clear.

During the first three months of 1942 Akers, Simon and Peierls all visited the United States where, notes an official résumé of Anglo-U.S. relations, 'the same complete freedom of investigation was allowed as had been accorded to the Americans in England'. In the expansion of the American effort which had followed Pegram and Urey's visit to Britain, a major effort was being mounted in Chicago to create a chain reaction with uranium and graphite, and it was now suggested by the British that Halban and his team should move to Chicago, thus centralizing the slow-neutron work. This was agreed by members of the American Technical Committee but turned down by Bush on the ground that it was impossible to employ any number of foreign-born nationals on secret U.S. projects.

The fact, reasonable enough in the circumstances, no doubt confirmed British caution and influenced Anderson when, on 23 March, he replied to Bush's proposal of some seven months earlier and turned down the idea of joint operations. Bush replied agreeably enough, proposing that when 'the pilot plant on either side of the water is ready for its tests, there should most certainly be visitors from the other side to observe its operation', while at a later stage it would be 'highly desirable that future action should be considered jointly'.

The first hint of the change to come arrived in a letter from Bush to Anderson in June. It described the major reorganization of the American effort in friendly terms but concluded with what must have seemed an ominous sentence: 'We will continue, on this matter as well as other phases of O.S.R.D. [Office of Scientific Research and Development] work, to adhere to the principle that confidential information will be made available to an individual only in so far as it is necessary for his proper functioning in connection with his assigned duties.'

This was one aspect of the controversial policy of compartmentalization favoured by Groves. 'In fact,' Akers later wrote to Perrin, almost in disbelief, 'he states that his intention is to divide the work into as many separate compartments as can be devised, and to fill each of these compartments with as many people as can possibly be used efficiently therein.

'He explains that only one person in each cell will be able to see over the top, as he expressed it, and that person will not be able to see into more than a minimum number of other cells.'

The attitude was rather different from the mutual trust which had been expected, and it was totally at odds with accepted scientific practice. However, it was being enforced by the Americans as much against their own nationals as against the British; furthermore, in the light of what is now known of Russian espionage activities inside the Manhattan Project, it seems possible that Groves was justified on a strictly profit-and-loss basis.

Even before the Army was given control, the Americans had adopted, after the attack on Pearl Harbor, a sensitiveness to security that was to have some peculiar results. One of the most curious came when Bush sought Einstein's aid. Many purely theoretical problems had to be solved before there could be hope of success with gaseous diffusion. Einstein was one man who could help with their solution, and the request for help was passed on to him by Dr Frank Aydelotte, director of the Institute for Advanced Study at Princeton, New Jersey, where Einstein had been working since 1933.

'As I told you over the telephone, Einstein was very much interested in your problem,' Aydelotte replied on 19 December to Bush in a letter which accompanied Einstein's handwritten answers to the problem.

> [He] has worked on it for a couple of days and produced the solution, which I enclose herewith. Einstein asks me to say that if there are other angles of the problem that you want him to develop or if you wish any parts of this amplified, you need only let him know and he will be glad to do anything in his power. I very much hope that you will make use of him in any way that occurs to you, because I know how deep is his satisfaction at doing anything which might be useful in the national effort.
>
> I hope you can read his handwriting. Neither he nor I felt free, in view of the necessary secrecy, to give the manuscript to anyone to copy. In this, as in all other respects, we shall be glad to do anything that will facilitate your work.

Einstein's calculations were passed on to Harold Urey. But a difficulty then arose. It was impossible for Einstein to tighten up the calculations if he was not given more information. But this, Bush confided to Aydelotte, was impossible. 'I wish very much that I could place the whole thing before him and take him fully into confidence,' he wrote, 'but this is utterly impossible in view of the attitude of people here in Washington who have studied into his whole history.' Einstein's pre-war pacifism was now coming home to roost; and the man who had signed a letter warning that a single nuclear weapon might destroy a whole port was prevented from knowing where gaseous diffusion fitted into the defence picture.

If the Americans felt it necessary to be cautious in their dealings with Einstein, it was understandable that they should have reservations in their dealings with the British. But this was hardly acceptable to Anderson who virtually ignored the semi-snub when he replied to Bush's note about confidential information. He had a delicate suggestion to make. He confessed that the work in Britain was going more slowly than had been hoped, mainly due to the other demands of the war. 'After full consideration of the position,' he went on, 'I have accordingly reached the conclusion that we should try to find some way of arranging for our pilot plant to be built in the U.S.A. This is also the view of the Prime Minister.' It would, he continued, be 'starting a fifth horse on equal terms in the joint race'. But the British should, he proposed, become members of Conant's committee. Halban, he added, should go to Canada where he would continue work on the slow-neutron research.

Bush politely declined this johnny-come-lately offer. 'I fear that you have a misconception at several points,' he somewhat frigidly replied. The Americans, also, he pointed out, must take care not to over-stretch themselves and they were already concentrating on every available method of producing a bomb for use in the war against the Axis powers.

From this genuine concentration on the war effort, and the war effort alone, there arose the tragic disagreements between the United States and Britain from which so much post-war insecurity developed. The Americans suspected, not without cause as it is now known, that Britain – whatever her spokesmen might aver – had the prospects of post-war power constantly at the back of their minds. In addition there was to be the open admission by Cherwell, at what Bush was to call 'a most extraordinary interview', that the British were considering the whole nuclear situation 'on an after-the-war military basis'. Britain would then have to keep ahead not only of her German enemies but of her Russian allies.

The crucial period of these deepening suspicions began during the early months of 1942 and was only ended in the late summer of 1943 by the virtual capitulation of the British. As Churchill himself was to say of a different matter, in history chronology is all, and it is the chronology of Anglo-U.S. nuclear negotiations during this period which lays bare the honest but conflicting aspirations of both sides.

Suspicions first arose over Britain's attitude to post-war nuclear power. 'Doubts about British motives had risen in the summer of 1942,' the official historians of the American project have stated.

> They took deep root during the discussions with Akers, whom Bush, Conant and Groves all judged to be an Imperial Chemical Industries man at heart, more interested in nuclear power plants than a bomb. The United States, Bush pointed out, was eager to transmit information when it would further the joint war effort. But it should not encourage Britain's post-war aspirations merely as an incident to developing the bomb, particularly when this threatened the security of the weapons program. If Britain expected the United States to yield findings produced at such great expense and effort for a purpose other than prosecuting the war, the question ought to be considered on its merits.

Groves was in no doubt about the position, and on the official U.S. report of a meeting in 10 Downing Street the following year between

Churchill, Cherwell, Anderson, Stimson and Bush, observed in a footnote: 'It was clear to me that Mr Akers was thinking primarily of commercial advantage to Britain after the war during his conferences in the fall of 1942.' Groves' suspicions were not without foundation. The previous year an internal minute in the Ministry of Aircraft Production had noted: 'If Halban goes to the U.S.A., it will be with the idea of developing processes arising from the work he has carried out in this country. Since we are passing to the U.S.A. all the knowledge gained over here on this matter, it has been suggested that steps ought to be taken to ensure that this country shall obtain an appropriate share of any benefits resulting from the immense industrial developments which will arise if the work is successful.' Shortly afterwards, David Pye stressed the same idea in a letter to Dr Charles Darwin, noting that while Halban's work had 'no direct application to the war effort, ... we are nevertheless anxious that it should be continued'. And as it was to be officially stated years later in Britain, while some of those working in Tube Alloys had gone to the United States to aid the bomb project, others '(notably a group of French refugee scientists) went to Canada where a combined research project was started with the object of developing civil applications of atomic energy'.

This approach no doubt showed through. The American reaction was rather brusque, but it might have been even more so had their extensive intelligence system gleaned the wording of a letter which Anderson sent to Malcolm MacDonald, Britain's High Commissioner in Canada, in August 1942. The Americans working in Chicago on the production of plutonium in a graphite-moderated pile were doing so because 'it is likely to prove the quickest method of obtaining the element in quantity for the military purpose.

'The British "94" team, on the other hand,' Anderson continued, 'are working on a system for the production of the element by using "heavy water" or even ordinary water, if uranium enriched with uranium 235 can be obtained. This is a longer-term project because it will be some time before either heavy water or enriched uranium is available. But it is generally agreed that it will in the end prove the more efficient method and the one which will eventually hold the field for the purpose of power production.'

But the Americans, already taking the lead, were not anxious to reveal to the British more than was necessary, and were quick at quoting security perils and strained resources for their actions. The British – with or without the unspoken blessing of I.C.I. – were anxious to secure their industrial position in the post-war world. This was the unhappy situation when, in the early autumn of 1942, it was finally decided that the French team should move to Canada.

The British had been verbally reassured more than once that the Chicago group was so occupied with work on a graphite-moderated pile that nothing would be put in the way of the French team if it moved to Canada. Earlier in the year the U.S. Government had signed an agreement with a private Canadian company to finance a plant at Trail, British Columbia, for the production of heavy water – a direct result of the French team's work in Cambridge – and it was accepted that some amicable arrangement would be agreed for disposition of the Trail plant's output. The difficulty of arriving at any such arrangement first became apparent when, in November 1942, Akers and Halban met Bush, Conant and Groves in Washington. It quickly became clear that the Americans were pushing ahead with their own heavy-water project as well as with all the possible ways of producing fissile uranium and were in a position to drive as hard a bargain as they wished for use of Trail's heavy water.

If all this was rather less trusting than might have been expected between devoted allies, an explanation was at hand. On 29 September 1942, Britain and Russia signed a solemn agreement for the exchange of technical information. Under its terms both countries were to 'furnish to each other on request all information including any necessary specifications, plans, etc. relating to weapons, devices, or processes which at present are, or in future may be, employed by them for the prosecution of the war against the common enemy.' There was, it is true, an escape clause allowing either country to withhold specific information, but in that case the reasons had to be given. The evidence as to who among the Americans became aware of the agreement during the succeeding weeks is contradictory but it would have been remarkable if neither Bush nor Conant had known of its existence if not of its details. Certainly Roosevelt was told of the agreement on 27 December. And on the following day he approved the draconian terms of a letter which Conant now sent to Dean Mackenzie, President of the National Research Council of Canada. So dictatorial were they that Conant felt it necessary to telephone Mackenzie while the letter was on its way. It would, he stated, read more sternly than intended.

Conant's letter, which outlined the conditions that for most practical purposes were to apply to the British, stated that the Du Pont Company was to build a number of American heavy-water plants, and a plant that would produce plutonium with the use of heavy water as a moderator. Secondly, Bush and he had been ordered to restrict exchange of information

on this whole subject by the application of the principle that we are to have complete interchange on design and construction of new weapons and equipment *only* if the recipient of the information is in a position to take advantage of it in this war. Such a principle is, of course, in the interest of secrecy. Since it is clear that neither your Government nor the English can produce elements '49' [plutonium] or '25' [uranium 235] on a time schedule which will permit of their use in this conflict, we have been directed to limit the interchange correspondingly.

There then came the nub of the matter as far as Halban's group was concerned.

We should very much like to have the group of scientists assembled in Canada carry on the fundamental scientific work for the use of heavy water so that Du Pont Company could base their designs on this experience. To this end we would be quite ready to release all the product of the Trail plant initially for the use of this group. We should expect that this Canadian group would direct their program along lines worked out in connection with the American engineers [Du Pont Company] and make all their results available to this designing group.

The application of the principle of limited exchange would mean that we should not be in a position to give to the Canadian group any information about the methods of extraction of element '49', nor the design of the plant for the use of heavy water for this purpose, nor the methods for preparing heavy water.

Under this arrangement and with exchange so restricted by order as I have stated, it would seem to us that it would be unwise to have any English or Canadian group of engineers associated with the enterprise, as this would certainly lead to conflict of authority and uncertainties as to procedure. The Canadian group would, of course, be free to interchange with Dr Compton's group at Chicago in so far as the use of heavy water in the chain reaction was concerned, and in regard to the chain reaction itself, but not in regard to the chemistry of element '49' or the separation.

Lest there be any lingering feeling that post-war commercial interests were involved, Conant ended:

I do not have to tell you how much we would appreciate the assistance of the Canadian group in this enterprise. It seems to us that the plans now adopted would provide for the maximum of assistance in what is, after all, a joint aim –

namely, the production of a weapon to be used against our common enemy in the shortest possible time under conditions of maximum security.

The possibility that the new attitude might successfully scupper British plans for diverting too much effort to post-war power production may well have crossed the minds of both Bush and Conant. But Dr Mackenzie, who in mid-January visited Conant in Washington, felt unable to support the bitter British denunciations which followed the Conant ultimatum. 'The army and the higher military police,' he wrote in his diary afterwards,

are quite convinced that secrecy is of extreme importance and unless they take the precautions stated there will be serious probability of leakage. General Groves says that anything that happens in a scientific laboratory or industrial plant finds its way very quickly to enemy countries. They [Bush and Conant] both assured me that the reason for excluding the United Kingdom from the large plant [Oak Ridge or Hanford?] which will be costing 200 millions has not anything to do with peace-time industrial interests, and Bush particularly is anxious that no deal be made which would give the isolation group in Congress an opportunity for adverse criticism. I gathered that they were particularly apprehensive of discussing all the details and know-how with the Montreal group, which is really not an Anglo-Saxon group, and that they felt there was no guarantee that the various nationals – French, Austrian, Russian, Czechoslovakian, German, Italian, etc. – could be guaranteed for any length of time. I think there is a great deal to be said for their point of view. We had a very pleasant and profitable discussion and they are extremely friendly to us.

Nevertheless, the impasse remained and the first months of 1943 witnessed an exchange of correspondence in which Anderson and an increasingly worried Churchill tried without much success to restore the earlier trusting co-operation.

On 11 January Anderson underlined to Churchill that the Americans were to pass over details of nuclear progress only if it would be of use in the war.

It appears that this principle is being interpreted to mean that information must be withheld from us over the greater part of the field of Tube Alloys. At the same time the United States authorities apparently expect us to continue to exchange information with them in regard to those parts of the project in which our work is further forward than theirs.

This development has come as a bombshell and is quite intolerable.

Churchill received Anderson's note as he was preparing to leave

London for the Casablanca Conference at which he, Roosevelt and their military leaders were to discuss the progress of the war. He raised the question of nuclear collaboration not only with Roosevelt but with Roosevelt's personal adviser, Harry Hopkins. But their talks were the prelude to seven months of wranglings and misunderstandings some of which would have been comic had they not left a legacy of grievance on both sides.

At Casablanca, Roosevelt somewhat airily assured Churchill that everything would be all right; without, however, knowing of the points that Bush and Conant were both raising. Churchill, possibly sensing this, took the precaution of discussing the matter with Hopkins who assured him that everything would be settled when the President had returned to Washington after the end of the conference on 24 January. Yet three weeks later, Churchill was writing to Hopkins: 'I should be very grateful for some news about this, as at present the American War Department is asking us to keep them informed of our experiments while refusing altogether any information about theirs.' A second note twelve days later, complaining that the American attitude 'entirely destroys' the idea of a joint effort, still failed to produce results, and in March Churchill prepared a somewhat drastic cable to Hopkins: 'Time is passing and no information is being exchanged. We have made some progress in the last three months and I shall shortly be forced to take decisions.' The wording was eventually toned down, possibly on the advice of Anderson, to 'collaboration appears to be at a standstill', and the empty threat about decisions was omitted.

Churchill's cable failed to produce any results, and at the end of March Hopkins's reticence was no doubt buttressed by a memorandum from Bush.

> The adopted policy is that information on this subject will be furnished to individuals, either in this country or Great Britain, who need it and can use it now in the furtherance of the war effort, but that, in the interests of security, information interchanged will be restricted to this definite objective.
>
> There is nothing new or unusual in such a policy. It is applied generally to military matters in this country and elsewhere. To step beyond it would mean to furnish information on secret military matters to individuals who wish it either because of general interest or because of its application to non-war or post-war matters. To do so would decrease security without advancing the war effort.

Meanwhile the British were taking what defensive actions they could. One opportunity arose when Urey and Fermi wished to confer with Halban, now continuing his work in Montreal with the Norwegian heavy water which had by this time survived the perils of an Atlantic crossing. The Montreal laboratory was therefore asked if Halban would visit New York for discussions. The Canadians said yes, but London said no; and since Halban was technically a British civil servant, it was the London order that had to be followed. The application had been made by Conant, and it was to Conant that Malcolm MacDonald explained that 'the authorities feel that whilst the present doubt about the basis of co-operation on the Tube Alloys project continues, such exchanges between Dr Urey and Dr Halban should be deferred'.

The action, however justified as a tactical move, was hardly the action of allies fighting a common war. But neither was that of the U.S. Army whom, the Canadians discovered, had under orders from Groves signed a series of secret contracts with the company producing most of Canada's uranium. 'For a time,' says the official Canadian history of its own nuclear enterprise, 'the Canadian Government was thrust into the indefensible and embarrassing position of not even being able to find out just what deals [the head of the company] and his associates had made with the Americans for Canadian ore and oxide.'

On 1 April Churchill repeated his concern to Hopkins, this time adding: 'That we should each work separately would be a sombre decision.' And to Anderson, later in the month, he stated that Britain could wait no longer 'and the necessary staff should be directed to draw out detailed plans for putting these projects into effect, with the necessary timetables and estimates'.

It is not clear how genuinely Churchill believed that Britain would be able to make her own nuclear weapons as the crisis of the war grew nearer. He did, however, believe that it was still possible for him to put pressure on Roosevelt, and an opportunity was offered during the Trident conference held in Washington in May. And here, indeed, Churchill once more appeared to have won over Roosevelt, cabling to Anderson after the conference was finished:

The President agreed that the exchange of information on Tube Alloys should be resumed and that the enterprise should be considered a joint one, to which both countries would contribute their best endeavours. I understood that his ruling would be based upon the fact that this weapon may well be developed in time for the present war and that it thus falls within the general agreement covering the interchange of research and invention secrets. Lord Cherwell to be informed.

But in Washington it was not only Roosevelt and Churchill who discussed the problems of nuclear energy. On 25 May Cherwell met Bush and Hopkins. They discussed the controversial matter of information exchange and Cherwell agreed that the information wanted by the British could not be utilized while the war lasted. 'The matter,' Bush said in a report of the meeting written the same day, 'finally came down to the point where he [Cherwell] admitted rather freely that the real reason they wished this information at this time was so that after the war they could then at that time go into manufacture and produce the weapon for themselves, so that they would depend upon us in this war for the weapon but would be prepared after this war to put themselves in a position to do the job promptly themselves.'

As far as Bush was concerned, this 'most extraordinary interview with Cherwell which had left me completely amazed at the British point of view' made it even more important that the British should be told no more than was absolutely necessary. Yet, astonishingly, it appears that in the press of events, neither Bush nor Hopkins gave the President details of the 'extraordinary interview'.

On the evening of the same day – only a few hours after Hopkins and Bush had been told by Cherwell of Britain's attitude – Roosevelt, Hopkins and Churchill dined together. It is not known whether Cherwell's 'extraordinary' statement had by this time been brought to Roosevelt's notice, or whether it was discussed over the dinner-table, but Cherwell wrote to Hopkins five days later indicating that the matter had been concluded satisfactorily while Churchill, in his message to Anderson after the conference was finished, appeared to spell it out in black and white.

However, Churchill, back in London, was given no indication of action in Washington. A cable to Hopkins merely produced a reply on 17 June saying: 'The matter of Tube Alloys is in hand and I think will be disposed of completely the first of the week.' It is not clear whether Hopkins had by this time told the President of the fateful Cherwell meeting. But Bush did so on 24 June. 'When I recounted that Cherwell had placed the whole affair on an after-the-war military basis, the President agreed that this was astounding. ... He said at one point ... he thought Cherwell was a rather queer-minded chap.'

Roosevelt was now in something of a dilemma. Both Bush and

Conant were against the renewed interchange of information which he had agreed with Churchill before hearing of the Cherwell revelation. By early July Churchill had still heard no more and on the 9th cabled to Roosevelt rather than to Hopkins, saying: 'My experts are standing by and I find it increasingly difficult to explain delay.'

Bush and Henry Stimson, the Secretary of War, were by this time preparing to leave Washington for London, where a variety of matters were to be discussed. After they had arrived, Roosevelt solved his problem. He had given his word to Churchill and on 20 July he told Bush: 'I wish, therefore, that you would renew, in an inclusive manner, the full exchange of information with the British Government regarding Tube Alloys.'

At least, that was the message he had sent. But it had to be coded before transmission to London. 'I thought,' Bush later recorded, 'I was instructed to review with the British the matter of interchange. On receiving the President's letter on returning to the U.S. I found that the wording was definitely "to renew full interchange".'

On 22 July (1943), Churchill, Anderson, Cherwell, Stimson and Bush – with Harvey Bundy as secretary – therefore sat down to discuss the question of Anglo-U.S. nuclear co-operation, with Bush, the American negotiator, believing that the complexity of 'review' rather than the simplicity of 'renew' was involved. Thus, as will be seen, the fateful 'Quebec Agreement' was to arise from a simple cable-coding error.

The previous day, 21 July, Anderson had warned Churchill of the dangers ahead. 'We cannot afford after the war to face the future without this weapon and rely entirely on America, should Russia or some other power develop it,' he minuted. 'However much, therefore, it may be tactically necessary to make use of the pretext of wartime collaboration, the Americans must realize that, failing collaboration, we shall be bound to divert manpower and materials from our work on radio, etc. in order to try to keep abreast.'

At Number Ten the following day the previous arguments were aired once again – the need for security, the need for collaboration, the need for husbanding of resources and the avoidance of duplication of effort: all matters which Bush could stress in his efforts to ensure that 'review' changed as little as possible. The situation was made more obscure as Churchill realized to his surprise that Stimson had not been told by the President of the Roosevelt–Churchill nuclear discussion during Trident and that Stimson 'was disturbed to know that I thought we had not been fairly treated'.

Eventually, after considerable argument, Churchill 'suggested that he would be in favour of an agreement between himself and the President of the United States having the following points ...'. The points formed the substance of what was to be known as the 'Quebec Agreement', not to be publicly revealed for nearly twelve years. The first, which was rephrased in the final text to call for a pooling of brains and resources, was that interchange should be 'to the end that the matter be a completely joint enterprise'. There followed clauses stating that neither side would use the weapon against the other and that neither should use it against third parties, or give information to third parties, without the consent of both. Finally, it would be agreed 'that the commercial or industrial uses of Great Britain should be limited in such manner as the President might consider fair and equitable in view of the large additional expense incurred by the United States.'

Various versions of the last clause were worked out during the next few days, and on one of them Churchill made its penultimate sentence read '... any post-war advantages of an industrial or commercial character shall be the property of the United States.' To this, Anderson commented: 'I do not, of course wish to be any less forthcoming than do you. I feel sure, however, that it is a mistake to talk about "property".'

In August, Churchill and Roosevelt once more met with their advisers, this time in Canada. On the 19th they met privately. Before them lay sheets of notepaper bearing the engraved inscription, 'The Citadel, Quebec'. To the typed 'Articles of Agreement Governing Collaboration between the Authorities of the U.S.A. and the U.K. in the matter of Tube Alloys', Roosevelt added the single word 'Approved' and the date. To it, he and Churchill added their signatures.

The document began by stating that the Tube Alloys project could be brought to fruition as early as possible 'if all available British and American brains and resources are pooled'. It then went on to outline five heads of agreement. Both countries agreed that they would not use 'this agency' against each other and that they would not use it against third parties without each other's consent. It was agreed that neither side would 'communicate any information about Tube Alloys to third parties except by mutual consent' – a clause which overlooked the fact that prior British obligations to the French already existed and which put Halban and the rest of the French team, whose loyalty to France had never been questioned, in a distinctly ambiguous position.

A Combined Policy Committee, consisting of Bush, Conant and

Stimson representing the United States; Field Marshal Sir John Dill and Colonel J. J. Llewellin representing Britain; and Hon. C. D. Howe representing Canada, was to be set up. Its function, 'subject to the control of the respective Governments', would be to guide the whole project. Interchange of information and ideas was to be complete between members of the Policy Committee and their immediate technical advisers, but in the field of scientific research and development, full and effective interchange was to be 'between those in the two countries engaged in the same sections of the field', a gentlemanly definition of General Groves's policy of compartmentalization.

However, the clause covering post-war developments still brusquely cut Britain down to size. This stated that 'in view of the heavy burden of production falling upon the United States as the result of a wise division of war effort, the British Government recognize that any post-war advantages of an industrial or commercial character shall be dealt with as between the United States and Great Britain on terms to be specified by the President of the United States to the Prime Minister of Great Britain. The Prime Minister expressly disclaims any interest in these industrial and commercial aspects beyond what may be considered by the President of the United States to be fair and just and in harmony with the economic welfare of the world.'

R. V. Jones, in charge of Britain's nuclear intelligence since the early days of the war, and one of the few to know of the secret agreement, later described his shock as the details arrived. 'It seemed to me that we had signed away our birthright in the post-war development of nuclear energy,' he wrote. And he immediately called on Cherwell to complain that he had given the Prime Minister such bad advice that the latter had signed Britain's rights away.

It now seems certain that Churchill had no alternative. Bush's fight for reviewal instead of renewal certainly made his task more difficult. Even so, there were factors which circumscribed what Roosevelt was able to do, as Bundy noted to Stimson in a comment on Bush and Conant. 'They are,' he said, 'trying to avoid at all costs the President being accused of dealing with hundreds of millions of taxpayers' money improvidently or acting for purposes beyond the winning of the war by turning over great power in the post-war world to the U.K. without adequate consideration or without submitting such a vital question for consideration and action by both Executive and Legislative authority.'

Churchill himself was confident that he had done as well as was possible for Britain. 'I am absolutely sure that we cannot get any better terms by ourselves than are set forth in my secret Agreement with the President,' he wrote to Cherwell nine months later.

It may be that in after years this may be judged to have been too confiding on our part. Only those who know the circumstances and moods prevailing beneath the Presidential level will be able to understand why I have made this Agreement. There is nothing more to do now but to carry on with it, and give the utmost possible aid. Our associations with the United States must be permanent, and I have no fear that they will maltreat us or cheat us. However I presume that we are free to start the manufacture here at any time we please. Nothing was said about this in the Agreement, and I cannot think they have any locus to object.

Whatever might happen in the future, Churchill was determined that the Quebec Agreement should be honoured. Early in 1944 it was decided that America, Britain and Canada should set up a Combined Development Trust to ensure supplies of uranium and during the negotiations he sent a worried enquiry to Cherwell and Anderson: 'Explain to me shortly,' he asked, 'What is the difference between the new document of trust and the agreement signed between me and the President of the United States? Nothing must be done which in the slightest degree detracts from the joint document signed by us.' And when Cherwell, protesting later that Britain was being denied information on the Americans' plutonium work, maintained that this made it impossible for the Americans to 'claim a [post-war] industrial monopoly', Churchill added in the margin: 'but see my agreement with the President'.

It was not only the British who were later to regret the results of the Quebec Agreement. In 1952, Conant wrote:

When the inside history of the negotiations between the three governments (U.S., Britain and Canada) on the manufacture of atomic bombs comes to be written perhaps a century from now, I believe it will be clear that troubles, misunderstandings, and bad feeling might have been avoided by greater frankness at the outset. If at least a committee of the Senate of the United States had been apprised of the U.S.–British wartime negotiations when they took place, the Bill that set up the Atomic Energy Commission of the United States [in 1946] might have taken a different form and Anglo-American relations after a rough passage might be better now. As a very humble observer from a distance of what occurred at high levels in 1943, I thought then and I still think that a treaty should have been drawn between the three nations involved, a treaty dealing with everything even distantly related to atomic energy.

However, when it was agreed that Britain's post-war nuclear advantages should be dealt with on terms which the American President should lay down, some doubt still remained about whether 'the bomb' would work, or whether the hope of power to be would ever be fulfilled. Those were the questions still to be answered by General Groves and the formidable teams of physicists now to be augmented by the men who had been working in Britain.

CHAPTER 8

Sowing the Wind

The first result of the Quebec Agreement was that nuclear research in Britain virtually ended for the duration of the war. A little work on industrial and technological problems was still carried on by I.C.I. but most of the scientific teams which had grown up since the early days of the M.A.U.D. Committee were disbanded and their members integrated into what was by now the huge American enterprise. In November most of the physicists set sail for the United States in the liner *Andes*. There had been difficulty in getting them to the Liverpool docks and they had finally arrived in black undertakers' cars, mobilized by Akers, with their luggage following in the hearse. If their departure fittingly symbolized the death of Britain's wartime nuclear effort, the event should not be regarded too chauvinistically. The British had more than played their part in the early days and even General Groves, a man not usually over-enthusiastic about the activities of non-Americans, was to write: 'I cannot escape the feeling that without active and continuing British interest there probably would have been no atomic bomb to drop on Hiroshima.'

Yet throughout 1942 and 1943 it had steadily become clearer that Britain, whether or not its leaders were willing to admit the fact, was quite unable to produce nuclear weapons during the war, even with the help of Canada. The daunting size of the enterprise and the finances required had only slowly been appreciated even by the best-informed physicists or the most competent of chemists and engineers. By 1943 there had been virtually no discussion of the almost intractable problem of assembling the weapon, notably that of bringing together two sub-critical masses at a speed which would avoid pre-detonation – the danger that the bomb would fizzle rather than explode. As to cost, the M.A.U.D. Committee had estimated that £8,510,000 would be needed to produce twenty-six bombs; by the summer of 1945 the Americans had spent more than $2,000,000,000 and had no more than three bombs available, although more would, it was hoped, be ready before the end of the year.

By the autumn of 1943 the release of nuclear energy rested entirely on the success of the huge American project since it was not only the British operation which had fallen by the wayside. In Germany, the increasing weight of Allied air-raids, the growing shortage of labour, and the concentration on the unmanned 'buzz-bomb' and the rocket of whatever manpower was available, all combined to rule out nuclear weapons, even had the German physicists been confident that they were on the right road, which was far from the case. Like the Americans up to the later months of 1941, they had conceived of practical nuclear energy only in terms of slow neutrons; they continued to do so and on hearing of Hiroshima many believed that the weapon must have been as big as a reactor.

In Russia, also, progress towards a nuclear weapon was still only at a very early stage. However, it was now beginning to move forward again, even if sluggishly, and in view of subsequent events it is worth while considering the change which was slowly taking place. In 1941 the German invasion had halted Russian nuclear research. The attack had brought to an end any use of the cyclotron in Leningrad and had stopped construction of a more powerful one in Moscow. Kurchatov had been taken off nuclear physics and directed to the Black Sea where he spent the next year and more on the problem of protecting Russian vessels from German mines. As the Germans continued to press more deeply into Russia, laboratories were uprooted and moved east to the Urals while all work was reorientated to meet the immediate threat. Peter Kapitza, the Russian physicist who had worked with Rutherford in the Cavendish in the early 1930s, and who had returned to Russia for good in 1934, is reported to have said: 'Our point was that scientific work which is not completed and produces no results during the war may even be harmful if it diverts our forces from work which is more urgently required.' It was a sensible argument which by 1943 was governing German policy, and much the same as Sir Henry Tizard had put forward in Britain in 1940.

But in Russia there had been a change which the Russians infer came about even before 1943. According to their version – given in Golovin's biography of Kurchatov – it was in 1942 that the physicist Flerov, visiting a university library, noted that neither American nor

British scientific journals were making any references to nuclear fission, and that the names of the more famous nuclear physicists which he expected to see, were not there. Flerov appears to have written to the State Defence Committee imploring its members that 'no time must be lost in making a uranium bomb'. He was then recalled from the Army for consultations with S. V. Kaftanov, the representative on the Defence Committee of science and higher education. There were, by this time, substantial reasons why the Committee should again consider nuclear research.

'By then the Soviet Government was in possession of information that urgent top-security work was in progress in Germany and the United States of America to create a new super-weapon,' Golovin claims. It may well have been the case. There had certainly been exchanges between Russian and German scientists following the Russo-German pact of August 1939, while it appears that Klaus Fuchs, and possibly others, were by 1943 feeding information to the Russians.

As a result of Flerov's letter, Kapitza and Ioffe were called to Moscow, and within a few days Kurchatov was also summoned. 'Three days later,' writes Golovin, 'having been appointed to head operations to make a uranium bomb, he returned, excited and pensive, to Kazan.'

The timing, whether in mid-May 1942 or early 1943, need not be taken too seriously. But the surrender of the German forces at Stalingrad in January 1943 confirmed that the military tide was turning. By the first months of the year it must have been evident that some effort might now be devoted to nuclear weapons, even if they could not be produced in time to help defeat the Germans. And subsequent events were to show that the Russians must have cleared away a good deal of the undergrowth covering nuclear problems during the last two or three years of the war.

They appear to have realized what had escaped the Germans: that fast neutrons would be needed for a bomb and that an early priority was to discover how best to separate the uranium isotopes and how to extract plutonium from a reactor. But this was largely theoretical undergrowth and it is clear from the statements – or admissions – in Golovin's biography, that between 1943 and 1945 the Russian effort was more on the scale of, and at the stage of, the M.A.U.D. Committee's work than that of the Manhattan Project. Thus in 1943 it was only in the United States that the scientific, technological, engineering and financial resources being deployed were sufficient to transform the scientists' nightmare into a practical weapon of war.

The formidable size of the commitment required had been underlined in June 1942. By that time it had become clear that there were five potential methods of producing fissile material, all equally likely to succeed. Uranium 235 could be produced by the gaseous-diffusion method for which the British had already built a pilot plant in North Wales. It could be produced in a centrifuge, and it could be produced by electromagnetic separation. Production of plutonium required a self-sustaining chain reaction in what was at first called a pile and later became known as a nuclear reactor; this required uranium and the use of a moderator which could be either heavy water or graphite. While all these five alternatives were considered capable of producing the fissile material required for a bomb, very little was known about their comparative efficiency, and each raised engineering and chemical problems that had never before been tackled, let alone solved. And it was still possible that the bomb might not work.

Yet the threat of Germany being first still appeared great, and long before signing the Quebec Agreement Roosevelt had decreed that work should be pressed ahead on all five methods – at a starting cost of some $500 million.

Colonel Groves – soon to be raised to General – was appointed to oversee the huge engineering enterprises now to be undertaken. His achievements have tended to be underrated. The problems that had to be overcome once the production of fissionable material was contemplated were not only engineering problems although these were forbidding enough. On one pile some 50,000 linear feet of joints had to be welded in positions that would be inaccessible once the plant started operating. Men sought for the task had to present work records going back fifteen years and even of those provisionally accepted less than twenty per cent then passed the necessary welding tests. Tens of thousands of graphite blocks, each produced to laboratory standards of purity, had to be machined to tolerances measured in thousandths of an inch. Many of the materials involved were new and exotic; very little was known about them and chances had to be taken that would have been unthinkable in any other circumstances.

One of the huge American piles was almost completed before it even was known whether enough uranium slugs would be produced to start operations. As the official historians of the enterprise have noted: 'No one could guess whether the slugs would withstand the prolonged exposure to high temperatures and intense radiation. Nor could anyone say confidently that the failure of one slug would not ruin the entire pile. Nothing but actual operation could provide the answer.' Yet it was the virtually unquantified hazard of radiation which introduced the greatest complication to much of the work. 'To build a pile,' it has been pointed out, 'the engineer had to know the effect of radiation on corrosion rates, on the properties of metals, on chemical reactions, on instruments and other equipment, and on man. He would need years to revise his handbooks. In the meantime, he could but resort to trial and error, leaving the systematic compilation of data to less critical times.'

The tenacity with which these problems were overcome demonstrates that the success of the Manhattan Project was due to much more than the accepted American ability for mass production. It also demanded some ever-present nagging spur. This was provided, in America as it had been earlier in Britain, by the threat that Germany might be first with nuclear weapons. It is ironic that, in practical terms, the threat never existed.

There were also political problems, some of them unexpected. Thus when Du Pont, the company which McGowan had been hopeful of drawing into Halban's nuclear-power scheme, was asked to build the chemical separation plant, the company was cautious. The board knew that the work could give them a head lead in post-war nuclear development, but they also appreciated what a key position in the Manhattan Project might mean in terms of post-war public relations. When the Nye Commission had investigated the arms industry in the 1930s, Du Pont had been named as one of the 'merchants of death' and its directors were anxious to prevent a repetition. It was therefore agreed that the company would take only a nominal one dollar fixed fee for the contract and that any patent rights which came from the work would automatically be handed over to the Government. Perhaps more significant, the Du Pont president, Walter S. Carpenter, Jr, maintained that the company would not take part in similar work after the war; production of nuclear weapons, he felt, should be controlled exclusively by the Government.

These massive enterprises, already swinging into operation as the British teams crossed the Atlantic in the autumn of 1943, were justified by one still secret but spectacular success; more fairly than many others, it can be used to mark the beginning of the nuclear age. It might, eventually, not be possible to bring two sub-critical blocks of fissile material into contact so that they produced the cataclysmic explosion forecast. But at least the self-sustaining nuclear chain reaction, which Halban had so nearly achieved in the winter of 1940, was now known to be possible.

When the Americans had at the beginning of 1942 started nuclear research in earnest, one of their first priorities had been to resolve this initial problem. Arthur Holly Compton, supervising the task at the University of Chicago's cryptically named 'Metallurgical Laboratory', brought in Fermi among others and early in 1942 set about the formidable task. Despite the huge amount of theoretical work carried out since Joliot-Curie had in 1939 reported a superfluity of neutrons in the fission process, an essential for a chain reaction, uncertainties still remained. Some of the neutrons released by fission might be captured by the uranium yet not produce further fission – non-fissioning capture; some might be absorbed in the graphite. Yet the neutrons might escape from the pile without being captured either by uranium or by the moderator. If the amount of uranium was increased, its volume grew more rapidly than the surface at which the fission-produced neutrons were able to escape, so it was quite clear that the larger the system, the smaller would be this third potential danger.

Shortage of materials limited what the Metallurgical Laboratory was able to do, and for many months Compton had to be satisfied with carrying out a series of experiments, each one of which provided data for the next. The proportion of uranium oxide to graphite was varied, oxides of different purities were used, and were processed into lumps of various shapes, sizes and degrees of compression. The lattice spacing was varied, the effect of surrounding the uranium oxide with beryllium and with paraffin was contrasted, and assemblies of similar lattice type but of different sizes were experimented with.

At first it was intended that the critical assembly should be built in the Argonne Forest outside Chicago. However, by October it was clear that the experiment could be carried out before the building at Argonne would be ready, and Fermi proposed that he could do it safely in Chicago.

'The only reason for doubt,' Compton wrote later, '... was that some new, unforeseen phenomenon might develop under the conditions of release of nuclear energy of such vastly greater power than anyone had previously handled. ... After all, the experiment would be performed in the midst of a great city. We did not see how a true nuclear explosion, such as that of an atomic bomb, could possibly occur. But the amount of potentially radioactive material present in the pile would be enormous and anything that would cause excessive ionizing radiation in such a location would be intolerable.'

Compton was an officer of the University of Chicago and according to the rules should have taken up the matter with President Hutchins. But Hutchins would inevitably object, and Compton believed that this would be wrong. 'So,' he has said, 'I assumed the responsibility myself. In a building under the west stands of the Stagg Athletic Field was a squash court. I told Fermi to use this room and go ahead with the critical experiment.' The plan had been to build a cubic lattice of normal uranium within a sphere of graphite, but the amount of pure uranium available, some 12,400 pounds (5,625 kg), was less than needed, so uranium oxide had also to be used. Despite this, it was calculated that the device would become critical earlier than forecast and the original design was changed; instead of a sphere, the construction was finished off in the form of an oblate spheroid, flattened at the top, rather like a conventional doorknob.

Two special teams were organized for construction of the pile. One machined the graphite, the other pressed the uranium-oxide powder, using a hydraulic press and specially made dies. It had been decided that the pile should be surrounded with a balloon-cloth tent so that it would be possible to operate it, if necessary, surrounded by carbon dioxide rather than by air, and within this tent the assembly began in mid-November 1942. Contemporary reactors – as piles are now called – demand plans drawn, and executed, to tolerances measured in ten-thousandths of an inch. The world's first nuclear pile was rather different.

'The frame supporting [it] was made of wooden timbers,' Herbert Anderson, one of Compton's staff, has written. 'Gus Knuth, the mill-wright, would be called in. We would show him by gestures what we wanted, he would take a few measurements, and soon the timbers would be in place. There were no detailed plans or blueprints for the frame or the pile. Each day we would report on the progress of the construction to Fermi, usually in his office in Eckhart Hall. There we would present our sketch of the layers we had assembled and indicate what we thought could be added on the following shifts.'

As layer was piled upon layer, the structure quickly became known as the pile, a useful word since – like the word 'tank' in the First World War – it effectively concealed its purpose and remained in use for some years. Running the entire height of the pile were ten deep slots. These were for the control rods, cadmium strips which absorbed neutrons and whose withdrawal or insertion into the pile thus governed the nuclear reaction inside it. One set of rods was automatically controlled from the balcony of the squash court. One single rod was manually operated. In addition, a hint of the powers which could be unleashed by the experiment, was an emergency safety rod. Attached to one end of it was a rope which ran through the pile and was weighted heavily at the other end. Before the crucial experiment started the rod was withdrawn and secured by another rope to the balcony. Standing by it throughout the experiment was one man with a heavy axe, ready to cut the rope and let the emergency rod fall into the pile should the mechanism controlling the other rods fail. Even this was not considered a sound enough precaution and a 'liquid-control' squad stood by, ready to flood the assembly with cadmium-salt solution if anything began to go disastrously wrong.

As the pile grew in size it became easier to predict when it could be made to go critical. As Anderson has said:

> Thus, we could tell that on the night between 1 and 2 December, during my shift, the fifty-seventh layer would be completed and the pile could be made critical. That night the construction proceeded as usual with all cadmium rods in place. When the fifty-seventh layer was completed, I called a halt to the work in accordance with the agreement we had reached in the meeting with Fermi that afternoon. All the cadmium rods but one were then removed and the neutron count taken, following the standard procedure which had been followed on the previous days. It was clear from the count that once the only remaining cadmium rod was removed, the pile would go critical. It was a great temptation for me to pull the final cadmium strip and be the first to make a pile chain react. But Fermi had anticipated this possibility. He had made me promise that I would make the measurement, record the result, insert all cadmium rods, lock them all in place, go to bed, and nothing more.

2 December 1942 dawned bitterly cold. The scientists assembled in the unheated west stands building, kept warm as best they could by moving about, then discovered in a locker a supply of old racoon-fur coats which were handed to the security guards standing outside the entrances.

Fermi, who arrived before 9.00 a.m., had a strict plan for the approach to criticality. At 9.45 a.m. he ordered the electrically operated control rods to be withdrawn. On the balcony the lights on the control apparatus showing the position of the rods began to come on. Then, from the counters, each resembling the face of a clock, with 'hands' indicating the neutron count, there came the clicking which revealed how many neutrons were being released. At 11.35 a.m. the automatic safety rod was withdrawn and the control rods were adjusted. The counters began to click faster.

Compton had been called to witness the final stages of the operation and, below, George Weil was manually operating the final cadmium rod.

The process was slow and deliberate and it was early afternoon before the critical moment was neared. Anderson has described the scene:

At first you could hear the sound of the neutron counter, clickety-clack, clickety-clack. Then the clicks came more and more rapidly, and after a while they began to merge into a roar; the counter couldn't follow any more. That was the moment to switch to a chart recorder. But when the switch was made, everyone watched in the sudden silence the mounting deflection of the recorder's pen. It was an awesome silence. Everyone realized the significance of that switch; we were in the high-intensity regime and the counters were unable to cope with the situation any more. Again and again, the scale of the recorder had to be changed to accommodate the neutron intensity which was increasing more and more rapidly.

Then suddenly Fermi raised his hand. 'The pile has gone critical,' he announced.

The men on the balcony watched as the instruments showed that, for the first time on earth, a nuclear fire was steadily burning.

Everyone wondered why the pile was not immediately closed down. But Fermi waited for a minute, and another. Then, as the tension mounted, he ordered: 'Zip in!' The emergency safety rod was released and dropped into the pile.

'No cheer went up,' says Anderson, 'but everyone had a sense of excitement. They had been witness to a great moment in history, Wigner was prepared with a bottle of Chianti wine to celebrate the occasion. We drank from paper cups and then began to say things to one another. But there were no words that could express adequately just what we felt.'

Compton returned to his office, telephoned Conant in Harvard, and in guarded words told him the news. 'The Italian navigator,' he said, 'has just landed in the New World. The earth was not as large as he had supposed [meaning that the pile of uranium and graphite needed to bring about the reaction was smaller than anticipated] so he arrived earlier than expected.'

'Were the natives friendly?' Conant asked.

'Everyone,' Compton replied, 'landed safe and happy.'

The success at Stagg Field was a reassurance. It confirmed the physicists in what they had been believing. Their theories had been right, at least as far as the chain reaction was concerned, and any remaining beliefs that the project involved an unjustifiable rather than a justifiable risk were weakened.

Whether the success encouraged the Russians as well is a speculation that is never likely to be resolved. It is true that two months after 'the Italian navigator ... landed in the New World', Field Marshal Paulus surrendered at Stalingrad, the limit of German penetration was visibly seen to have been reached, and the Russians started preparations for the series of offensives which were to bring them to the banks of the Elbe. This alone may have accounted for the renewed interest which from now onwards was given in the Soviet Union to nuclear research, to the work that even if it did not impinge on the war would provide Russia with a new source of power during the years of post-war reconstruction. Yet if the one secret of the bomb was whether it would or would not explode, the more basic secret of nuclear research was whether a chain reaction could be created by nuclear fission. Once the answer was known to be yes, the incentive for further research was obvious.

It may have been mere coincidence but a month after Fermi had shown a self-sustaining chain reaction to be possible the Soviet Purchasing Commission in the United States notified the War Production Board that it wished to purchase eight tons of uranium oxide and the same amount of uranyl nitrate salts, materials later considered enough to duplicate the Chicago pile. It is also curious that the first Russian pile, the P.S.R., put into operation in 1947, so closely replicated the first pile built as a result of the Chicago experiments. Diameter, lattice spacing, loading and rod diameter – all vital statistics – were all similar.

Although the only evidence of a chain reaction was from the small pilot plant in Chicago, and although the early experiments there produced only half a milligram of plutonium – roughly enough to cover a pinhead – the Americans quickly pressed ahead with their plans. Within less than a year, the world's first power-producing pile went critical at Clinton with a fuelling of only some 30 tons of uranium, less than had been thought necessary. There was a week's shutdown for testing and then, with an increase to 36 tons, power was being delivered at 500 kilowatts, roughly half the plant's capacity. Before the end of 1943 it had produced some 500 milligrams of plutonium, and from then onwards was to produce steadily increasing amounts.

Elsewhere, progress continued at much the same steady rate, and it was soon obvious to most members of the British team that while they could be usefully employed on small self-contained items of research, while their ideas were frequently of help and would usually be listened to, they were not essential to the joint enterprise in any meaningful way. The British contingent was spread throughout the various Manhattan Project establishments, some going to Chicago and some to Los Alamos, the extraordinary laboratory in New Mexico where the mechanisms for the world's first nuclear weapons were eventually perfected. Once the possibility of a chain reaction had been proved in Chicago, once the first trickle of uranium 235 and of plutonium from the production plants became more than a promise, one problem began to assume increased importance. It was the problem which had gibbered on the sidelines of the physicists' world ever since nuclear weapons had appeared to be a practical possibility: would it be possible to bring together the two sub-critical parts of a weapon so quickly that the result was an explosion rather than a fizzle? In the early days it had been the physics of the problem that had appeared most important; but these doubts had been removed, only to be replaced by questions of a more mundane character. They were typified by a matter which Thomson had raised with Simon even before the M.A.U.D. Committee had completed its work.

> I have been calculating the temperature which would be reached by the portions of the bomb when they hit after being shot together, and taking a speed of 6,000 feet per second it comes to about 5,000 degrees absolute, which is rather formidable. I have made a calculation to see whether the resulting vapour pressure would be so great as to blow the halves apart before they could explode, assuming that for any reason the initiation of the explosion was delayed for a fraction of a second.

It was to settle such matters, and the even greater complexities of weapon design which later became apparent, that the Los Alamos Laboratory had been planned in New Mexico. The site was that of an isolated school, reached by a tortuous road from Santa Fé which climbed to the top of a high mesa and had to be largely rebuilt to carry the traffic to and from the world's first centre for nuclear weapons research. It had been chosen by General Groves and by J. Robert Oppenheimer, a professor at the California Institute of Technology who had been picked to head the laboratory. A New Yorker, Oppenheimer's postgraduate research in England had brought him into touch with Thomson, Rutherford and Born, and by 1943 he was a physicist with a formidable record. 'More than any other man,' Hans Bethe was to say after his death, 'he was responsible for raising American theoretical physics from a provincial adjunct of Europe to world leadership.'

Los Alamos, arena for a constant if subdued contest between Groves the military commander and Oppenheimer the scientist, each admiring yet suspicious of the other's abilities, was a unique concentration of human beings. Their differences were highlighted by Groves after the war when he admitted that the scientists did not agree with his policy of compartmentalization.

They were not in sympathy with the security requirements. They felt that they were unreasonable. I never held this against them, because I knew that their whole lives from the time they entered college almost had been based on the dissemination of knowledge. Here, to be put in a strange environment where the requirement was not dissemination, but not talking about it, was a terrible upset. While I was always on the other side of the fence, I was never suprised when one of them broke the rules.

The riddle of 'what made Los Alamos different' has been answered at least in part by Alice Kimball Smith, the historian of the physicists' community in the 1940s.

The isolation, the sun-drenched mesas and sparkling air, the cosmopolitan tone, the frantic pace of work capped suddenly by the horror of Hiroshima – all these are tossed into the hopper. And always, in some form, the influence of Robert Oppenheimer. It was a special kind of person, suggests one of his former students, who responded to Oppie's influence, and this made the place special. Perhaps this was no less true of those who had met him for the first time as recruiter for a half-crazy project to win the war. Then there was his oft-cited leadership of the laboratory – his refusal to compartmentalize information, and his own sense of style in science. And this influence permeated the strange community mushrooming among the pines and canyons. Wives knew that his insistence on a few amenities had left trees outside and put fireplaces inside their mirror-image apartments. Or perhaps they had seen his tall figure emerge from the darkness by a door where tragedy had struck. A little aloof but still a warm and comforting presence.

At Los Alamos, in the view of Hans Bethe, a development took place which 'changed everything; it took scientists into politics'. At first glance the claim appears extravagant. Sir Isaac Newton was, among other things, adviser to the British Admiralty. The great Lavoisier was head of the French State Arsenal. During the American Civil War, the Federal Government's Navy Department appointed a Permanent Commission of Scientists that included Joseph Henry and Charles Henry Davis. Michael Faraday was scientific adviser to the British War Office, while Rutherford had in the First World War headed an international body that produced the anti-submarine device, ASDIC. More recently, Bush and Conant had been two men whose careers suggested that Los Alamos hardly altered the position, and much the same could be said for Cherwell in Britain.

Yet even the most politically conscious of these men tended to conform to the judgement of one egregious British Minister that 'scientists should be on tap but not on top'. They knew their place and kept to it; they gave advice but left it to the politicians to take it or leave it. But at Los Alamos a subtle transformation began to take place. The scale of the subject concerned, the size of its impact on the future of the world, were both so great that a daring possibility began to be considered: scientists might not wish to be on top, but unless their views on the political implications of their work were at least seriously considered, they might no longer be on tap.

This feeling grew slowly as, throughout 1944, the Los Alamos teams wrestled with a growing number of problems. Even late in the enterprise the critical mass of uranium 235 was not known with the accuracy required and only when significant quantities of the isotope began to arrive was it possible to test the various calculations which had been made. One method was to build small almost spherical assemblies, a dangerous process since if the critical size was exceeded by even a small amount the result would be a burst of lethal radiation. 'On one occasion,' Otto Frisch has written,

I was making such an assembly (nicknamed Lady Godiva because there was no neutron-reflecting material round it), and just as we were getting close to critical size the student who was assisting me pulled out the neutron counter which he said seemed unreliable. I leaned over, calling out to him to put it back, and from the corner of my eye I saw that the neon lamps on the scaler had stopped flickering and seemed to glow continually. Hastily I removed a few pieces of uranium 235, and the lamps returned to their flickering. Obviously, the assembly had briefly become critical because my body – as I leaned over – reflected neutrons back into it. By measuring the radioactivity of some of the uranium 235 bricks afterwards, we could calculate that the reaction had been growing by a factor of 100 every second! As it happened I had received only about one standard daily dose in those two seconds but it would have been a lethal dose if I had hesitated for two seconds longer.

Other men were not so lucky. One young scientist who dropped a block of reflector material at the wrong moment died from radiation injuries two weeks later even though he had swept the material aside without delay.

As the work progressed, fresh problems presented themselves. It had at first been hoped that no test explosion would be necessary; whatever the pros and cons, the expected scarcity of both uranium 235 and of plutonium meant that the 'waste' of a test bomb should be avoided if possible. However, long before the end of 1944 it was clear that such a test was essential. More important, the early view that a gun-type weapon would be suitable for both a uranium and a plutonium bomb had to be abandoned. As a result, work had to be carried out simultaneously on two types of weapon. In the first, the uranium bomb, a sub-critical mass of uranium 235 would be shot into another sub-critical mass at the target end of the device. When plutonium was the fissile material, however, an entirely different method had to be employed. In this, a sub-critical sphere of plutonium would be surrounded by encircling explosive charges; when these charges were exploded the fissile material would be compressed by the implosion. The surface area of the fissile material would become less, the sub-critical mass would become critical, and the nuclear explosion would take place. Work was carried out simultaneously on both types of weapon although it was still uncertain whether sufficient quantities of uranium or of plutonium would be available before the end of the war. On a dozen lesser matters, huge sums had to be spent on materials or processes before it was known whether they would ever be needed. Thus a remarkable feature of the Manhattan Project was that so many interlinked operations had to be carried on before any one of them had been brought to a satisfactory conclusion.

The year 1944 saw the numbers involved still growing, an increasingly optimistic sign that most things were progressing on the right lines. But there was still no certainty as to when the first bombs would be ready, and there were still setbacks: as late as December Groves was reluctantly forced to report that one implosion system on which everyone had relied now had to be ruled out.

CHAPTER 9

Letting Matters Drift

From the first days of Tube Alloys and the Manhattan Project, the minor leakages of information which worried the Western Allies were based on the fear that aid might be given to the Germans. Not until late in 1944, when Allied forces occupied Stuttgart, was it certain that Heisenberg and his colleagues had hardly started along the road to production of an atomic bomb. More important, and a pointer towards the great crisis which lay ahead in the post-war world, was the fear behind two major controversies which erupted in 1944 and which threatened to restore the bitter suspicions which had been partially removed by the Quebec Agreement. These concerned potential leakages of information not to the Germans but to the Russians whose Red Army was, like Housman's mercenaries, still holding the sky suspended, and was at great cost grinding the Wehrmacht into the ground.

All centred on a simple question. How much, if anything, should one tell one's Russian ally about the new power which it was confidently being predicted would soon divide the old world from the new? As in most political or military dilemmas, the moral issues were quickly and quietly overlooked. What mattered was whether an open admission to the Russians of the great schemes afoot would or would not help to make a more secure world when the war had been won. 'The secret' of the bomb, that catch phrase of the scientifically illiterate, was not involved; even the most liberally minded of those in the West betrayed little intention of telling the Russians technical details. Far more important for the future, it was instead suggested, would be an honest statement that the Western Allies were at work on a nuclear project and that they hoped for post-war collaboration in some form of international control.

Any such control would have to have the fully committed support of the Americans and on 27 April 1944, Anderson proposed to Churchill that the Prime Minister should cable Roosevelt pointing out that the nuclear efforts were coming to fruition: 'But their very success,' he continued, 'makes me wonder whether any plans for future world security which do not take account of the implications and possibilities of this tremendous development must not be quite unreal. I do not know whether you have any ideas on this. It seems to me to require deep thought.' Churchill's reply was unaccommodating. 'I do not think any such telegram is necessary,' he minuted; 'nor do I wish to widen the circle who are informed.'

Any doubt that Anderson was realistically advocating some communication with the Russians is dispersed by a further note from him to Churchill:

> As soon as we get into discussion with the Americans, one point which we shall have to settle quickly with them is whether, and if so when, we should jointly say anything to the Russians. If we jointly decide to work for international control, there is much to be said for communicating to the Russians in the near future the bare fact that we expect, by a given date, to have this devastating weapon; and for inviting them to collaborate with us in preparing a scheme for international control. If we tell them nothing, they will learn sooner or later what is afoot and may then be less disposed to co-operate. At the same time, there would seem to be little risk of the Russians, if they chose to be unco-operative, being assisted in the development of their own plans by a communication of the kind suggested.

Churchill's reaction was to ring the word 'collaborate' and put in the margin beside it 'on no account'. Beside the last sentence he wrote 'no'. This was the more surprising since Anderson had co-opted one of the few men whose advice Churchill frequently took. 'This minute,' he had concluded, 'has been prepared in consultation with Lord Cherwell, who agrees.' To which Churchill added the four words: 'I do not agree.'

Anderson and Cherwell also conscripted on to their side Field Marshal Smuts, the South African Prime Minister, having obtained Churchill's consent to speak to him about the nuclear project. Smuts was an elder statesman and was often consulted by the War Cabinet. According to his own account, he had known of the work at least since the summer of 1943, perhaps through the search for uranium then carried out in South Africa. Although the Field Marshal felt that Stalin should not be told at the moment, the letter which he wrote in his own hand to Churchill on 15 June makes his position clear.

> Something will have to be done about [nuclear energy's] control, but exactly what is at present far from clear. While it may be wise to keep the secret to ourselves for the moment, it will not long remain a secret, and its disclosure after the war may start the most destructive competition in the world. It would therefore be advisable for you and the President once more to consider this matter, and especially the question whether Stalin should be taken into the secret. There must of course be the fullest trust and confidence between you as a condition precedent of any such disclosure. But the matter cannot be allowed to drift indefinitely.

Only a few months later Roosevelt himself was to tell the Canadian Prime Minister, Mackenzie King, that he 'thought the time had come to tell [the Russians] how far the developments had gone. Churchill was opposed to doing this. Churchill is considering the possible commercial use later.' Mackenzie King's own views are revealed by the next words in his diary: 'I said it seemed to me that if the Russians discovered later that some things had been held back from them, it would be unfortunate.'

It has been suggested that King may have misunderstood Roosevelt's exact meaning, and it is certainly true that during the last days of 1944 Roosevelt 'thought he agreed' with Stimson that it was not yet time to tell the Russians anything. However, the President's fluctuating views on what should be told when, were views on a problem that still did not demand any urgent solution. There were enough pressing matters to be dealt with, and until it became clear that the atomic bomb was a practical weapon the final solution could wait; until then, it was natural that Roosevelt, anxious for a post-war rapprochement with the Russians yet regularly appalled by their actions, should express contradictory opinions. The most that can be said with certainty is that he did not uncompromisingly rule out, as Churchill ruled out, any simple statement to the Russians that the Western Allies were at work on such a weapon.

One reason for this was no doubt Roosevelt's earlier awareness that the Russians knew of the Manhattan Project even though there was no evidence about how much they knew. Before the end of 1944 – and possibly some while before – Stimson told Roosevelt that he was 'troubled by the possible effect' of trying to conceal from the Russians the basic fact that they already knew. However, Churchill was given much the same information early in 1945 when he was told by Anderson that a combined Anglo-American Committee on Tube Alloys intelligence had reported that 'Russian interest in the project is confirmed by recent scientific publications and the American T.A. organization has detailed evidence regarding Russian attempts to learn about the T.A. programme in the U.S.A.'

Against this background the controversy over Niels Bohr which erupted in 1944 is significant and ominous, highlighting Churchill's unwillingness to take the advice of his experts on a matter which was gravely to affect the post-war world. In 1942 Bohr had been visited by Werner Heisenberg; the two men subsequently gave very different accounts of the meeting, Heisenberg maintaining that he wanted to discuss the moral issue of working on nuclear weapons, Bohr believing that Heisenberg's aim was to pump him for information. Whatever the truth of the matter, and the two versions are not mutually exclusive, it is nevertheless true that at that time Bohr had no knowledge of what progress, if any, had been made by the Allies in producing a nuclear weapon.

Early in 1943 he received through the Danish Resistance a message from England. Shortly afterwards there arrived a bunch of keys. By following instructions, Bohr found that a minute hole, half a millimetre either way, had been bored in one of the keys. Inside was a microfilm on which there appeared what seemed to be a single dot. Studied under the microscope, it was revealed to be a letter the British Intelligence Service had persuaded Chadwick to write, inviting Bohr to come to England. 'There is no scientist in the world who would be more acceptable both to our university people and to the general public,' he said. He added that Bohr might be able to help on certain matters, but made no mention of what these were.

Bohr refused since he felt it necessary 'to assist in the protection of the exiled scientists who have sought refuge here'. But he guessed what Chadwick's other matters had been, adding: 'Above all, I have to the best of my judgement convinced myself that, in spite of all future prospects, any immediate use of the latest marvellous discoveries of atomic physics is impracticable.'

So Bohr, among the greatest of all living physicists, cut off from all that had been happening in the United States and Britain since the summer of 1940, still ruled out the immediate prospect of using nuclear energy. After all, as he still saw it, the separation of these elusive seven atoms of uranium 235 in every 1,000 of natural uranium would be a task which would never succeed in producing the large amounts necessary. Like Frisch and Peierls before February 1940, he still thought the amounts to be huge. This was spelt out a few months later in a second message to Chadwick in which he reaffirmed his views – but on the assumption that it would not be possible to separate uranium 235 on a large enough scale.

Then, in September 1943, Bohr was warned that he and his family were likely to be arrested. The same night he was taken by the Danish Resistance across the Kattegat into Sweden. The British had been kept informed and after only a few days in Stockholm Bohr was invited to England by Cherwell. Shortly afterwards he was flown to Britain where, carefully protected and in complete anonymity, he was briefed by Cherwell and Anderson on the astounding progress made towards nuclear weapons during the previous two years.

Bohr's reaction appears to have been very different from that of most physicists, most politicians, and most of the military. While they, with few exceptions, concentrated on the impact which nuclear fission might now have on the current war, Bohr took a longer view. If the Allies were, as he was told, on the verge of producing nuclear weapons, then of course success would end the war. There was really nothing to discuss about that. But what happened next? The Russians would immediately increase their work on similar weapons. There seemed to be, in Bohr's opinion, the choice of trying to lay the foundation for telling the Russians now that work was in hand – but not giving them details – or of risking post-war suspicion and the consequent near certainty of a nuclear arms race.

He had been in London only a few days before he was invited to America where Groves realized that his advice might be of outstanding use. Bohr accepted at once, ostensibly so that he could help the project on its way. Later he was to say he had accepted in order to press his views on America's policy-makers. 'They didn't need my help in making the atom bomb,' he added. What they did need, he was certain, was help in dealing with the situation after the war.

Before leaving Britain for America, Bohr had been told that if in need of guidance on any problem while in the United States he should approach Lord Halifax, the British Ambassador in Washington. It was not long before he did so, asking how the gravity of the situation could be forced upon the Americans. Halifax knew that the only approach must be through Roosevelt and by a lucky chance Bohr had already renewed acquaintance with his old friend, Felix Frankfurter, not only a distinguished Supreme Court justice but an unofficial adviser to and friend of Roosevelt. Frankfurter himself had already been approached by a number of American scientists, worried about the prospects opened up by nuclear weapons. When he and Bohr met in the Supreme Court building the two men were therefore able to talk about 'X', the bomb, without either betraying any confidences. Frankfurter, impressed by Bohr as 'a man weighed down with a conscience and with an almost overwhelming solicitude for the dangers to our people', was glad to intervene. He was glad that he had done so when, during a long interview with Roosevelt, the President told him that the atomic bomb problem 'worried him to death'. Roosevelt then told Frankfurter that he could pass on to Bohr the news that he, Roosevelt, would be happy to explore the situation with Churchill.

'My father,' Aage Bohr later wrote, 'was entrusted with the message that Roosevelt would welcome suggestions from Churchill as to how the matter should be approached. Halifax considered this development to be so important that he thought my father should go to London immediately, and Anderson agreed.' Bohr returned to London in April 1944, comparatively confident: he had carried out at least part of his self-imposed mission and believed that he would now be successful in putting his views to Churchill.

At this point, Bohr's luck changed; so, it might be suggested without too great a dramatization, did that of the human race. On arrival back in Britain Bohr was told that a letter was awaiting him at the Russian Embassy whose members, whatever precautions had been taken by the British, were aware of his presence in London. It was handed to him by the Counsellor who, Bohr later wrote, 'in a most encouraging manner stressed the promises for the future understanding between nations entailed in scientific collaboration. Although, of course, the project was not mentioned in this conversation I got nevertheless the impression that the Soviet officials were very interested in the effort in America about the success of which some rumours may have reached the Soviet Union.'

The letter was from his old friend Peter Kapitza, who said that the Russians had learned of Bohr's escape. They were worried about his safety and would be happy if he wished to settle in Russia with his family. Every facility would be put at his disposal, and both he and the rest of Kapitza's colleagues would be honoured. The letter was, in some ways, uncannily like Chadwick's invitation to Britain and the offers that were by this time being made by the Americans.

Bohr showed the letter to Anderson and to the British intelligence authorities. The latter proposed that he reply in a friendly tone, and then vetted and approved his letter. Dated 29 April it began: 'I do not know how to thank you for your letter ...'. Bohr then stated that his plans were unsettled; mentioned that he had visited America, while giving no hint of the reason; and then ingeniously explained his escape from Denmark.

Notwithstanding my urgent desire in some modest way to try to help in the war efforts of the United Nations, I felt it my duty to stay in Denmark as long as I had any possibility of supporting the spiritual resistance against the invaders and assisting in the protection of the many refugee scientists who after 1933 had escaped to Denmark and found work there. When however, last September I learned that they all, besides a large number of Danes, like my brother and myself, were to be arrested and taken to Germany my family and I had the great luck of escaping at the last moment to Sweden, among the many others who due to the unity of the whole Danish population succeeded in counterfoiling the most elaborate measures of the Gestapo.

Bohr delivered personally to the Russian Embassy in Kensington the letter which for all its kind words was in reality a firm no. Then he continued his efforts to see Churchill.

Anderson was by this time trying to arrange a meeting while Cherwell was also pressing on Churchill the need for action. 'Mackenzie King spoke to me on Sunday about Tube Alloys,' he minuted on 10 May.

(You may remember that you asked me to discuss it with him at Washington last year.) He also feels that it must play a vital part in any security arrangements for the post-war world.

Should it be desired to try to prevent anyone else developing these weapons, this will involve periodic inspections not only of enemy but of Allied countries, which will have to start soon after the armistice. (Members of the American Army now controlling it would, I gather, like to insist on inspecting us 'with the rest'.) If, as there is reason to think, the President hopes some other way can be imagined of preventing a competition in this form of super-armament, even more difficult problems will have to be solved.

I must confess that I think plans and preparations for the post-war world or even the peace conference are utterly illusory as long as this crucial factor is left out of account. And all the information I get from America tends to show that it will be more difficult to make head [sic] against the desires of the American generals in control if we wait till after the election.

As you have no time to consider these questions yourself, would it perhaps be agreeable to you to have the position explained to Field Marshal Smuts, who could advise you as to whether any action should be taken or whether it is safe to let the matter drift?

Much the same message was being given by Sir Henry Dale, the President of the Royal Society, and a member of Britain's Scientific Advisory Committee to the Cabinet, in a letter to Churchill.

I cannot avoid the conviction that science is even now approaching the realization of a project which may bring either disaster or benefit, on a scale hitherto unimaginable, to the future of mankind. The devastating weapon, of which the early realization is now almost in view, must apparently put the power of world mastery into hands having a one-sided control of its use. This position will have been created by the work of leading men of science of the U.S.A. and Britain, largely on the basis of Bohr's theoretical work and recently with his active participation; but these men of science cannot concern themselves with its tremendous political implications. It is impossible, nevertheless, for a man of science who has been allowed to see what is happening and what may be involved, to neglect any opportunity of furthering the timely consideration of these implications by the only two men in whose power it may yet lie to take effective action – yourself and President Roosevelt. It is my serious belief that it may be in your power, even in the next six months, to take decisions which will determine the future course of human history. It is in that belief that I dare to ask you, even now, to give Professor Bohr the opportunity of brief access to you.

Eventually Churchill gave in. He agreed to meet Bohr on 16 May. Bohr spent three days drafting and discussing what he would say. He was then taken to Churchill by Cherwell.

The meeting was disastrous. 'It was terrible,' Bohr said afterwards. 'He scolded us like two schoolboys!'

One account of the meeting that was to have such enormous repercussions for the future, has been given by R. V. Jones. It misfired from the start, he has said.

Churchill was in a bad mood, and he berated Cherwell for not having arranged the interview in a more regular manner. He then said that he knew why Cherwell had done it – it was to reproach him about the Quebec Agreement. This of course was quite untrue, but it meant that Bohr's 'set piece' talk was thrown right out of gear. Bohr, who used to say that accuracy and clarity were complementary (and so a short statement could never be precise), was not easy to hear, and all that Churchill seemed to gather was that he was worried about the likely state of the post-war world and that he wanted to tell the Russians about progress towards the bomb. As regards the post-war world Churchill told him: 'I cannot see what you are talking about. After all this new bomb is just going to be bigger than our present bombs. It involves no difference in the principles of war. And as for any post-war problems there are none that cannot be amicably settled between me and my friend, President Roosevelt.'

At the time Churchill was hard pressed by a multitude of more urgent matters. The Allied invasion of Europe on which so much rested was less than three weeks away. It seemed essential to him that any possibility of offending the Americans must be ruled out: and hovering in the background there were the stringent terms of the Quebec Agreement in which he had signed away Britain's options.

But Bohr refused to give up. Early in June he returned to the United States, and in Washington he and his son Aage drafted for Roosevelt a seven-page memorandum which again stressed the need for some approach to the Russians before the bomb was used. This was handed to Frankfurter who gave Roosevelt an oral summary, discussed it twice with the President, and eventually persuaded the President to see Bohr.

The meeting took place on 26 August, after Roosevelt had received the Bohr memorandum from Frankfurter, and lasted an hour. The central issue, developed in what Frankfurter called Bohr's 'able but quaint English' has been well summarized by the annotator of the Frankfurter–Roosevelt correspondence, Max Freedman.

> Continued secrecy would poison American-Russian relations and endanger the prospects for peace. Russia would have reason to think that she had suffered a gigantic and unexampled double-cross if the United States, while talking of the partnership of war, worked in unity with Britain but kept its entire atomic program an absolute secret from the Soviet Union. Perhaps a policy of disclosure would not produce the open world. But the failure to deal frankly with Russia would almost certainly produce a closed world, a world of division and discord, an armed world capable for the first time in history of reducing itself to a charred and poisonous rubble.

Roosevelt appeared to be impressed. Certainly he pointed out that he would be meeting Churchill the following month, and he left Bohr with the firm impression that he would stress to the Prime Minister the need for some communication to the Russians. To Bohr it must have seemed that at almost the last moment of the last hour a progress to disaster was to be avoided. What happened was something very different.

On 19 September Churchill discussed the nuclear problem with Roosevelt at Hyde Park, Roosevelt's home town in the south-east of

New York state. Neither appears to have left any first-hand account of their conversations. Yet the outcome was spelt out in an *aide-mémoire* signed by both. It said:

1. The suggestion that the world should be informed regarding Tube Alloys, with a view to an international agreement regarding its control and use, is not accepted. The matter should continue to be regarded as of the utmost secrecy; but when a 'bomb' is finally available, it might perhaps, after mature consideration, be used against the Japanese, who should be warned that this bombardment will be repeated until they surrender. 2. Full collaboration between the United States and the British Government in developing Tube Alloys for military and commercial purposes should continue after the defeat of Japan unless and until terminated by joint agreement. 3. Enquiries should be made regarding the activities of Professor Bohr and steps taken to ensure that he is responsible for no leakage of information, particularly to the Russians.

The gratuitous insult to Bohr was belatedly taken back nine months later when Field Marshal Sir Henry Maitland Wilson had joined the Combined Policy Committee in Washington. Instructed to send Stimson a copy of the Hyde Park *aide-mémoire* he added, 'The last paragraph of the enclosed document arose from a misunderstanding which was subsequently cleared up by Lord Cherwell and Dr Bush.'

But there was more to it than the Hyde Park *aide-mémoire*. The day after it was signed, Churchill sent a note to Cherwell which suggests how greatly he had influenced Roosevelt.

The President and I are much worried about Professor Bohr. How did he come into this business? He is a great advocate of publicity. He made an unauthorized disclosure to Chief Justice [sic] Frankfurter who startled the President by telling him he knew all the details. He says he is in close correspondence with a Russian professor, an old friend of his in Russia to whom he has written about the matter and may be writing still. The Russian professor has urged him to go to Russia in order to discuss matters. What is all this about? It seems to me Bohr ought to be confined or at any rate made to see that he is very near the edge of mortal crimes. I had not visualized any of this before, though I did not like the man when you showed him to me, with his hair all over his head, at Downing Street. Let me have by return your views about this man. I do not like it at all.

Churchill's extraordinary reaction is explained if not excused by the inability of any man to ride half a dozen horses at the same time. Tube Alloys was only one of a score of subjects with which he had to concern himself and he could not be well briefed on all of them. This had been made clear earlier in the year when he questioned Cherwell about an American hint of 'a German bomb which emitted some liquid starting radioactivity over an area which might be two miles square, causing nausea and death and making the area unapproachable. ...

'All this seemed very fruity,' he went on. 'I do not know whether he is mixing up the possible after-effects of an explosion on the lines of Anderson's affair – (I have forgotten the code-name).'

In September 1944, Churchill was deeply concerned with the progress of the Allied advance across Europe. He was dismayed by the perfidy of the Russians who were refusing to aid the Polish rising in Warsaw, and his reaction to a situation on which he had been only partially and inadequately briefed was unfortunate but explicable.

More complex is the *bouleversement* of Roosevelt's attitude to Bohr. It has been claimed that in May 1944, he was anxious that the Russians should be told nothing at all and that during his meetings with Bohr he was 'following his frequent practice of genial deception'. By contrast, an equally responsible historian of Roosevelt's policies has maintained that he genuinely wished to continue promoting Soviet-Western accord. 'Aware through Bohr himself that the Soviets had invited him to live and work in Russia, Roosevelt undoubtedly assumed that his conversation with Bohr would become known to them. The talk, therefore, was his way of telling Moscow that he was ready to entertain a Soviet role in ultimate control of atomic power, or that post-war co-operation on other matters could be a prelude to shared control of the bomb.'

It seems likely that neither deception nor Byzantine manoeuvring provides the explanation. More probably and simply Roosevelt's long-term hope for a post-war East-West understanding was first given support by Bohr's pleadings and then dispersed by the force of Churchill's oratory. '[He had] scented danger in Bohr's suggestion that the scientific community might help in promoting the process of political co-operation with Russia, and he infected Roosevelt with his suspicions of Bohr's purposes ...', Max Freedman has commented. 'Bohr was declared off limits. He was placed under secret scrutiny. For a time the most detailed watch was kept on his movements. Under Churchill's influence, Roosevelt's trust in Bohr was replaced by suspicion. Even Stimson, who agreed generally with Bohr, could not see him.'

Yet it seems possible that the repercussions of the incident affected even Bohr, who quickly heard of Churchill's views, since less than three months later he was to nip in the bud another, and certainly less responsible, attempt to start a dialogue with the Russians.

This time the move came from Einstein who had been visited in Princeton by an old colleague, Otto Stern, a consultant to the Manhattan Project. Stern warned Einstein of the post-war problems of nuclear control and on 12 December Einstein, almost in desperation it appears, wrote to Bohr at the Danish Legation in Washington. Surely, he asked, men such as Bohr himself, Cherwell in London, Kapitza in Russia, could bring combined pressure on their political leaders 'to bring about an internationalization of military power'.

Bohr, fearing that Einstein might take matters into his own hands, hurried to Princeton without delay, and convinced him of the perils of stepping out of line. Einstein, he reported to Washington, now 'quite realized the situation and would not only abstain from any action himself, but would also ... impress on the friends with whom he had talked about the matter, the undesirability of all discussions which might complicate the delicate task of the statesmen.' The poacher 'on the edge of mortal crimes' had turned gamekeeper.

The controversy over Niels Bohr should never have arisen. His proposal was not that the Russians should be given 'the secret' but merely that they should be told of what, it was by now known, they had already discovered for themselves – that the Americans were building the world's first nuclear weapons. The revelation might or might not have helped post-war negotiations between East and West; it is difficult to see how it could have made the Russian acquisition of nuclear weapons any easier.

However, a comparable argument about the French in Montreal which also came to a climax towards the end of 1944 was on a very different level of complexity. There were five Frenchmen involved – Dr Halban, Dr Kowarski, Professor Auger, Dr Guéron and Dr Goldschmidt. All knew, though to differing degrees, something of the work of the Manhattan Project. The circumstances under which they worked were, however, very varied. Halban and Kowarski were technically British civil servants, although in addition Halban had come to an arrangement with the British under which he would do his best to secure for Britain the non-French patent rights the French team had in 1939 assigned to the Centre National de la Recherche Scientifique; in return, the French were to have the British rights in the patents taken out after Halban and Kowarski had arrived in Britain. Dr Guéron was employed only by the Free French while Auger and Goldschmidt were in between, having been employed by Britain's D.S.I.R. on the wartime basis of one month's notice on either side. However, according to de Gaulle, Guéron, Auger and Goldschmidt had all started nuclear work 'with [his] authorization'.

This was enough to make the situation rather difficult once the liberation of France had begun in the wake of the Normandy landings since any of the Frenchmen might well wish to return home. In the case of Halban and Kowarski, the question of their visiting France after the liberation had been raised during the patent negotiations.

But there were two other complicating factors. One was the secrecy agreement signed by Churchill in Quebec, since the French were not aware of the Agreement and the Americans had not been made aware of Britain's prior obligations to the French. The danger of the French talking out of turn to their own physicists was compounded by the awkward fact that the Frenchman in whom they were most likely to confide was Joliot-Curie, a self-confessed Communist and soon to become, by reason of his work in the French Resistance, not only a famous physicist but a national hero.

Little thought had been given to the potential dangers and it appears to have been almost by chance that Cherwell discovered during a visit to Montreal in October 1944, that Auger had already returned to France and that Halban was planning to visit London. Without delay he cabled to Anderson saying that as far as Auger was concerned,

> only a promise which he might disregard if he thought his country's interest so required, stands in the way of his disclosing any or all of the considerable volume of information he possesses to anyone he pleases.
>
> I had always believed that foreigners at Montreal, most of whom are in possession of vital information, were naturalized but I find that many of them are not. I can quite understand American reluctance to impart knowledge to these people if they are at liberty to return home and place their services at the disposal of anyone they think fit.

Until a policy had been decided, he recommended, the French should be prevented from leaving either Canada or Britain – Britain no doubt being added in the hope that even if Halban reached London he would not cross the Channel. Five days later he followed up his first cable with another.

> Joliot, who has great influence on our French members, will press them for

information and may declare it is their duty to him as representative of science in France to give him information. (b) It is certain that the Fighting French organization has some knowledge of project. It is quite possible de Gaulle or some accredited agent of French Government may command them in the interest of their country to disclose their information in which event they have no choice but to obey.

However, while Cherwell and Chadwick in Washington emphasized the need to propitiate the Americans, a different view was taken in London. There were, Anderson told Churchill, two reasons which made it desirable in Britain's interest that Halban should not only visit Paris but also meet Joliot-Curie. '(a) I wanted to try to clinch the patent arrangements outlined in our contract with Halban, and (b) I considered that it would be to our advantage if Halban, as he was willing to do, gave Joliot such an account of the continuation of the work initiated in Paris as would lead the latter to feel that T.A. was not a matter of immediate or vital importance to France.'

Halban went, and he talked with Joliot-Curie, but not before a potential diplomatic disaster had been only narrowly avoided. With Halban in London, Groves firmly instructed John Winant, the U.S. Ambassador, that Halban was not to visit France. Anderson intervened and Winant, awaiting clarification from Washington, was persuaded to agree that the Frenchman who had played such a key part in the whole nuclear enterprise should be allowed to visit his own country after four years' absence.

Halban scrupulously obeyed his instructions to say nothing about certain matters to either Joliot-Curie or Auger, but exactly what his instructions were has never been clarified. Dr Spencer Weart, who has carried out a detailed study of the French nuclear scientists' activities, believes that Halban probably told Joliot-Curie that large-scale uranium isotope separation was feasible, that nuclear reactors worked and that these could produce fissionable plutonium. The danger, if danger it was, could not really be confined, and within a few weeks Anderson was warning Churchill that 'we should all have it in mind that Professor Joliot may very well either ask point-blank where the French stand *vis-à-vis* the United States or, despairing of any prospect of a fair deal from them, advise General de Gaulle to take precipitate action in another direction'; a warning against which Churchill noted: 'If there is real danger of this, Joliot should be forcibly but comfortably detained for some months.'

But all such precautions were in practice too late. In July 1944, General de Gaulle had visited Ottawa. Knowing of the coming event Auger, Goldschmidt and Guéron decided that de Gaulle should be told at least that nuclear weapons would probably be found practicable before the end of the war. 'It was Guéron,' Goldschmidt later wrote, 'the only one of us de Gaulle had previously met, who had the honour of giving our message in a little chamber hidden at the end of a corridor which the General, forewarned, visited for three precious minutes.' This was only a beginning.

Anderson himself met Joliot-Curie early in 1945 and was told 'that the French had approached the Russians about [the atomic bomb]; they had answered that they were interested but would not give any information.' Threats of proliferation were thus on the way even before the first bomb had been tested.

The strong feelings – indeed, the vehemence – which are revealed in Churchill's correspondence on both Bohr and the French reflect the Prime Minister's growing realization that nuclear weapons would provide the key to post-war power. But he remained constantly mindful of the Quebec Agreement. Following the four-power meeting at Yalta early in 1945 he wrote to Anderson and Cherwell: 'On board the *Quincy* yesterday I mentioned to the President, in the presence of Mr Hopkins, that we should be going ahead with our own work on T.A., and I read out the paragraph marked on the Paymaster-General's [Cherwell's] paper.' This paragraph, which Cherwell had included in a note to the Prime Minister on 26 January, read as follows: 'I think the best way of putting it would be to say that, after the war, we shall want to do work here on a scale commensurate with our resources; indeed, we should like, even earlier, to do more to pull our weight and shall wish to discuss with the Americans fairly soon what work we can best undertake to this end without impairing our contribution to the production in America of a weapon for use in this war.'

After listening to this masterly description, the President, in Churchill's words, 'made no objection of any kind'.

The Prime Minister's feeling that nuclear fission was so important that knowledge should be handled on a 'need-to-know' basis, was illustrated early in 1945 by the case of Sir Henry Tizard whose subcommittee of a subcommittee had started the British investigations five years previously.

The British Chiefs of Staff had been asked by the Ministers of Production and of Home Security for advice on post-war defence and it had been proposed that Tizard should chair a Joint Technical Warfare Committee which would provide the necessary advice. Even the Chiefs of Staff had been given only the vaguest information about Tube Alloys and when Anderson asked that they should be put 'fully into the picture', Churchill replied as follows.

> The Chiefs of Staff have already been informed and there is no reason why they should not have the whole matter laid before them. There is surely no necessity for them to go into the technical details which are a life-study in themselves. General Ismay should also be informed, but I am entirely opposed to any other person in the Defence system who does not already know, being made a party. As to Sir Henry Tizard, he should not be made a party at the present time.
>
> 2. It would really be better to wait a month till we have had our talks, before taking any decided new step.

The proposal that the Chiefs of Staff should be given more information on a matter about which they were to be advised than the man who was to give them the advice was a little unusual, and General Ismay pointed out to the Prime Minister that he had a choice of three alternatives, each objectionable. The least so, he felt, was for Churchill to stick to his ruling even though that might mean Tizard's resignation.

However, instead of resigning Tizard tried to find a way round the difficulty. 'Personally,' he was told by Ismay, 'he [Anderson] would see no objection to your putting questions to the British physicists engaged on the special project, some of whom happen to be in this country for a short time now. He thinks, however, that you would be well advised to make it clear to the Prime Minister that you do not intend to make any allusion to the topic in your main report; and that your best line would be to indicate that at this stage you are only seeking background.'

The idea of omitting from his report any mention of what might be the most important item of all was hardly agreeable to Tizard, who persevered. He had no luck. 'Pray see my previous minute,' Churchill replied to Ismay, 'and reply in that sense on my behalf to Sir Henry Tizard. He surely has lots of things to get on with without plunging into this exceptionally secret matter. It may well be that in a few years or even months this secret can no longer be kept. One must always realize that for every one of these scientists who is informed there is a little group around him who also hear the news.' So Ismay, who almost exactly five years earlier had appealed to Tizard for the ton of uranium oxide needed by George Thomson, now had to instruct Sir Henry that his report on post-war defence must refer to the possibilities of nuclear weapons in only vague terms.

By the spring of 1945, as Tizard was prevented from plunging into the 'exceptionally secret matter' which he himself had helped to start, another question began to crystallize.

Both in Britain and the United States nuclear research had been almost frantically pushed ahead by the fear that the Germans might win the race for the atomic bomb. But in the autumn of 1944 an American team under Presidential control had found proof in Germany that the Germans had failed to solve even the preliminary problems involved in making a nuclear weapon. The results of this team – code-named Alsos, the Greek for groves – were conclusive. They were swiftly sent to Washington. And from this date the operational use of the atomic bomb inevitably became a more controversial subject. It now seemed likely that the war with Germany would be ended before the first bombs were ready. But Japan, whose nuclear capacity was known to be non-existent, was already being beaten to her knees by relentless and decreasingly opposed air raids as well as by submarine attacks which had almost stopped her essential seaborne supplies. If Japan capitulated before the bomb was even tested, what then? Might it not be possible to conceal the very essentials of the Manhattan Project, not only from the Russians but from all except the small number of British – and French – who knew of it?

First, however, as the huge American plants began to produce respectable quantities of fissile material, as Oppenheimer began to solve the hitherto intractable problems of efficient implosion and explosion, there had to be some assurance that the bomb would, in fact, really work. Apart from anything else this would be a justification for spending on it a sum already approaching 2,000 million dollars.

CHAPTER 10

A Trace of Indecision

As preparations went ahead for the crucial test, for the experiment which would confirm, once and for all, that a nuclear weapon would explode rather than fizzle, a new and in some ways illogical questioning began to rustle through the laboratories and plants where the bomb was being made possible. Three years previously, as scientists had willingly uprooted themselves and embarked on an enterprise of which they knew no more than the barest details, there had been few doubts. Among those joining the Manhattan Project there were, it is true, a few who did so only after an agonizing struggle with pacifist consciences. More normal was the reaction of Alice Kimball Smith, the historian of the scientists' protest movement. 'I remember vividly,' she has written, 'the succinct explanation that covered our own migration to an undesignated spot in the western mountains: the project would almost certainly end the war and afterwards promised almost limitless benefits to mankind. Amidst the fears and uncertainties of the winter of 1942 and 1943, so easily blurred and forgotten in later knowledge of victory, what more could one ask?'

Another incentive was hinted at by Oppenheimer. 'Almost everyone knew [the Project] was an unparalleled opportunity to bring to bear the basic knowledge and art of science for the benefit of his country. Almost everyone knew that this job, if it were achieved, would be a part of history. This sense of excitement, of devotion and of patriotism in the end prevailed.'

For Oppenheimer, significantly, 'the excitement' came first. Would it really be possible to release the almost unimaginable energy locked within the nucleus? Here was an opportunity to find out; to reject it would have been as difficult as for Adam to reject the apple or Doctor Faustus to reject the offer of Mephistopheles. Yet if the sober patriotism of workers like Alice Kimball Smith and the scientific curiosity of Oppenheimer tended to submerge any twinges of conscience about weapons of mass-destruction, there was another factor which even from 1939 had been far more significant. Einstein and Szilard were both refugees from totalitarian Germany; so were Edward Teller, Eugene Wigner and James Franck, while Fermi had escaped from fascist Italy. These men knew, far better than their colleagues in America could ever know, what subjugation to the Nazis would mean for the rest of the world. They knew, also, that nuclear fission had been discovered in Berlin and, although they knew little about German progress, the fear that Hitler might eventually possess the world's first nuclear weapons pressed forever on their minds. James Franck, who knew nothing of the Alsos team's discoveries, emphasized that 'to the last day of the European war, we were living in constant apprehension as to their possible achievements'.

But all that was soon in the past. The final collapse of the Third Reich early in May confirmed what the Alsos mission already knew: that throughout the entire war – although for reasons which were not yet clear – there had been no practical possibility of the Germans building a nuclear weapon. Thus the great fear had been removed just as the chance of successfully killing civilians by the tens of thousands began to change from a distant possibility to a likely event of the day after tomorrow. The prospect, like that of Dr Johnson's man to be hanged in a fortnight, concentrated wonderfully the minds of many physicists who had dedicated themselves to the job when success seemed distinctly unlikely.

There now arose in some men's minds a trace of doubt. Would it perhaps not be better if schemes for releasing this new power, pursued in the project throughout the past three years solely as a means of destruction, were never brought to fruition? Since the first years of the century the power within the atom had from time to time appeared before scientific eyes as a mirage, a vision, a Holy Grail, a promise or a threat according to individual opinion. But while it had appeared as only a distant prospect it had been easy to give the search unqualified support; now, almost at the end of the trail, things began to look slightly different.

By the spring of 1945 two interlinked questions forced themselves to the fore. The great riddle no longer concerned the release or non-release of a mind-boggling amount of energy; that, give or take the odd doubt or two, was answered. But, as Szilard asked himself: 'What is the purpose of continuing the development of the bomb, and how would the bomb be used if the war with Japan has not ended by the time we have the first bomb?' More bluntly, would it be necessary to prove the weapon's worth by incinerating the population of a major city or would a demonstration of its power be enough to bring about surrender? This question was intimately linked with the subject which the politicians had been debating for some time. Should the Russians be told of the Manhattan Project before or after the weapon had been demonstrated to exist, and how much, if anything, should they be told?

These questions were more political than military yet it appears inescapable that the scientists were more aware of their significance than most of the politicians or of most military men. Perhaps this was inevitable. Admiral Leahy, Roosevelt's Chief of Staff, still looked on nuclear weapons as 'a professor's dream'. Truman, who in April 1945 was to succeed Roosevelt, was as totally ignorant of the nuclear work as was Clement Attlee who succeeded Churchill as British Prime Minister three months later. Only a few men on either side of the Atlantic appeared to have had the slightest realization that once the bomb had been shown to work, one world would have ended and another begun.

The first tentative suggestion that it might not be necessary to use the bomb in anger came from President Roosevelt when, on 21 September 1944, he met Bush, Admiral Leahy and a British representative on atomic energy. Roosevelt, according to the account which Bush gave to Conant two days later 'raised the question of whether ... [the bomb] should actually be used against the Japanese or whether it should be used only as a threat with full-scale experimentation in this country.' The suggestion came, Bush believed, 'in connection with Bohr's apparent urging that a threat be employed against Germany, which would of course, I think, be futile'. The question was not yet urgent and was left in the air.

It was revived two months later when Alexander Sachs – the Einstein–Roosevelt intermediary of 1939 – read a memorandum to the President in which he had outlined his proposals. Following a successful test in the United States, this went,

> there should be arranged (a) a rehearsal demonstration before a body including internationally recognized scientists from all Allied countries and, in addition, neutral countries, supplemented by representatives of the major [religious] faiths; (b) that a report on the nature and portent of the atomic weapon be prepared by the scientists and other representative figures; (c) that, thereafter, a

warning be issued by the United States and its allies in the Project to our major enemies in the war, Germany and Japan, that atomic bombing would be applied to a selected area within a designated time limit for the evacuation of human and animal life; and, finally, (d) in the wake of such realization of the efficacy of atomic bombing, an ultimatum demand for immediate surrender by the enemies be issued, in the certainty that failure to comply would subject their countries and peoples to atomic annihilation.

Alexander Sachs was a curious foster-father to what is sometimes condemned as an over-humanitarian idea since he has gone on record as always believing that 'the real warmongering, combined with defeatism, is done by the pacifists'. Roosevelt's reaction is unknown although Sachs, predictably enough, claimed that it was favourable. This may well have been the case and the, admittedly scarce and sometimes contradictory, evidence suggests that if Roosevelt had lived even four months longer, then history might have followed another path.

Throughout the first months of 1945, as the Manhattan Project moved towards its goal, and as the defeat of Germany was correctly estimated to be only weeks away, there was an increase in the doubts felt among many of the scientists involved. Leo Szilard, whose efforts had first alerted Roosevelt in 1939, was the first to act. Fearful of the conditions he had done so much to create, Szilard now saw the threat of a post-war nuclear arms race as the main danger to the future. He saw no point in discussing the matter with Groves, or Conant or Bush; because of security regulations there was no one at what he called 'intermediate level' to whom he could turn. He could have approached Stimson but seems to have been determined that the Secretary of State should be bypassed. James Franck who, like Szilard, feared a post-war nuclear arms race, tried to change his views, warning him: 'You may be cleverer than I but, believe me, I am wiser.' Typically, Szilard replied: 'Sir, I agree with one half of your statement.'

However, Szilard believed he had one strong card to play. In 1939 Einstein had opened a way to the President. Perhaps he could do so again. Szilard saw Einstein. He asked for and received a letter of introduction to Roosevelt which, he explained to Einstein, was needed for a matter he was not able to disclose. And he then sent the note of introduction to Mrs Roosevelt with whom he had previously been in correspondence.

Szilard had prepared a memorandum for Roosevelt and, on hearing from Mrs Roosevelt that the President was willing to see him on 8

May, showed the memorandum to Arthur Holly Compton in Chicago, half-expecting to be rebuffed. But Compton reacted with the words: 'I hope that you will get the President to read this.' Szilard, elated, went back to his own room. 'I hadn't been in my office for five minutes,' he later wrote, 'when there was a knock on the door and Compton's assistant came in, telling me that he had just heard over the radio that President Roosevelt had died.'

Szilard arrived in Washington shortly afterwards, hoping to see the new President, but Truman diverted him to James Byrnes, Director of War Mobilization and soon to become Secretary of State. Byrnes was not impressed by Szilard. 'As the Einstein letter had indicated he would,' he later wrote, 'Szilard complained that he and some of his associates did not know enough about the policy of the government with regard to the use of the bomb. He felt that scientists, including himself, should discuss the matter with the Cabinet, which I did not feel desirable. His general demeanour and his desire to participate in policy-making made an unfavourable impression on me, but his associates were neither as aggressive nor apparently as dissatisfied.'

Byrnes appears to have believed that most scientists had a weakness at least for the Left and possibly for the Russians, a belief encouraged by General Groves and one which conditioned much of the official response to the growing scientific agitation in the first half of 1945. However, the exchanges between Byrnes and Szilard were hardly on a mutually informed level. 'When,' Szilard later wrote, 'I spoke of my concern that Russia might become an atomic power – and might become an atomic power soon, if we were to demonstrate the power of the bomb and use it against Japan – his reply was "General Groves tells me there is no uranium in Russia".' Since the whole atomic project had been fostered by fears that the Germans were mining the rich uranium deposits of Czechoslovakia, by this time becoming available to the Russians, Szilard began to despair.

On returning to Chicago he found that Groves had strongly objected to his visit to Truman and to his handing of the memorandum to Byrnes. In an attempt to calm things down, Dr Compton now set up under James Franck a Committee on Social and Political Implications of Nuclear Weapons. The outcome was the Franck Report, a sober and prescient forecast of what would happen if certain mistakes were made by America in revealing the almost unimaginable power which she now commanded. Most of the mistakes were made and most of Franck's Cassandra-like forecasts were to be justified.

He began by pointing out that the British and the French both knew as much about the basic physics of the bomb as did the Americans, and that nuclear fission had itself originated in Germany. 'In Russia, too,' he went on, 'the basic facts and implications of nuclear power were well understood in 1940, and the experience of Russian scientists in nuclear research is entirely sufficient to enable them to retrace our steps within a few years, even if we should make every attempt to conceal them.'

As to raw materials, it was true that most of the known reserves of uranium were controlled by Canada, Belgium or Britain. Yet the Czech deposits which had so worried Szilard and Einstein when they had been under German control in 1939 were now available to the Russians. '[They are] known to be mining radium on [their] own territory; and even if we do not know the size of the deposits discovered so far in the U.S.S.R., the probability that no large reserves of uranium will be found in a country which covers one-fifth of the land area of the earth (and whose sphere of influence takes in additional territory), is too small to serve as a basis for security.'

The United States, it was true, could, at the time of writing, make more and better bombs than any other nation. But, Franck went on, this was a double-edged weapon to wield, since 'such a quantitative advantage in reserves of bottled destructive power will not make us safe from sudden attack. Just because a potential enemy will be afraid of being "outnumbered and outgunned", the temptation for him may be overwhelming to attempt a sudden unprovoked blow – particularly if he should suspect us of harbouring aggressive intentions against his security or his sphere of influence. In no other type of warfare does the advantage lie so heavily with the aggressor.'

Franck knew as well as most physicists, and better than most politicians or Servicemen, that the one 'secret' of the bomb was whether it would explode as forecast. Therefore, he warned, 'the race for nuclear armaments will be on in earnest not later than the morning after our first demonstration of the existence of nuclear weapons'. It would, moreover, be a race in which the Americans could expect nothing but distrust from her competitors. 'It may be very difficult,' Franck pointed out, 'to persuade the world that a nation which was capable of secretly preparing and suddenly releasing a weapon as indiscriminate as the rocket bomb and a thousand times more destructive, is to be trusted in its proclaimed desire of having such weapons abolished by international agreement.' If such a race began, he went on, there would be only one defence for any developed country, such as the United States, with its great centres of population: that would be a massive dispersal of industry and populations – a gigantic operation which has not been attempted even a third of a century later under the threat of the intercontinental H-bomb-tipped rocket.

The only solution Franck envisaged was the one which had been tentatively suggested, then dropped, some months earlier: a demonstration to the Japanese of what the new weapon could do to their country. 'This may sound fantastic,' he admitted, 'but in nuclear weapons we have something entirely new in order of magnitude of destructive power, and if we want to capitalize fully on the advantage their possession gives us, we must use new and imaginative methods.'

One other proposal circulating among the scientist, that the weapon should be secretly tested, but never demonstrated to the world and never used in war, failed to gain even minimal support. Thus the chance of locking back the contents of Pandora's box and pretending that it had never been opened, an idea doomed to failure, was abandoned almost as soon as it was put forward.

The idea of a demonstration lingered on, but even today it is impossible to speculate with any certainty on what the results would have been. The most ambitious of the proposals for a demonstration echoed Alexander Sachs's idea of the previous November. Military and political observers not only from the Western Allies but from Russia and from Japan would be invited to witness a demonstration in the United States, and Japan would then be given the choice of capitulation or devastation by the new weapon. Alternatively, it would be possible to serve notice on Japan that a new weapon would be dropped at a specific time on a named target within her territories. One school of thought believed that an uninhabited stretch of the Japanese mainland should be designated as the target. Another felt that a region such as the Truk Islands in the Pacific, heavily defended but containing no civilians, would be a more suitable choice.

However attractive such options were to anyone with even a trace of humanity, the practical risks they involved appeared overwhelming and insurmountable. The first, which applied to any form of demonstration, was that the bomb might fail to explode. By the spring of 1945 every theoretical detail had been checked, counterchecked and counterchecked once again. Yet the possibility of failure could not be entirely ruled out, and failure before an invited audience that included the enemy would have been a psychological disaster. Moreover only three bombs would be available even by summer so there was little chance of counteracting failure by a further attempt.

Secondly, a point affecting a test in the United States or on an uninhabited part of the Japanese mainland, it was by no means clear what the visual effects of an atomic explosion would be in such circumstances. Oppenheimer later remarked, as he looked at the barren desert landscape of Alamogordo after the bomb test in July, that there was comparatively little to impress questioning observers. A third danger, which applied to a test on enemy-held territory, was that the Japanese, once warned of time and place, would bring on to the target as many Allied prisoners of war as possible.

These were some of the questions which in the spring of 1945 increasingly exercised many men who had either initiated or supported the nuclear project in very different circumstances, had contemplated the use of a nuclear weapon against a different enemy, and who had contemplated even that only because no other course appeared to be possible.

Now, with the test of the bomb only three months away, the scientific conscience at last produced a response from official quarters. At the end of April Niels Bohr – his status possibly enhanced by the absurdity of Churchill's attack – again met Vannevar Bush. With Bush he left two memoranda which outlined his fears of a post-war nuclear arms race. Bush, much impressed, forwarded the memoranda to Harvey Bundy, Stimson's assistant; Bundy showed them to George Harrison, Groves' deputy, and Harrison proposed to Stimson that a small committee should be set up 'to study and report on the whole problem of temporary war controls and publicity, and to survey and make recommendation on post-war research, development and controls, and the legislation necessary to effectuate them.' This brief did not call for any discussion of how the bomb was to be used. However, the members of the Interim Committee as Stimson called it, not only discussed the question but came to decisions which were strictly followed. On the all-important subject of informing the Russians it collectively changed its mind and, as the evidence showed, had its final decision watered down to ineffectiveness, possibly by the powerful influence of Churchill on Truman during a private meeting at Potsdam.

The seven-man Interim Committee, chaired by Stimson, included Bush, James B. Conant and Karl T. Compton as well as representatives from the Navy and State Department, and James Byrnes who sat as

Truman's personal representative, and in effect took over the duties of a Military Advisory Committee that had been set up two years previously. On Conant's advice there was also an advisory scientific panel whose members were Arthur Holly Compton, Fermi, Lawrence and Oppenheimer.

The Interim Committee met four times in May. On the first three occasions its members discussed the history of the Manhattan Project, the amount of information which should be released to the public when the bomb had been used and, in general terms, the problems of post-war international control. Only on the 30th was an attempt made to grasp the two most prickly nettles – the question of what, if anything, should be revealed to the Russians, and the question of how the bomb should be used.

The meeting was a large one, including the members of the scientific advisory panel as well as General Marshall, General Groves and Harvey Bundy. At least some of those present were astounded when Oppenheimer, reviewing the current state of the nuclear art and the immediate prospects, spoke of three stages. The first, about to be brought to fruition, was manufacture of a bomb having the explosive force of anything from 2,000 to 20,000 tons of T.N.T. In the second stage the figures were estimated to be from 50,000 to 100,000 tons, while it was considered possible that a third-stage weapon might be equivalent to between 10 million and 100 million tons of T.N.T. With the prospect of a single bomb equivalent to 100 million tons of T.N.T. being brought from nightmare to possible reality, even the most stout-hearted of his listeners began to wonder what they were doing. 'As I heard these scientists and industrialists predict the destructive power of the weapon, I was thoroughly frightened,' admitted James Byrnes.

It was possibly with these figures still in mind that Arthur Compton subsequently intervened with the suggestion that had tentatively been talked about many months earlier and had then been dropped. 'At the luncheon following the morning meeting, I was seated at Mr Stimson's left,' he has written. 'In the course of the conversation I asked the Secretary whether it might not be possible to arrange a non-military demonstration of the bomb in such a manner that the Japanese would be so impressed that they would see the uselessness of continuing the war.'

Stimson favoured such an idea, and earlier in the month (16th) had told Truman: 'The reputation of the United States for fair play and humanitarianism is the world's biggest asset for peace in the coming decades. I believe the same rule of sparing the civilian population should be applied, as far as possible, to the use of any new weapons.'

But the qualification of 'as far as possible' remained even after a long afternoon's discussion and, state the minutes of the meeting, 'the Secretary expressed the conclusion, on which there was general agreement, that we could not give the Japanese any warning; that we could not concentrate on a civilian area; but that we should seek to make a profound psychological impression on as many of the inhabitants as possible. At the suggestion of Dr Conant the Secretary agreed that the most desirable target would be a vital war plant employing a large number of workers and closely surrounded by workers' houses.' The decision had a similarity with the Deputy Chief of the Air Staff's directive of 14 February 1942, that bombing was 'to be focused on the morale of the enemy's civilian population and in particular, of the industrial workers'; it enabled the maximum number of civilians to be killed with the maximum moral rectitude.

As to whether the Russians should be told, Mr Byrnes was strongly against. General Marshall appeared favourable, even going so far as to ask whether it might not be desirable to invite two Russian scientists to watch the test explosion soon to be carried out in New Mexico. General Groves appears to have made no comment; he may well have been speechless at the idea. No decision was taken, and it was merely agreed that on Byrnes's suggestion every effort should be made 'to better our political relations with Russia'.

The position on this vital matter was soon to be clarified. Stimson had asked, after the discussion of 31 May, that the science panel should formally advise on the use of the bomb. On 16 June they reported, and with obvious reluctance, that 'we can propose no technical demonstration likely to bring an end to the war; we see no acceptable alternative to direct military use'. At the same time, however, the four scientists were well aware of what this would mean internationally. They therefore recommended 'that before the weapons are used not only Britain, but also Russia, France and China be advised that we have made considerable progress in our work on atomic weapons, that these may be ready to use during the present war, and that we would welcome suggestions as to how we can co-operate in making this development contribute to improved international relations.'

The Interim Committee now made up its collective mind, and on 21 June recorded its view that

> there would be considerable advantage, if suitable opportunity arose, in having

the President advise the Russians that we were working on this weapon with every prospect of success and that we expected to use it against Japan.

The President might say further that he hoped this matter might be discussed some time in the future in terms of ensuring that the weapon would become an aid to peace.

The recommendation was clear enough. The President was being advised to reveal the existence of American work on a nuclear weapon; and, if words meant anything at all, they meant that he was being advised to make clear that the subject of his, presumably formal, statement to the Russians was not just a more powerful explosive but something which would dominate the post-war world.

However, there were contrary forces still at work, as was shown while the Interim Committee was making up its mind. Some weeks earlier a group of distinguished American and British scientists had been invited to attend the 200th anniversary session of the Soviet Academy of Sciences in Moscow during the second half of June. Arrangements were completed without trouble but on 13 June a number of the American scientists were told in the United States that they would not be allowed to leave for Russia. Simultaneously, the same news was given to eight members of the 29-man British delegation in London.

The scientists banned from the journey were those involved in the nuclear enterprise. They had been invited because of their distinction in various branches of physics, and it would have needed little imagination on the part of the Russians correctly to assess what they had been doing. As Irving Langmuir put it: 'I believe that these attempts to maintain secrecy resulted in giving to the Russians the very information which the Army most wished to keep from them. Any sensible Russian scientist knowing of these facts would have believed that we were developing an atom bomb and were keeping it secret from the Russians.'

This was no more than the blunt truth. Three months earlier Churchill, in a letter to Anthony Eden, the British Foreign Secretary, dealing with the French, had written. 'I shall certainly continue to urge the President not to make or permit the slightest disclosure to France or Russia.' And he had then added a sentence which was edited out before the Minute was sent: 'Even six months will make a difference should it come to a show-down with Russia, or indeed with de Gaulle.'

CHAPTER 11

Over the Brink

By the beginning of July 1945 the future appeared to have been decided. Before the end of the month the first bombs would be ready. One of them would, with luck, be tested in mid-month, before the 17th when the President was to meet Stalin and Churchill in Potsdam. Truman would tell the Russian leader of progress on the atomic bomb. If the Japanese were still fighting there was little doubt that the bomb would settle the war. Then America and her allies, hopefully including Russia, would set about building a permanent peace.

Preparations for the all-important test, now code-named 'Trinity', had been continuing for some weeks. As far back as early May Kenneth Bainbridge, who had heard the deliberations of the M.A.U.D. Committee some four years previously, had fired 100 tons of explosives spiked with fission products from the Hanford plant in order to test observation methods and routines. Groves had approved a test site at the Alamogordo Air Base, some 120 miles from Los Alamos, and had perfected a first-class communication network by which details of the test could be sent to Washington without fear of leakage. Although the remoteness of the test site was expected to satisfy security requirements, a cover-story describing an unexpected explosion of stored materials at the air base had been prepared for public consumption.

The plutonium for the first bomb arrived at Los Alamos early in July. It was fashioned into an implosion-type device and on the afternoon of 12 July was driven out from Los Alamos to Alamogordo. A few hours later a convoy of vehicles carrying the non-nuclear components drove out to the test site where a 100-foot metal tower built for the occasion now rose from the barren desert landscape. The following day was spent assembling the device in a tent at the foot of the tower; by evening everything was ready except for insertion of the detonating device. The next morning the bomb was lifted by hoist to a platform at the top of the tower. Here the detonators were installed, then the mechanism by which they would be fired. By five in the afternoon all was ready.

The extensive instrumentation that was to record the world's first nuclear explosion was checked and checked again. General Groves, keyed up for the climax towards which he had been working for more than three years, finalized his complicated set of arrangements. One continuing worry was the danger of radioactive fall-out, a danger of which he had, almost incredibly, been made aware only a few months earlier. 'The hazard that I feared the most,' he later admitted, 'was that of radioactive fall-out on the areas over which the radioactive cloud would pass. This had not been considered for too many months as it was only at the turn of the year that Joseph Hirschfelder had brought up the possibility that this might be a serious problem. I learned later that the possibility of this danger had been indicated in the British Maud report, but I had been unaware of the existence of this report.'

Groves' late awareness of the fall-out danger merely highlighted the question marks which even in July 1945 still hung over the enterprise. Frisch's calculations in what was now the prehistoric past of Christmas 1938 had seemed certain enough. So had all the others that followed in Britain and France and America. The flickering needles on the graphs in Chicago on that historic afternoon of 2 December 1942 had surely confirmed everything that the physicists had forecast. But 'the boiler' was not 'the bomb'. Doubts remained, and in some cases the doubts were still hopes that the huge energies locked within the nucleus might never be released.

There were still other doubts, often jokingly made, much as the lesser doubts of earlier wars had been revealed by jokes and limericks not to be taken too seriously. On the eve of the test, now set for 4½ hours before dawn on the 16th, Fermi greatly annoyed Groves by offering to take wagers on whether the tests would ignite the atmosphere; and, if it did so, whether the explosion would merely destroy New Mexico or destroy the entire world. Fermi was not entirely serious; but his attitude did indicate the residual uncertainty that now, after so many years of dedication and intellectual struggle, still remained. Others took a less apocalyptic view. About a hundred scientists at Los Alamos wagered a dollar each on the power that the test would reveal. Oppenheimer guessed 300 tons, and most of the rest put the figure at less than 500. Rabi, finding that most of the lower figures in the lottery had been taken, picked 18,000.

Late on the 15th Groves made his final dispositions; but he was still worried about the weather and the radioactive cloud that had been forced so late in the day into his calculations. 'It was extremely important,' he wrote, 'that the wind direction be satisfactory, because we did not want the cloud, if one developed, to pass over any populated areas until its radioactive contents were thoroughly dissipated. It was essential that it did not pass over any town too large to be evacuated. The city about which we were most concerned was Amarillo, some 300 miles away, but there were others large enough to cause us worry. The wind direction had to be correct to within a few degrees.'

But every hour increased the danger of something going wrong with the complicated instrumentation. All the key men involved in the test were by this time strung up for the event, and shortly before midnight Groves decided that the test should go ahead. Zero hour was changed two hours to 4.00 a.m.

The decision was made at the control shelter built in the desert 10,000 yards south of the tower, and for the next few hours Groves remained here, talking with Oppenheimer and from time to time leaving the shelter with him to inspect the weather.

It was a cloudy night, with light rain falling, only a few stars visible, and at times the rumble of distant thunder. At 3.00 a.m. Groves and Oppenheimer noted that the stars were gleaming more brightly and decided to postpone the test, first for another sixty minutes and finally for a total of ninety. This brought zero hour to 5.30 a.m., one hour before dawn. No further delays were practicable.

At 5.00 a.m. the five men making last-minute adjustments at the tower were ordered away, and each drove his separate vehicle out of the danger zone. Huge floodlights were switched on to illuminate the tower and the squat object on its uppermost platform.

Ten minutes later, as Groves drove back to the main base shelter, 17,000 yards from the tower, the final countdown started.

'As the time intervals grew smaller,' Brigadier General Thomas F. Farrell, Groves' second in command, wrote later, 'and changed from minutes to seconds, the tension increased by leaps and bounds. Everyone in that room knew the awful potentialities of the thing that they thought was about to happen. The scientists felt that their figuring must be right and that the bomb had to go off, but there was in everyone's mind a strong measure of doubt.'

Of all those present Oppenheimer was the most tense, absolutely silent, supporting himself by holding on to a post as the voice over the radio counted down from 'ten – nine – eight ...'

As the word 'Now' came into the silent room the desert was illuminated by what Groves was to call 'a lighting effect ... equal to several suns in mid-day', a searing burning light that was to be seen nearly 200 miles away. 'It was golden, purple, violet, grey and blue,' Farrell later wrote. 'It lighted every peak, crevasse and ridge of the nearby mountain range with a clarity and beauty that cannot be described but must be seen to be imagined. It was that beauty the great poets dreamed about but describe most poorly and inadequately.' Chadwick felt much the same, noting in his diary that 'a great blinding light lit up the sky and earth as if God himself had appeared among us ... there came the report of the explosion, sudden and sharp as if the skies had cracked ... a vision from the Book of Revelation.' Oppenheimer remembered later that a few people laughed, a few people cried, most people were silent. 'There floated through my mind,' he went on, 'a line from the *Bhagavad-Gita* in which Krishna is trying to persuade the Prince that he should do his duty: "I am become death, the shatterer of worlds." I think we all had this feeling more or less.'

But some others reacted differently. 'Finally,' wrote Kenneth Bainbridge, 'I could remove the goggles and watch the ball of fire rise rapidly. It was surrounded by a huge cloud of transparent, purplish air produced in part by the radiations from the bomb and its fission products. No one who saw it could ever forget it, a foul and awesome display.' Of those who did see it, one was as far as 150 miles away to the west, a civilian who commented: 'It was just like the sun had come up and then suddenly gone down again.'

At the base shelter, some nine miles from the bomb, Fermi dropped from his outstretched hand a group of torn paper strips. Some were blown several feet by the blast wave and Fermi had only to measure the distance before he could calculate the blast. His figures were remarkably close to the more detailed, instrumented, results. Groves remembered how Blondin had crossed Niagara Falls on a tightrope and noted that his personal tightrope had been three years long. Greisen exclaimed with surprise: 'My God, it worked,' while Kistiakowsky, who a few days earlier had bet a month's salary against ten dollars that the gadget would work, put his arm round Oppenheimer's shoulder and said: 'Oppie, you owe me ten dollars.' Bush silently shook hands with Groves. So did Conant. Groves commented: 'This is the end of traditional warfare.'

As data were collected from the instruments during the next few hours it became clear that the bomb had been even more powerful than expected. Whereas the British M.A.U.D. report, giving the world's first estimate of what a nuclear explosion would mean, had suggested an energy yield of about 1,800 tons of T.N.T., Groves was now able to inform Stimson that at Alamogordo the yield had been between 15,000 and 20,000 tons. The steel tower to which the bomb had been strapped had been entirely evaporated by the heat. Even 1,500 feet away, a stout four-inch iron pipe, 16 feet high and set in concrete, had completely disappeared. At the site of the explosion there remained a crater 1,200 feet across.

Most impressive of all to Groves was the impact of the explosion on a massive 220 ton steel test cylinder half a mile away. This had been securely anchored in concrete and was comparable to a steel building bay to be found in a twenty-storey skyscraper. But the blast had torn the cylinder from its foundations, ripped it apart and laid it flat on the ground. 'The effects ... indicate that, at that distance, unshielded permanent steel and masonry buildings would have been destroyed,' noted Groves. 'I no longer consider the Pentagon a safe shelter from such a bomb ... None of us had expected [the construction] to be damaged.' And Groves, the Army engineer responsible for building the Pentagon, had previously thought it invulnerable.

On the face of it, a deep sigh of relief was justified. Two thousand million dollars had not been poured down the drain. The scientists had been shown to be right and the politicians had been shown to be right in their trust of the scientists. For all practical purposes it could be taken for granted that the war with Japan would soon be over, that there would no longer be any need for the long-dreaded frontal assault on the beaches of Japan.

Yet the situation was not quite so simple. As Trinity was providing its own unique warning of the wrath to come, Truman, Churchill and Stalin, together with their staffs and advisers were already gathering at Potsdam, a few miles from Berlin, for the Conference, opening on 17 July, called to settle plans for the occupation of Germany. The timing was at least partly fortuitous, since Trinity could always have been postponed, but it was extremely convenient. Yet if certain problems had been removed, others had sprung up in their place. The first, omnipresent as the Western delegations moved about the quarters of the Cecilienhof Palace, provisioned, administered and guarded by the formidable apparatus of their Eastern allies, was still the problem of what the Russians should be told.

Since these Eastern allies had at fearful cost been engaging the bulk of the Wehrmacht for some four years it might have been expected that the good news would be shared with them. But with the advance of the Red Army into Europe the West's suspicions had been justified, and by the time of Potsdam Russian intransigence, and her open determination to swallow up as much of Eastern Europe as the West would allow, had hardened the feelings of those who had to deal with this prickly problem. It is true that the Interim Committee, set up specifically to advise the President on the atomic bomb, had unambiguously recommended that 'the President advise the Russians that we were working on this weapon with every prospect of success and that we expected to use it against Japan'. But at Quebec Roosevelt and Churchill had signed a solemn agreement that neither Britain nor America would 'communicate any information about Tube Alloys to third parties except by mutual consent', and far from consenting Churchill was still showing determined opposition to any such idea.

Secondly, with the Japanese now on their knees, was it necessary to use the uranium bomb at all; was it, perhaps, even wise to do so, since many scientists had always maintained that there was really only one 'secret' about atomic weapons – the riddle, unsolved until the day of Trinity, of whether they would work? And if the weapon was to be used, how, and exactly on whom, was it to be used, a matter on which there had already been much, and as some felt quite needless debate? These two questions, as it was only later to be realized, were together to affect not only the ending of the war with Japan but the prospects for nuclear energy, the chances of international control, and possibly even the future of the human race.

Churchill's first inkling that the Americans had 'changed their view that nothing should be said until the weapon had been used', had come on 29 June in a minute from Anderson who noted that he had just received from Field Marshal Wilson, Britain's representative on the Combined Policy Committee, 'some highly confidential details of the outline plan for the first use of the weapon by the Americans against Japan'. It was, Anderson went on, 'obviously desirable that you and the President should speak with one voice' if the Russians were to be informed. Two weeks later, a more specific briefing had been prepared for the Prime Minister by Cherwell. Truman, he said, would no doubt raise the question of what was to be said to the Russians about the atom during 'Terminal', the code-name for the Potsdam Conference. There were, said Cherwell, in his typically lucid way, two alternative courses:

(1) Not to say anything and face the fact that when, a few weeks after TERMINAL, concealment is no longer possible Stalin may claim that our lack of candour in not mentioning such an important matter justifies him in going back on any engagements he may have made.

(2) To say, in broad general terms, that we have done a great deal of work on T.A. and hope shortly to make an operational test, but to refuse to say anything further. This might spoil the atmosphere at TERMINAL if the Russians demanded further information. But they have no right to any and I am sure the Americans will not give it.

The arguments are fairly closely balanced. But as the Americans have contributed such an overwhelming proportion of the effort, I do not think it would be easy to oppose Truman strongly, if he seems anxious to take the second course.

In any event the issue may be forced by Stalin asking point-blank what the position is.

Truman was urged to take the second course a few days later. On 3 July Stimson briefed him on the coming conference. The President should, he wrote in his diary, 'if he found that he thought that Stalin was on good terms with him', tell the Generalissimo we were nearly ready with this new weapon, that we would use it against Japan, and 'that we proposed to then talk it over with Stalin afterwards, with the purpose of having it make the world peaceful and safe rather than to destroy civilization. If he pressed for details and facts, Truman was simply to tell him that we were not yet prepared to give them. The President listened attentively and then said that he understood and he thought that was the best way to do it.'

Stimson's proposal for Truman's handling of the situation was more than a suggestion from Minister to President. Indeed, when he chaired a crucial meeting of the Combined Policy Committee the following day, 4 July, the proposal was written into the minutes. First, Field Marshal Wilson formally announced that the British Government concurred in the use of the bomb against Japan. Then, after the operational aspects had been discussed, Stimson explained how Truman was to handle Stalin.

Truman and Churchill arrived in Potsdam in mid-July, each quite convinced that his own tactics were the best; Truman willing to follow Stimson's proposal and tell Stalin that they were approaching success with nuclear weapons, Churchill even more reluctant to tell the Russians than he had been to tell his own Chiefs of Staff.

The first news of the Alamogordo explosion arrived in Potsdam from Harrison at 7.30 on the evening of the 16th. 'Operated on this morning,' said the message to Stimson. 'Diagnosis not yet complete but results seem satisfactory and already exceed expectations. Local press release necessary as interest extends great distance. Dr Groves pleased. He returns tomorrow. I will keep you posted.' Whether this would have deceived either the Russians or the Japanese had either intercepted and decoded the message is a moot point; any doubt would probably have been removed by the second message which arrived on the morning of the 18th. 'Doctor has just returned most enthusiastic and confident that the little boy is as husky as his big brother,' this went. 'The light in his eyes discernible from here to Highold [Stimson's home on Long Island] and I could have heard his screams from here to my farm [in Upperville, Virginia].'

Only the following morning, as he left a brief meeting with Stimson, did Churchill receive the news. 'As he [Churchill] walked down to the gate,' Stimson wrote in his diary, 'I told him of Harrison's message. He had not heard from his own people about the matter. He was intensely interested and greatly cheered up but was strongly inclined against any disclosure [to the Russians]. I argued against this to some length.'

But the following day Churchill had the chance of pressing his views on Truman whom he invited to an intimate lunch in his personal quarters. No others were present, and Churchill's only account of the vital meeting appears to be a summarized note of the conversation which was at the time given the remarkably small circulation of only four copies – to Anthony Eden, the Foreign Secretary; to Anderson; to Cherwell; and to General Ismay for the Chiefs of Staff Committee.

The following subjects were touched upon:

1. T.A.

The President showed me the telegrams about the recent experiment, and asked what I thought should be done about telling the Russians. He seemed determined to do this, but asked about the timing, and said he thought that the end of the Conference would be best. I replied that if he were resolved to tell, it might well be better to hang it on to the experiment, which was a new fact on which he and we had only just had knowledge. Therefore he would have a good answer to any question, 'Why did you not tell us this before?' He seemed impressed with this idea, and will consider it.

On behalf of His Majesty's Government I did not resist his proposed disclosure of the simple fact that we have this weapon. He reiterated his resolve at all costs to refuse to divulge any particulars.

It appears clear from Churchill's account that Truman had arrived in Potsdam determined to tell the Russians not merely that the Western Allies had a powerful new weapon but that it was a nuclear weapon, with all that that would imply. There is no indication in the account that Churchill had been able to erode that determination; yet the reader is left with the firm impression that the Prime Minister was most anxious that if Truman remained 'resolved to tell', then the telling should be at least watered down, and the more water the better.

The following day Cherwell met Stimson and Bundy, and it is difficult to believe that he had not been briefed to discover how determined Stimson was in his intentions. 'At twelve o'clock,' says Stimson's enigmatic diary entry, 'Lord Cherwell called and he and Bundy and I sat out under the trees and talked S-1 [the U.S. code for the bomb project]. He was very reasonable on the subject of notification to the Russians, feeling about as doubtful as we.'

But if doubts were increasing they seem to have been at least partly removed by the full account of Trinity that now arrived in Potsdam. Truman quickly told Churchill who, according to Harvey Bundy, enthusiastically commented: 'Stimson, what was gunpowder? Trivial. What was electricity? Meaningless. This atomic bomb is the Second Coming in Wrath.'

From this, as from other evidence, it appears that the effects of the bomb surprised the Western leaders at Potsdam as much as they had surprised many scientists at Alamogordo. There was good reason for this. Groves had formally told Marshall at the end of 1944 that the first bomb would be equivalent to about 500 tons of conventional explosive, and that the second would be about twice as powerful. And as late as the middle of May 1945, he was to write, 'The responsible heads at Los Alamos felt that the explosive force of the first implosion-type bombs would fall somewhere between 700 and 1500 tons.'

This was sufficiently different from the ten-ton block-busters then in service to stick in the mind and there was probably more than a hint of suspicion about Groves' estimates for the test explosion of between 15,000 and 20,000 tons. The figure of 500 tons could be grasped by the imaginations of ordinary men; the effect of 15,000 tons was beyond most of those at Potsdam, with the probable exception of Cherwell.

However, with the Russians continuing their intransigence about

Marie and Pierre Curie who in December 1898 *discovered radium while investigating the phenomenon they had called 'radioactivity'. This was the process in which uranium had been found to give off alpha, beta and gamma rays*

Physics Research Students at the Cavendish Laboratory, Cambridge, June 1898. *Left to right, back row: S. W. Richardson, J. Henry. Middle row: F. B. H. Wade, G. A. Shakespear, C. T. R. Wilson, Ernest Rutherford, W. Craig-Henderson, J. H. Vincent, G. B. Bryan. Front row: J. C. McClelland, C. Child, Paul Langevin, Prof. J. J. Thomson, J. Zeleny, R. S. Willows, H. A. Wilson, J. Townsend*

Ernest Rutherford in his laboratory in the Macdonald Physics Building of McGill University, Montreal, 1905

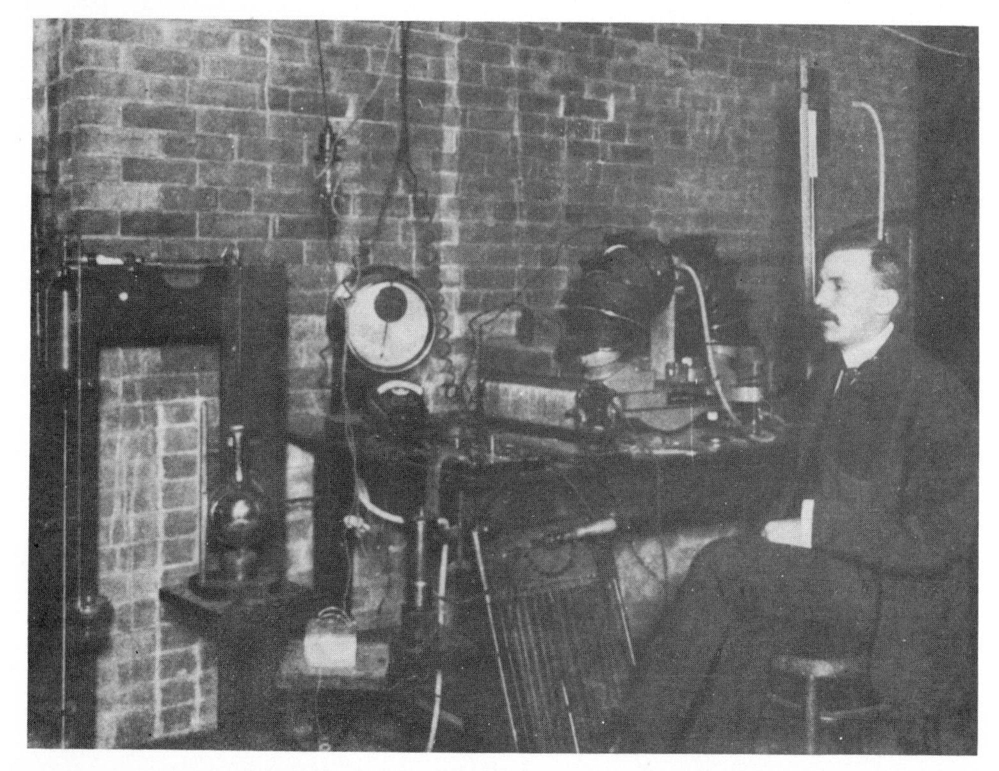

A corner of Rutherford's laboratory in Manchester in 1919. *A reconstruction showing his nuclear disruption apparatus, with microscope attached*

Rutherford's laboratory in the Cavendish Laboratory, Cambridge, about 1920

Francis Aston in the Cavendish Laboratory, Cambridge, with one of the mass spectrographs which enabled him to show that most stable elements were mixtures of isotopes differing in mass but not in chemical properties

John Cockcroft (left) and George Gamow discussing problems whose solution led to the artificial transmutation of the atom by Cockcroft and Walton in 1932

A reconstruction of the apparatus which led to James Chadwick's discovery of the neutron in 1932. *George Crowe, seen with the apparatus, was at the time an assistant to Rutherford and Chadwick in the Cavendish Laboratory*

The apparatus with which Cockcroft and Walton artificially transmuted the atom in Cambridge in 1932

Niels Bohr working in his study in Copenhagen

Physics Research Students at the Cavendish, June 1932. Left to right, back row: N. S. Alexander, P. Wright, A. G. Hill, J. L. Pawsey, G. Occhialini, H. Miller. Third row: W. E. Duncanson, E. C. Childs, T. G. P. Tarrant, J. McCougall, R. C. Evans, E. S. Shire, E. L. C. White, F. H. Nicoll, R. M. Chaudhri, B. V. Bowden, W. B. Lewis. Second row: P. C. Ho, C. B. Mohr, H. W. S. Massey, M. L. Oliphant, E. T. S. Walton, C. E. Wynn-Williams, J. K. Roberts, N. Feather, Miss Davies, Miss Sparshott, J. P. Gott. Front row: J. A. Ratcliffe, P. Kapitza, J. Chadwick, R. Ladenberg, Prof. Sir J. J. Thomson, Prof. Lord Rutherford, Prof. C. T. R. Wilson, F. W. Aston, C. D. Ellis, P. M. S. Blackett, J. D. Cockcroft

post-war Europe, the new weapon was clearly a sword which could be rattled in the scabbard if necessary. '[Churchill] now not only was not worried about giving the Russians information on the matter,' Stimson wrote in his diary after an hour's meeting on 22 July with Bundy, Churchill and Cherwell, 'but was rather inclined to use it as an argument in our favour in the negotiations. The sentiment of the four of us was unanimous in thinking that it was advisable to tell the Russians at least that we were working on that subject and intended to use it if and when it was successfully finished.' According to Field Marshal Lord Alanbrooke (then Sir Alan Brooke) Churchill had been 'carried away' by the report of Trinity: 'He was already seeing himself capable of eliminating all the Russian centres of industry and population without taking into account any of the connected problems, such as delivery of the bomb, production of bombs, possibility of Russians also possessing such bombs, etc.' On the 23rd after a session with Truman, Stimson noted: 'We had a brief discussion about Stalin's recent expansions and he confirmed what I have heard. But he told me that the U.S. was standing firm and he was apparently relying greatly upon the information as to [the bomb].'

Thus it now appeared that the Interim Committee's recommendation, reinforced by Stimson's advice which had been written into the minutes of the Combined Policy Committee, had been strengthened still further by the success of Trinity. Not only should the Russians be told of the nuclear weapon which the West now had in its armoury, but the news could be used 'as an argument in our favour in the negotiations'.

Yet nothing of the sort was to happen. The West's only reference to the bomb, a reference that carefully omitted any indication of what it really was, came after the plenary session on 24 July. As those who had attended the session chatted informally in twos and threes before dispersing, Truman casually approached Stalin, moving up to the Russian leader without his interpreter in order to appear as casual as possible.

Stalin may have wondered what was coming. He also was casual, commenting that the Japanese had been making peace overtures through their Ambassador in Moscow. Truman was already fully aware of these through the breaking of the Japanese diplomatic code, achieved some four years previously; nevertheless, he looked pleasantly surprised. Then it was his turn. 'I casually mentioned to Stalin that we had a new weapon of special destructive force,' he later wrote. 'The Russian Premier showed no unusual interest. All he said was that he was glad to hear it and hoped we would make "good use of it against the Japanese".'

As the meeting finally broke up Truman and Churchill left the building and found themselves side by side, waiting for their cars. 'How did it go?' the Prime Minister asked. 'He never asked a question,' Truman replied. 'This was the end of the story so far as the Potsdam Conference was concerned,' Churchill later wrote. 'No further reference to the matter was made by or to the Soviet delegation.'

It was known to Americans and British that the Russians had been seeking information on the Manhattan Project, and both had feared that Stalin might ask point-blank about the bomb. Thus it is possible that Truman expected his statement to be followed by questions which would open up the subject and allow him to follow his briefing from the Interim Committee and from Stimson. Stalin allowed him no such chance. The two men drifted apart, a course that their two countries were now to follow with increasing speed.

It is still not certain how much significance Stalin attached to Truman's single sentence. At the time it was rather innocently believed that whatever Russian intelligence might know about the Manhattan Project, Stalin knew nothing. According to Churchill he 'seemed to be delighted. A new bomb! Of extraordinary power! Probably decisive on the whole Japanese war! What a bit of luck! This was my impression at the moment, and I am sure that he had no idea of the significance of what he was being told.' James Byrnes, also watching intently, took the same view although expecting the Russians to ask within the next day or so for details of the wonderful new weapon. Later revelations about Russian espionage networks inside the Allies' nuclear projects in the United States and Canada have suggested that Stalin was dissimulating, creating a doubly ironic situation in which Truman had concealed his knowledge of the Japanese peace moves and Stalin had concealed his knowledge of the bomb. This view is supported by the evidence of Marshal Zhukov, a not entirely unbiased witness, who in his memoirs says: 'On returning to his quarters after this meeting Stalin, in my presence, told Molotov about his conversation with Truman. ... "Let them. We'll have to have a talk with Kurchatov and get him to speed things up." I realized they were talking about research on the atomic bomb.'

An entirely different story is given by another Russian witness, General Shtemenko, who in his detailed account of *The General Staff at War, 1941–45* says: 'Antonov [Stalin's interpreter] told me later that Stalin had informed him of the Americans' possession of a bomb of very great destructive power. But neither Antonov, nor apparently Stalin himself, had gathered from the conversation with Truman the impression that this was a weapon on entirely new principles. In any case, no additional instructions were given to the General Staff.'

Exactly why Truman failed to implement the Interim Committee's recommendation and exactly how much Stalin gathered from the brief exchange are both likely to remain subjects for speculation. But Truman may well have thought that the obliteration of a Japanese target would have more effect on the Russians if they had no advance news of the American success. He obviously wished to avoid being pressed by Stalin for details. He may also have been affected by Churchill's swift change of opinion. It seems clear that during the two men's meeting on the 19th Churchill had only reluctantly agreed that anything at all should be said to the Russians. Yet, in Field Marshal Alanbrooke's words, he had been 'carried away' by the full report of Trinity and immediately saw it as a means of putting pressure on the Russians. Truman may well have thought that a middle course was the course to take.

On balance, it seems likely that Stalin would have suspected that Truman was referring to a nuclear weapon. But given the briefness of the exchange and the ease with which the nuances of meaning can get lost in translation, it would be remarkable if he had read into Truman's words – before consultation with others – either the fact that the crucial test had been successfully carried out or that the new weapon was so vastly different from conventional explosives. To Stalin at Potsdam the significance of Truman's statement, if it had any significance at all, was merely confirmation that the Western Allies did not trust the Russians, not that the Russians' position in Europe had been dramatically changed.

That was made clear on 6 August in Hiroshima and in Nagasaki three days later. Stalin's reaction after the first attack was reported by the U.S. Ambassador in Moscow Averell Harriman who met Stalin on the 8th. 'Stalin was – or pretended to be – impressed,' says a memorandum on their meeting,

> the Soviet scientists, he remarked, had told him that it was a very hard problem to work out. Harriman said that if the Allies could keep it and apply it for peaceful purposes, it would be a good thing. Stalin assented, saying that it

would mean the end of war and aggressors – but the secret would have to be well kept. He then went on to tell the Ambassador that the Red Army had found one laboratory in Germany where the Germans had been working on the same problem without result. If they had found out how to make the weapon Hitler would never have surrendered. Even England, Stalin continued, although it had excellent physicists, had gotten nowhere with its research in this field.

Stalin's immediate reaction appears to have been fear that a revived Germany might eventually find out how to produce nuclear weapons. Before the end of the year the implications had begun to sink in and were described in a long message to Ernest Bevin, the British Foreign Secretary in the Labour Government, from Sir Archibald Clark-Kerr, the British Ambassador in Moscow. He noted how the Russians had nearly been defeated by the German invasion of 1941 and how, after immense losses, they had gained not only victory but what they believed to be national security.

Then, plump, came the atomic bomb. At a blow the balance which had now seemed set and steady was rudely shaken. Russia was balked by the West when everything seemed to be within her grasp. The three hundred divisions were shorn of much of their value. About all this the Kremlin was silent but such was the common talk of the people. But their disappointment was tempered by the belief inspired by such echoes of the foreign press as were allowed to reach them that their Western comrades in arms would surely share the bomb with them. That some such expectation as this was shared by the Kremlin became evident in due course. But as time went on and no move came from the West, disappointment turned into irritation and, when the bomb seemed to them to become an instrument of policy, into spleen. It was clear that the West did not trust them. This seemed to justify, and it quickened, all their old suspicions. It was a humiliation also and the thought of this stirred up memories of the past.

This, with what is known now of Clark-Kerr's political views, can certainly be taken as special pleading. But within a few weeks of Potsdam the Russians had launched a crash programme to catch up with the Americans. And before the end of the year Thomas P. Whitney, an attaché in the U.S. Embassy in Moscow, could warn the Secretary of State: 'The U.S.S.R. is out to get the atomic bomb. This has been officially stated. The meager evidence available indicates that great efforts are being made and that super-priority will be given to the enterprise.'

Thus the Potsdam Conference had become the starting-pistol for the nuclear arms race. It would probably have started whatever happened and it is certainly true that later events were to show that both Churchill and Stalin were to be justified in their deepest suspicions of each other. Churchill was right in thinking that it would be unwise to tell the Russians any more than they could find out by espionage against their allies; Stalin was correct in believing that the West would tell the Russians no more than they had to. As for Truman, he had been steered into an awkward position; he now had to maintain that the few casual words which aroused 'no unusual interest' did implement the recommendation that he inform the Russians of a nuclear weapon that could 'become an aid to peace'.

The outcome of this last-minute attempt to tell the West's allies what they already knew might have been different had Roosevelt died four months later than he had or had the British General Election, which unseated Churchill and put Clement Attlee in the saddle before the Potsdam Conference had finished, taken place a month or so earlier.

A few months after Potsdam, Attlee met the Canadian Premier, Mackenzie King, who spoke of his discussion with Roosevelt earlier in the year on what should be revealed about the Manhattan Project. '[Attlee] said,' King recorded in his diary, 'he thought it was a mistake not to have told the Russians. I said that President Roosevelt had said the same thing to me. He thought Stalin ought to be told that I had agreed with him.' With the later revelations of the Russian spy ring in Canada, King was to change his views and to praise Churchill's obduracy rather than Roosevelt's idealism. Nevertheless, it appears extremely likely that had Roosevelt and Attlee rather than Truman and Churchill taken the vital decision, the Interim Committee's advice would not have been watered down. It is by no means certain that the history of the post-war years would have been very different had the Western powers at Potsdam followed the advice of Roosevelt, Anderson, Smuts and, to a certain extent even of Cherwell, a man somewhat allergic to communism; but it is difficult to see what would have been lost. As Professor Bernstein has summed up at the end of his long dissection of the policies followed by Roosevelt and then Truman, 'The wartime policy on atomic energy represented one of a number of missed opportunities at achieving limited agreements and at testing the prospects for Soviet-American co-operation on a vital matter.'

Equally controversial was the decision not to halt plans for the atomic bombing of Japan even though Japan was already trying to end the war. The long-term results of this decision were to be even more important for the world than the destruction of two great cities. That nuclear energy could be released and controlled had been demonstrated in Chicago; that it was the world's most powerful explosive had been demonstrated at Alamogordo. But these facts were still known to only a handful of people, and the inability even of the scientists to understand what they were handling is shown by their estimate of deaths in Hiroshima: 10,000 against the reality of 70,000. Even in the second half of July 1945 it would still have been possible to keep shielded from the mass of mankind the nature of the genie that had been let out of the bottle. The Chicago success would have remained a new and interesting, if formidably impractical, way of generating electricity. 'The bomb' could have sunk back to being little more than Churchill's 'more powerful explosive'. A lot more than the lives of some 100,000 Japanese therefore rested on the decision still to be taken.

In the chronology of the next two weeks, a period during which mankind crossed a watershed into the reality of the atomic age, lies a clearer understanding of the traumatic events of late July and early August 1945. Truman had casually told Stalin of the 'new weapon of unusual destructive force' on 24 July.

It now appears that he had vestigial misgivings, and that he was under some misapprehension about the targets which had been chosen. Within hours of making the pseudo-revelation to Stalin he had ordered the War Department in Washington to drop the first bomb on Japan as soon after 3 August as the weather permitted. The following day, 25 July, he wrote in his diary:

> We have discovered the most terrible bomb in the history of the world. It may be the fire destruction prophesied in the Euphrates Valley era, after Noah and his fabulous ark.
>
> This weapon is to be used against Japan between now and August 10. I have told the Secretary of War, Mr Stimson, to use it so that military objectives and soldiers and sailors are the target and not women and children. Even if the Japs are savages, ruthless, merciless and fanatic, we as the leader of the world for the common welfare cannot drop this terrible bomb on the old capital or the new [a reference to Kyoto and Tokyo].
>
> The target will be a purely military one and we will issue a warning statement asking the Japs to surrender and save lives. I'm sure that they will not do that, but we will have given them a chance. It is certainly a good thing for the world that Hitler's crowd or Stalin's did not discover this atomic bomb. It seems to be the most terrible thing ever discovered, but it can be made the most useful.

Plans for the atomic bombing had already been discussed in great detail and to put the decision at Potsdam into context it is necessary briefly to describe how, as the plans developed, Japan's will to fight on was being slowly but remorselessly battered out of existence.

Even when it had been decided that there was no alternative to the atomic bombing, without warning, of a target in Japan there had been great discussion of what the city should be. A Target Committee consisting of civilian scientists and Army Air Force officers had been set up in late April 1945, even before the formation of the Interim Committee, and by July had been studying for three months the multiple problems that faced them. The object of the attack would not be to kill as many humans as possible or even to destroy as much property as possible, although both might be subsumed in the main aim of the exercise: to impress on the Japanese Government that loss of the war was inevitable and that the longer they failed to give in, then the greater would be their country's sufferings. The sight and sound of the new weapon was therefore a factor to be considered along with destruction – which could, for all practical purposes in the then state of Japan's defences, be produced with conventional weapons wherever the U.S. Army Air Forces wished.

It was also felt necessary to discover, from the first bombing, what the scale of destruction would be, since all that could be inferred from the test would be theoretical extrapolations. Therefore the target had to be one not previously attacked – if it were, then it would be difficult to separate new damage from old – and it had to be of such a size that it would include the whole area of expected damage.

Uncertainty about what would impress the Japanese Government led to the abandonment of more than one ingenious idea. It was proposed by Lewis Strauss (who was a special assistant to the Secretary of the Navy) that a bomb detonated above a forest some forty miles from Tokyo would burn or bowl over tens of thousands of Japanese redwoods as though they were matchsticks; but it was finally decided that the Japanese would not be greatly impressed. Various targets around Tokyo Bay were proposed, and it was suggested that if the bomb only detonated on impact an immense tidal wave might be created. But all targets in the Tokyo area were eventually ruled out: if the Emperor, living on the outskirts of the city, were even injured, let alone killed, the shocked Japanese might carry on the war with increased ferocity.

Kyoto, the ancient capital of Japan, with a population of more than a million, the centre of a huge industrial complex, was one city on the list. But Stimson, hearing from an enthusiastic Orientalist about the beauty of the old imperial residences, decided to consult a book showing the glories of the city. Kyoto was struck from the target list, and later requests that it should be put back were firmly turned down. However, it was not only idealism that saved the ancient glories of Japan. Truman agreed with the exclusion of Kyoto from the target list, according to Stimson, because 'the bitterness ... caused by such a wanton act might make it impossible during the long post-war period to reconcile the Japanese to us in that area rather than to the Russians. It might thus, I pointed out,' Stimson added in his diary, 'be the means of preventing what our policy demanded, namely a sympathetic Japan to the United States in case there should be any aggression by Russia in Manchuria.'

Hiroshima took the place of Kyoto and operational plans were perfected for the movement of the unfused bomb from Los Alamos to Tinian, a small island in the Western Pacific, and for its delivery, when ordered, by the special crews who had been training for the task for some months. In only one important way did Hiroshima fail to fill the bill of requirements. It was 500 miles from the seat of the Japanese Government in Tokyo, and the destruction wrought by the bomb was so great that communications with the capital were completely severed. The Government learned of the disaster which had struck the city only three days after the bombing.

As a fuller evaluation of the Trinity results took place after 16 July it seemed clear that a touch of the nuclear weapon would bring the Japanese to their knees. However, the Japanese were already on their knees although this was known only to a relatively small number of high Washington officials.

Five years previously the American cipher- and code-breakers had achieved one of the greatest coups of the war – the war in which the Americans were not then involved. They had broken the Japanese 'Purple' code – so-called despite the fact that it was a cipher – used by the Japanese for the most secret diplomatic messages transmitted from Tokyo to Japanese embassies throughout the world. 'Breaking' the code was not as all-embracing as it sounds and for some while only some messages, or even some parts of some messages, could be decoded in Washington. By the summer of 1945, however, the Americans had mastered the intricacies of the system: with comparatively few exceptions, they were reading, 'in clear', messages transmitted by the Japanese Foreign Office almost as soon as those to whom the messages were sent. The fact that the enemy was anxious to end the war some while before the destruction of Hiroshima and Nagasaki was therefore known, in detail, to the American Government.

As early as 22 June the Emperor told the six most important members of the Supreme Council for the Direction of the War that he wished his Ministers to seek peace through the Soviet Union. On 12 July, the Foreign Minister, Shigenori Togo, informed the Japanese Ambassador in Moscow, Naotake Sato, that a swift end to the war was wanted, and that he was sending a special envoy to Moscow in the hope of obtaining this. The only stumbling block now appeared to be the formula of 'unconditional surrender' which the Allies were still demanding – a formula which it was correctly thought on both sides was intended to bring about the abdication of the Emperor.

By the start of the Potsdam Conference the Japanese knew just how increasingly desperate their situation had become. On 15 July, the day before the Trinity test, and as the Allied Ministers were preparing for Terminal, the American Secretary of the Navy, James Forrestal, received a further batch of decoded messages from the Japanese Foreign Minister to Moscow. 'The gist of his final message,' Forrestal noted, 'was that it was clear that Japan was thoroughly and completely defeated and that the only course open was quick and definite action recognizing such fact.'

This was still the situation as the Potsdam Conference began. By 24 July, when Truman made his casual semi-revelation to Stalin and gave orders for the bombing as soon after 3 August as possible, the only change had been a steady deterioration in the already perilous situation of the Japanese. On the following day, 25 July, Sato was ordered by Tokyo to meet Molotov, the Soviet Foreign Minister, at any place the latter might suggest, and to inform him that while the Japanese still did not consider unconditional surrender to be acceptable, complete collapse was imminent and they had no objection to a peace based on the Atlantic Charter.

It is not known whether this message was immediately transmitted to Truman in Potsdam; possibly not, since it did little more than underline the known fact that the Japanese were desperately anxious to end the war. This anxiety would, it was hoped, be translated into action by the proclamation broadcast to Tokyo from Potsdam on the 26th under the names of Truman, Chiang Kai-shek and Churchill. A solemn warning of the wrath to come, it yet gave no hint of the new weapon 'of unusual destructive force'. Perhaps more important – since nothing less than a physical demonstration could reveal the power of nuclear weapons – the proclamation made no reference to the position of the Emperor once the war had been ended. Stimson felt that the lack of any reference to the Emperor's status was a mistake, noting in his diary his belief that it was 'important and might be just the thing that would make or mar their acceptance'. And he told President Truman he 'hoped that the President would watch carefully so that the Japanese might be reassured verbally through diplomatic channels if it was found that they were hanging fire on that one point. [Truman] said that he had that in mind, and that he would take care of it.' The penultimate sentence of the Declaration called for 'the unconditional surrender' of all the Japanese armed forces. This was not to be accepted and the Japanese finally agreed to the Potsdam terms only if the sovereignty of the Empire was retained. Eighteen months later Truman was to state of the bomb: 'The Japanese were given fair warning, and were offered terms which they finally accepted, well in advance of the dropping of the bomb.' Some semantic conjuring seems necessary before the statement can be squared with the facts.

The day after the Potsdam declaration had been issued, Mackenzie King noted in his diary; 'I feel we are approaching a moment of terror to mankind, for it means that, under the stress of war, men have at last not only found but created the Frankenstein which conceivably could destroy the human race.'

The following day the Japanese effectively rejected the Potsdam Declaration. The crucial word used was '*moku-satu*', first translated by the American monitors as 'ignore' and later officially interpreted as 'reject', although the more normal translation is 'kill with silence'.

The official Japanese response was quite at variance with the diplomatic instructions being monitored by the Americans, and on 2 August the Japanese Foreign Minister made yet another urgent demand to his Ambassador in Moscow that the Russians should be asked to mediate. 'There are only a few days left in which to make arrangements to end the war,' this went. '... Since the loss of one day relative to this present matter may result in a thousand years of regret, it is requested that you immediately have a talk with Molotov.'

However, the Russians were not unduly anxious to act as mediators and by 2 August Truman was already returning from Potsdam to Washington. Just how many of the final messages revealing Japan's desperate situation reached him are not known, but William Friedman, the cryptologist whose team had broken the Purple Code, later went on record as saying: 'If only I had had a channel of communication to the President I would have recommended that he did not drop the bomb – since the war would be over within a week.'

In Britain, meanwhile, a memorandum was drawn up by a number of scientists intimately concerned in Tube Alloys. It pointed out that there was no possibility of Japan having a nuclear weapon, outlined the disadvantages of the Allies dropping one, and suggested that its use would be comparable to using a sledgehammer to crack a nut. Arrangements were made for the memorandum to be submitted to the Government through Lord Hankey.

In the Pacific the operation continued as planned. The bomb had been ready on Tinian since 1 August and all now rested on the weather. Only two bombs were yet available, one armed with uranium, the other armed, as the Trinity test weapon had been, with plutonium. Visual aiming was considered essential and it was only at 02.45 on 6 August that the 'Enola Gay', piloted by Colonel Paul Tibbetts, commander of the specially formed 509th Composite Group, lifted off the Tinian runway. Three weather planes had taken off earlier, the task of each being to reconnoitre one of the three alternative targets. Two observation planes followed.

At 8.15 Tibbetts was told that Hiroshima was relatively cloud-free. An hour later he was over the target. At 31,600 feet he released the 'Thin Man', the long gun-type uranium bomb. It exploded at 1,000 feet over the city, killing 70,000, injuring 51,000, destroying more than 70,000 buildings and making homeless more than 170,000 people.

News of what had happened at Hiroshima filtered through only slowly to Tokyo and only the bare details of the disaster were known in Tokyo by the 8th. But in the afternoon the Japanese Ambassador in Moscow, Naotake Sato, called on Molotov in the Kremlin, hoping that he would now at last be able to enlist the Russians as mediators. He was, instead, handed a note stating that from the next day Russia would be at war with Japan.

It is part of the mythology about the nuclear attacks on Japan that they were intended to pre-empt Russia's entry into the war; no longer required on the battlefield, it is postulated that she would, if kept neutral until the Japanese surrender, have no role in the post-war settlement. The chronology does raise suspicions while Philip Morrison, one of those working on the weapon, has gone on record as saying: 'I can testify personally that a date near 10 August was a mysterious final date which we, who had the daily technical job of readying the bomb, had to meet at whatever cost in risk or money or good development policy.'

Yet the Americans, in the shape of James Byrnes, found an ingenious method of helping Russia into the war in time. At Potsdam Molotov had suggested that the United States, Great Britain and the other Allies should formally ask the Soviet Government to declare war on Japan. 'The request,' Byrnes has written, 'presented us with a problem. The Soviet Union had a non-aggression pact with the Japanese. ... We did not believe the United States Government should be placed in the position of asking another government to violate its agreement without good and sufficient reason.' And while the Russians had a few months earlier warned the Japanese that they intended to abrogate the non-aggression treaty, that treaty was still in force.

However, ingenuity came to the rescue. The Charter of the United Nations, recently signed by the Soviet Union in San Francisco, demanded consultation on 'joint action on behalf of the community of nations'. Article 103, moreover, provided that 'in the event of a conflict between the obligations of the Members of the United Nations under the present Charter and their obligations under any other international agreement, their obligations under the present Charter shall prevail.' The letter of the Charter was quickly invoked to break the non-aggression pact and early on 9 August Russian troops crossed the Manchurian frontier in an attack on the Japanese Kwantung army.

Later on the 9th a second atomic bomb, this time an implosion-type weapon with a charge of plutonium, was dropped on the great port of Nagasaki. In the wake of the bomb there went an unconventional message – to confirm to the Japanese what they had probably deduced already from the Hiroshima explosion. On Tinian three of the American physicists preparing the bomb had decided for themselves that they had a special duty. Before the war they had worked in the United States with a distinguished Japanese physicist, Dr Ryokichi Sagane. Now they concocted a message addressed to him, made three copies, taped them to three instrument boxes, and arranged that these should be parachuted over the target.

> We are sending this as a personal message to urge you that you use your influence as a reputable nuclear physicist to convince the Japanese General Staff of the terrible consequences which will be suffered by your people if you continue in this war.

You have known for several years that an atomic bomb could be built if a nation were willing to pay the enormous cost of preparing the necessary material. Now that you have seen that we have constructed the production plants, there can be no doubt in your mind that all the output of these factories, working twenty-four hours a day, will be exploded on your homeland.

Within the space of three weeks, we have proof-fired one bomb in the American desert, exploded one in Hiroshima, and fired the third this morning.

We implore you to confirm these facts to your leaders, and to do your utmost to stop the destruction and waste of life which can only result in the total annihilation of all your cities if continued. As scientists we deplore the use to which a beautiful discovery has been put, but we can assure you that unless Japan surrenders at once, this reign of atomic bombs will increase manifold in fury.

From three former colleagues. [Luis Alvarez, Robert Serber, Philip Morrison]

In Nagasaki nearly 40,000 were killed and 25,000 injured. As at Hiroshima, the town was almost totally destroyed. The following day the Japanese offered to surrender on the Potsdam terms – but with one proviso: the sovereignty of the Emperor must be retained. On the 11th the Allies sent their reassurances; while the Emperor would have to support the Allied commander in enforcing the surrender terms, there was, after all the fine words, no hint that his role would otherwise be seriously changed. Later, Stimson, according to his biographer and collaborator, did 'believe that history might find that the United States, by its delay in stating its position, had prolonged the war'. But the Russians had at least been shown what any aggression on their part would be up against, a fortuitous warning of the state of the world in the summer of 1945.

It is significant of the mistrust felt by the Americans not only for the Russians but for the Japanese – and an important factor in justification for the bombings – that despite the Allied retreat on this point, it was felt necessary that conventional bombing should continue. The day after Nagasaki, Stimson suggested that all raids should be stopped. He cited, says Forrestal, 'The growing feeling of apprehension and misgiving as to the effect of the atomic bomb even in our own country.' But negotiations were drawn out by the Japanese until 14 August, and on that day 828 Superfortresses made their heaviest assault of the war on Japan. Had more fissile material been available, it seems likely that one plane would have made the third nuclear attack of the war.

The fact that America knew by the end of July that Japan was desperately suing for peace does support, if superficially, the view that some 100,000 human beings were incinerated for reasons more devious than the simple wish to end the war as soon as possible. Chronology suggests as first alternative candidate the belief that atomic bombs would suddenly produce peace and prevent the Russians coming in at the last minute to claim their seat at the victor's table.

It must also have been obvious even to the most innocent that a display of the weapon's power would tend to inhibit Russian ambitions. It has never been effectively denied that the Western Allies' bombing of Dresden in February 1945 was influenced by a wish to demonstrate the strength of Allied air power, and after the war it was widely believed that but for the threat of the bomb the Russian advance would have continued westwards across Europe. The theory is incapable of proof or disproof, but it is significant that Leo Szilard, reporting his meeting with Byrnes in the spring of 1945, subsequently wrote: 'Mr Byrnes did not argue that it was necessary to use the bomb against the cities of Japan in order to win the war ... Mr Byrnes's ... view [was] that our possessing and demonstrating the bomb would make Russia more manageable in Europe.'

It has also been pointed out that Congress would eventually ask what had happened to the $2 billion spent on nuclear research, and that wartime use of the weapon would help justify the expenditure. The scientists' desire to confirm that the weapon was operationally viable, and that the uranium version utilizing explosion was as effective as the plutonium test version which utilized implosion, is also cited as influencing the bomb's use.

It would be difficult to maintain that these factors were not present. Yet the justification for using at least the first of the two bombs, and possibly the second, rests firmly on the sombre doubt that the Japanese might not surrender as expected. On battlefields such as Iwo-Jima, Japanese troops had shown their fanatical resolution to fight on against all odds, to the last man, whatever the dictates of order and common sense. Stimson who comes well out of the records as a man trying to be decent, was to declare: 'As we understood it in July, there was a very strong possibility that the Japanese Government might determine upon resistance to the end, in all the areas of the Far East under its control.' And Karl Compton, at General MacArthur's headquarters by mid-August, was told by a Japanese that if the bombs had not been dropped, 'We would have kept on fighting until all Japanese were killed, but we would not have been defeated.' Stimson may have been poorly briefed and Compton may have been reporting only the obvious reply of a proud people. But there were to be other signs that it was wise to consider the decoded messages with caution. Even after the Emperor had ordered surrender, the Army tried to stage in Tokyo a rising against it. In Washington it must have seemed possible, if not likely, that without the shock of two bombs the Peace Party in Japan might have lacked sufficient support, the Government might have failed to enforce any order to lay down arms, and a defeated country might have gone on to commit national hara-kiri – and to exact an appalling penalty as it did so.

The situation in Japan, it was eventually discovered, had been complex. 'It is apparent,' the U.S. Strategic Bombing Survey was to report, 'that the effect of the atomic bombings on the confidence of the Japanese civilian population was remarkably localized. Outside of the large cities, it was subordinate to other demoralizing experiences.' The authors of the Survey were in no doubt about the implications, in effect supporting the Navy view that blockade and bombardment would have done the job before the end of 1945. 'Based on a detailed investigation of all the facts,' it reported, 'and supported by the testimony of the surviving Japanese leaders involved, it is the Survey's opinion that certainly prior to 31 December 1945, and in all probability prior to 1 November 1945, Japan would have surrendered even if the atomic bombs had not been dropped, even if Russia had not entered the war and even if no invasion had been planned or contemplated.'

Yet little of this evidence which buttressed the plausibility of the intercepted Japanese peace overtures was known in Washington or in Potsdam. Thus ignorance was one reason for the world's first nuclear bombings. Another was the momentum which the operation had gathered during the preceding years. A third was that in the press of wartime circumstances none of those with their fingers on the button had the time fully to consider the physical consequences of their action, its political implications or its potential impact on the future.

'When we first began to develop atomic energy, the United States was in no way committed to employ atomic weapons against any other power,' General Groves has written.

With the activation of the Manhattan Project, however, the situation began to change. Our work was extremely costly, both in money and in its interference with the rest of the war effort. As time went on, and as we poured more and more money and effort into the project, the government became increasingly committed to the ultimate use of the bomb, and while it has often been said that we undertook the development of this terrible weapon so that Hitler would not get it first, the fact remains that the original decision to make the project an all-out effort was based upon using it to end the war.

Much the same attitude was taken by Churchill who has written that 'there was never a moment's discussion as to whether the atomic bomb should be used or not. The historic fact remains, and must be judged in the aftertime, that the decision whether or not to use the atomic bomb to compel the surrender of Japan was never an issue. ... There was unanimous, automatic, unquestioned agreement around our table; nor did I ever hear the slightest suggestion that we should do otherwise.'

The recriminations came later, when it had been demonstrated that a nuclear weapon was something more than a more powerful explosive. Admiral Leahy was quite outspoken in saying to Mackenzie King, 'that he did not think it ever should have been used as the United States had used it, against defenceless women and children which was the case in the two cities that were bombed'. General Eisenhower went on record as having told Stimson of his

> grave misgivings [about the use of the bomb], first on the basis of my belief that Japan was already defeated and that dropping the bomb was completely unnecessary, and secondly because I thought that our country should avoid shocking world opinion by the use of a weapon whose employment was, I thought, no longer mandatory as a measure to save American lives. It was my belief that Japan was, at that very moment, seeking some way to surrender with a minimum loss of 'face'. The Secretary was deeply perturbed by my attitude, almost angrily refuting the reasons I gave for my quick conclusions.

Even Churchill, a man attuned to the grim necessities of war, appears to have had a post-operational twinge of doubt. Speaking to Mackenzie King in May 1946 about the two nuclear bombings, he said, according to King's diary, that 'he was surprised that the second attack came so soon after the first. They had, however, given very ample warning', a statement that can be disputed. 'The way Churchill put it when discussing it with me was he expected that he would have to account to God as he had to his own conscience for the decision made which involved killing women and children and in such numbers,' King continued. 'That God would ask him why he had done this and that he would reply he had seen the terrors of war.' Churchill went on to say that he thought that what he had done was right and 'said something to the equivalent of welcoming a chance to be judged in the light of omnipotent knowledge'.

Irène Joliot-Curie (daughter of Madame Curie) and her husband Frédéric Joliot-Curie whose discovery of artificial radioactivity in 1934 paved the way for the post-war production of radioactive isotopes, now widely used in medicine, industry and agriculture

Otto Hahn who with Fritz Strassman split the nucleus of the uranium atom in December 1938, thus demonstrating that the use of nuclear energy might be possible

The apparatus with which Hahn and Strassman split the nucleus of the uranium atom in December 1938

Declassified SECRET

F2

Meeting No 1

Meeting of sub-committee on U bomb of C.S.S.A.W. held at Royal Society 2.30 P.M. April 10. 1940.

Present Prof. G. P. Thomson chair
Prof M. L. Oliphant
Prof. J. D. Cockcroft
Lt. Allier of Ministère d'Armament France } Present during first part only
Dr P. B. Moon

Lt Allier made a statement as to the efforts which the Germans were believed to be making to obtain information about work done in France on the U bomb, and to obtain heavy water. He asked that all details should be kept strictly secret and not mentioned outside the meeting.

The possibility of separating isotopes was then discussed and it was agreed that the prospects were sufficiently good to justify small scale experiments on uranium hexafluoride. It was decided to ask Dr Whitlaw Gray to provide a specimen for test. Prof Oliphant was asked to find a chemist with suitable experience to do the preliminary work at Birmingham.

It was agreed that Dr Frisch should be informed of the importance of avoiding any possible leakage of news, in view of the interest shown by the Germans. It was agreed that he should be informed that his proposal was being considered but should not be told details. There was however no objection to his continuing scientific works on his own lines.

G P Thomson

G. P. Thomson's hand-written minutes of the first meeting of the sub-committee (of the Committee for the Scientific Study of Air Warfare) on the U-bomb, later the M.A.U.D. Committee, held in London, 10 April 1940

James Chadwick (left), senior British scientist on the atomic bomb project in the United States and General Groves, head of the American Manhattan Project which built the first bombs

Enrico Fermi, the Italian physicist under whose direction the world's first atomic pile was constructed in Chicago. It became self-sustaining, and thus demonstrated the feasibility of nuclear power, on 2 *December* 1942

The world's first atomic pile under construction in Chicago in December 1942. The pile consisted of alternate layers of solid graphite blocks and of graphite blocks containing either uranium metal or uranium oxide. Here a layer of partly-completed solid graphite blocks is shown above a lower layer of blocks containing uranium

The mushroom cloud rising above the Americans' Bravo test of 1 *March* 1954. *The first of six explosions in the Castle series, it released as much energy as* 15 *million tons of T.N.T. An unexpected change of wind brought the radioactive fall-out from the explosion over the Japanese fishing vessel* Lucky Dragon

Members of the Montreal Laboratory in 1944. Hans Halban is standing eighth from left, and John Cockcroft eleventh from left

Four British members of the British Tube Alloys team after they had received the American Medal of Freedom for their work in the wartime Manhattan Project. Left to right: Dr Penney, Dr Frisch, Professor Peierls and Professor Cockcroft

In a different world, Edward Teller, who at Los Alamos had been largely preoccupied with a weapon still only talked about behind closed doors – the hydrogen bomb – was to ask: 'Could we have avoided the tragedy of Hiroshima? Could we have started the atomic age with clean hands?' and was to reply: 'No one knows. No one can find out.' Forrestal, the Navy Secretary, who had a closer view than most men of the decoded messages between Tokyo and Moscow which showed that Japan was anxious to surrender, was later to raise the question as a major strategic decision of the war. Did the bombing 'turn out', he asked, 'to have been a sound decision or not, in view of the exchange of dispatches between Sato in Moscow and the Japanese Foreign Office which, by mid-July 1945, clearly indicated the hopelessness of the Japanese situation?'

Ironically enough, Field Marshal Smuts, and Niels Bohr, two of the men most unhappy about the Western Allies' failure to inform the Russians of the nuclear project, had few if any doubts about the wisdom of bombing Hiroshima and Nagasaki. 'I daresay,' Smuts wrote to his friend M. C. Gillett, a week after Nagasaki, 'this sudden collapse was largely due to the atomic bomb, the effect of which, both physical and psychological, must have been shattering. From that point of view its use has been justified. But it has been even more justified from the point of view of preventing war in future. ... We are now forewarned of what is coming if war is not ended for good.' Bohr revealed his feelings to Felix Frankfurter in November 1946, the day after he had met Herbert Hoover, the former American President who said he thought it was a crime to have dropped the bomb. 'Of course that is not what Bohr thinks,' Frankfurter wrote in his diary. 'Bohr fully realizes that the decision was taken after the most careful consideration and as an exercise of judgement by men who had the responsibility of making it.'

Although Smuts's forecast was to be over-optimistic, and Bohr's possibly over-simplistic, one result of the bombs on Japan was to be as ironical as the outcome of the saturation raids on Germany, designed to halt German war production. The raids on Germany signally failed in their main aim. What they did, however, was to suck a huge segment of enemy industry into production of night-fighters. Thus in the summer of 1944 the Germans lacked the daylight fighter-bombers which, almost alone, could have halted the Anglo-U.S. invasion of Normandy in its stride and made the outcome of the war a very different affair.

The unplanned and unexpected result of the bombing offensive against Germany was to have a parallel. Six years after Potsdam it was seriously being debated in Washington whether nuclear weapons should be used in Korea, where the Communist North Korean forces were driving all before them. It seems probable that if the effects of nuclear attack had been as unknown then as they were before 6 August, the world's first fission bombs would have been used in Korea. But by that time the Russians would have been able to respond accordingly.

Yet with all this in a future whose shape few could yet discern, the first weeks of peace was a period for second thoughts. Thus Oppenheimer, one of the few men, if not the only man, who had had the knowledge and influence to change the verdict of the Interim Committee, and suggest that an approach could be made to the Russians with no chance of loss and perhaps a chance of strengthening future hopes, was to greet an Army award to Los Alamos with unexpected words: 'If atomic bombs are to be added as new weapons to the arsenals of a warring world, or to the arsenals of nations preparing for war,' he said, 'then the time will come when mankind will curse the names of Los Alamos and Hiroshima.'

PART THREE

On to the Darkling Plain

CHAPTER 12

'An Entirely New Situation'

Even by the middle of August 1945 it had become obvious to most men and women that the world was facing what a report from a Commission of the British Council of Churches was to call 'an entirely new situation in human history'. On the 6th, President Truman had issued from Washington a sober statement explaining that the single bomb dropped on Hiroshima had the explosive force of 20,000 tons of T.N.T.

In London the new Prime Minister, Clement Attlee, issued a similar statement, prepared earlier on the orders of Churchill who had been uncertain how much it was wise to say of Anglo-U.S. disagreements and agreements. When the announcement was being planned, Cherwell warned:

> I think it very undesirable to refer specifically to either of these agreements [the Quebec Agreement and the Hyde Park *aide-mémoire*] at the present time as this would only lead to demands for publication which would be very embarrassing to the Americans and perhaps even to us. They have given a very fair description of the arrangements in Mr Stimson's statement and, I understand from Bundy, are anxious that this particular aspect should be handled with considerable care as very difficult political issues may be involved.

The Quebec Agreement in particular was something that neither the British nor the Americans wished to bring into the open. In the autumn of 1945 a British M.P., Captain Blackburn, raised the question of a secret agreement but failed to prise any information from either the Labour Government or from Churchill, now Leader of the Opposition. At the same time President Truman, asked in Washington if Roosevelt and Churchill had 'reached a secret agreement at Quebec on the peace-time use of the atom bomb' had replied: 'I do not think that is true.' And finally to end speculation Churchill, quoting Truman's statement in the House of Commons, added: 'Subject to anything that the Foreign Secretary may say, I strongly advise the House for the present to leave the question where it now lies.'

In the United States, the Americans had followed Attlee's brief statement in August 1945 with the 144-page Smythe Report, a history of the United States nuclear enterprise and of the bomb's manufacture, subjects which Truman had been dissuaded from even mentioning to the Russians a mere sixteen days earlier. Certain details included in the duplicated copies of the Report issued to the press were deleted on security grounds before publication, possibly on the recommendation of Anderson and Cherwell in London. However, it was the duplicated copy which was translated and soon published in Moscow; thus for many years ordinary Americans had to consult the copy in the Library of Congress if they were to know as much about the bomb as the Russians had already been told. The British, unaware that the Smythe Report was to be issued in Washington so quickly, published as a White Paper the modest *Statements Relating to the Atomic Bomb*, a masterly survey written by Michael Perrin in a single 24-hour session. For fifteen years it remained the only official indication of the part played by Britain in the birth of the bomb.

These statements came as a shock which grew as their meaning seeped out into common understanding and as reports from the two incinerated Japanese cities began to reach the outside world. The shock was all the greater since the secrecy in which the nuclear project had been successfully cloaked had prevented any indication of things to come. At the end of the Middle Ages men had been prepared, at least by discussion, before Galileo maintained that the earth was not the centre of the universe. In the nineteenth century Darwin's firm statement that man's claim to special creation was a figment of faith came long after Lyell and others had begun to ask awkward questions. Even Freud's demonstration that man's intellect was at the mercy of his repressed emotions was the capstone rather than the foundation of a whole edifice of ideas. But now the fairy-tale of limitless power had in a single destructive flash, without proper warning, been transformed into a threat which by the day after tomorrow might wipe out civilized life as it was known.

The surprise was almost complete. In February 1943, Pope Pius XII had, it is true, spoken to the Pontifical Academy of Science about the advances of atomic physics and had warned about the dangers of nuclear weapons. In the United States, the Department of the Interior's *Minerals Yearbook* for 1943 had commented that the uranium industry was being 'greatly stimulated by a government program having materials priority over all other mineral procurement, but most of the facts were buried in War Department secrecy'. In London a British Member of Parliament, Dr Haden Guest, had told the House of Commons that Mr Churchill had been spending huge sums on investigating a new weapon, and added: 'I do not know whether it is £10 million or £20 million or £30 million. I hope that the Leader of the House will inform the Prime Minister that I have made this statement and ask him to let the House know what authority there is for his conducting experiments of that kind and spending that amount of money without Parliament knowing anything about it.' But these, and a few other potentially dangerous leaks, were never enlarged upon and in neither the United States nor Britain did anyone outside the nuclear project have any inkling of what was afoot.

Even at the top, the need-to-know policy had been rigidly enforced. On Roosevelt's death, the new President had to be hurriedly initiated into the potentialities of nuclear fission. In Britain, where security was directed as pointedly towards the Opposition as towards the enemy, the unseating of Churchill in July 1945 meant that Clement Attlee had to be hurriedly briefed on a subject about which he had, even as Deputy Prime Minister, been kept in almost total ignorance. At Cabinet level only four ministers knew of Tube Alloys, while until the first months of 1945 the heads of the Royal Navy, the Army and the Royal Air Force had been treated as enquiring children to whom nuclear weapons were of no concern.

This secrecy had been enforced to keep the Russians as ignorant as the Germans, Italians, and Japanese. After August 1945, with the Axis powers defeated and the image of the Russian allies only slowly being tarnished in most men's eyes, the need for secrecy became explicable mainly in terms that neither the American nor the British Government was particularly anxious to spell out. Thus from the end of the war onwards nuclear secrecy became a subject in which Rutherford's 'science is international, and I trust will ever remain so' was perpetually being challenged by political expediency and military fears that Governments were not particularly anxious to explain to the ordinary people.

In London the first shot in what was to be an unending guerrilla war was fired by Sir Henry Dale, not only President of the Royal Society and one of those most scandalized by the ban on Allied scientists visiting Moscow in June 1945, but also a member of the Scientific Advisory Committee of the Cabinet which in 1941 had recommended that Britain should try to build her own bomb. Two days after the destruction of Hiroshima, Sir Henry wrote to *The Times* saying that scientists would guard the secrets involved in nuclear weapons until the war with Japan was over. 'Then,' he continued, 'I believe they will all wish to be done with it for ever. We have tolerated much, and would tolerate anything, to ensure the victory for freedom; but when the victory has been won we shall want the freedom.' The letter was both a warning of the way in which the bombing had scraped across the knuckles of the human conscience and a warning of the rift in human attitudes that nuclear weapons were increasingly to widen during the years ahead.

In the United States, Dale's appeal was echoed by Samuel Allison, the physicist who had tolled out the countdown for the Trinity explosion and who by the start of September 1945 had been appointed the new director of the University of Chicago's Institute for Nuclear Studies. At a luncheon on 1 September, Allison forthrightly spoke out against the continuing censorship being imposed on those who had worked on the Manhattan Project. 'We are determined to return to free research as before the war,' he said, and followed it up with a sentence that quickly went from coast to coast. If exchange of scientific information was stopped by the military, he warned, then research workers would leave the field of atomic research and 'devote themselves to studying the colors of butterfly wings'.

Overall, the reaction of scientists in the United States was much the same as in Europe, and to a certain extent the same was true of ordinary laymen. Londoners who had experienced what 600 tons of bombs a night could achieve during the blitz, Americans who had taken part in the Anglo-U.S. air offensive against Germany, and Germans who had been de-housed by it, all knew that the equivalent of 20,000 tons from a single bomb ended the history of war between great powers as it had so far been fought. Even before they understood the dangers of fall-out, even before they became aware of nuclear war's genetic dangers, most people sensed in their bones that the world would never again be the same.

The questions that they raised in the first weeks of the peace were partly moral, partly military, partly political. Immediately pertinent was the question of whether the killing of more than 100,000 men, women and children, mainly civilians, in Japan was so very different from killing tens of thousands in a single night's fire-bombing of Tokyo or killing 45,000 in ten nights' 'conventional' bombing of Hamburg. The attacks on Hiroshima and Nagasaki did, as the report from the British Council of Churches put it, involve 'a great extension of the practice of indiscriminate massacre'; but by 1945 indiscriminate massacre had gained a respectability it had lacked even a few years earlier. Thus honest men could come to very different decisions.

Monsignor Fulton Sheen quoted, in St Patrick's Cathedral, New York, the warning against nuclear weapons given by the Pope at the Pontifical Academy of Science in February 1943 and commented: 'This counsel was not taken. This moral voice was unheeded.' The Hiroshima bombing was against the moral law, he said, before adding, 'We have invited retaliation for that particular form of violence.' But there was also the Rev. John A. Siemes who had been in Hiroshima on 6 August, and who in an eyewitness account published in the *Jesuit Mission* came to this conclusion: 'It seems logical to me that he who supports total war in principle cannot complain of war against civilians. The crux of the matter is whether total war in its present form is justifiable, even when it serves a just purpose. Does it not have material and spiritual evil as its consequences which far exceed whatever good that might result? When will our moralists give us a clear answer to this question?'

Against a background of such argument it was asked whether President Truman's statement that America was holding the secret of the bomb in sacred trust for all mankind was anything more than a cosmetic aimed at salving a nation's conscience. This was of course a subject on which the American argument was likely to be different from the British. It was, after all, American lives rather than British which appeared to have been saved by the bombings since the United States were to provide the bulk of the forces invading Japan. And the British, who had begun the war with a pledge to avoid civilian targets and ended it by de-housing the civilian German population, had to choose their objections with care.

There were other questions which were asked in the aftermath of Hiroshima with a different emphasis on either side of the Atlantic. With the German V-2 rocket not yet developed into the intercontinental ballistic missile, the United States could still – with the threat of the suitcase bomb also still in the future – regard the Atlantic and Pacific oceans as keeping her secure from nuclear attack. The great plants at Hanford and Oak Ridge were producing fissile material on a production-line basis and with delivery systems available in the shape of Superfortresses or the planes already planned to supersede them, the prospects for the *Pax Americana* looked as healthy as they had once looked for Hitler's Third Reich.

By contrast, Britain's position in the entirely new situation that the bomb had created was distinctly vulnerable, strategically because of her geography and industrially because of the Quebec Agreement which appeared to limit her exploitation of peace-time nuclear energy to what President Truman would allow. Clement Attlee, who had succeeded Churchill midway through the Potsdam Conference, quickly tried to alter this state of affairs, proposing in September 1945 that he and Truman should meet. 'Responsible statesmen of the great Powers', he said, 'are faced with decisions vital not merely to the increase of human happiness but to the very survival of civilization.' Therefore, he went on, 'while realizing to the full the importance of devising means to prevent as far as possible the power to produce this new weapon getting into other hands, my mind is increasingly directed to considering the kind of relationship between nations which the existence of such an instrument of destruction demands.'

Truman responded and early in November met both Attlee and the Canadian Premier, Mackenzie King, in Washington. Attlee's hopes were set out in a memorandum which he discussed with the Cabinet shortly before leaving for the United States. In this he emphasized that there was no defence against nuclear weapons, and that 'Britain [was] peculiarly vulnerable to attack by atomic bomb owing to her geographical position and her concentration of population'. He was against giving Russia information until relations had improved; he wholeheartedly supported the United Nations; and he made a final important point: 'I should tell the President that we are naturally interested in the development of atomic energy, both as a means of self-defence, and as a source of industrial power. I should endeavour to regain our freedom by securing a reasonable interpretation if not the abolition of Clause 4 of the Quebec Agreement which laid it down that the President should specify the terms upon which any post-war commercial advantages were to be dealt with.'

There was some very tough bargaining in Washington, but it appeared from the communiqué, issued after considerable difficulty had been overcome, that Attlee had not come away entirely empty-handed. Those signing it admitted that there was no defence against atomic weapons and, perhaps more significantly, that in their use 'no single nation can in fact have a monopoly'. But if this was a reluctant agreement that the Russians would eventually have the bomb, there was no need to hurry them on their way, either to the bomb or to nuclear power.

> We are not convinced that the spreading of the specialized information regarding the practical application of atomic energy before it is possible to devise effective, reciprocal, and enforceable safeguards acceptable to all nations would contribute to a constructive solution of the problem of the atomic bomb. On the contrary we think it might have the opposite effect. We are, however, prepared to share, on a reciprocal basis with other of the United Nations, detailed information concerning the practical industrial application of atomic energy just as soon as effective enforceable safeguards against its use for destructive purposes can be devised.

The best way of implementing such a scheme, it was proposed, would be for the United Nations to set up a Commission. This would, among other things, make specific proposals for 'extending between all nations the exchange of basic scientific information for peaceful ends', and 'for the elimination from national armaments of atomic weapons and of all other major weapons adaptable to mass destruction'. The first of these laudable ideas, recorded in a memorandum drawn up by Anderson and Groves as 'full and effective co-operation in the field of basic scientific research among the three countries', was to be made illegal by America's McMahon Act in less than a year; the second was to be subverted early in 1947 by Britain's secret decision to build her own nuclear weapons.

The question of whether the Russians should be given even more technical information than had been contained in the Smythe Report was directly linked to the attempts at international control foreshadowed in the communiqué issued after the November 1945 meeting in Washington, started in the United Nations in 1946 and ending without success in 1948. By that time, however, the entire question of control was being changed by the process of proliferation. The only two countries who in 1945 could conceivably be considered as potential nuclear powers, even if mini-powers, were Britain and France. In both, the population was war-weary, the exchequer was nearly empty, and enthusiasm was minimal for devoting to nuclear rearmament the massive money and manpower which would be necessary. However, the statesmen in both countries had already decided to support Cherwell's view that it was unwise to put oneself at the mercy of one's neighbours.

Britain was first off the mark. Even before the end of 1943 it was being taken for granted by Anderson, Cherwell, Churchill, and the small handful of others who knew of Tube Alloys, that if the atomic bomb were a success, then Britain would have to build her own when the war was over. Defence was not the only factor involved. General de Gaulle's later statement that no country without the bomb could call itself a great power was already being taken for granted. The question of whether an impoverished Britain would have the money, men or materials for the job was not given too much thought.

The implicit assumption had become explicit as early as May 1944 when the British Technical Committee, set up in 1941 when the Tube Alloys organization came into being, met to advise Sir John Anderson on Britain's nuclear future. The programme its members then recommended 'would shorten as much as possible the time required to produce in the United Kingdom after the war a militarily significant number of bombs, say ten'. There was considerable opposition, particularly from Sir Henry Dale, who felt that a nuclear weapons programme would suck in scientific, educational and labour resources urgently needed for reconstruction. Sir Henry Tizard and P. M. S. Blackett were among others who believed that such a programme would cripple Britain's chances of post-war recovery.

However, there was also the possibility of nuclear power. Whatever problems still had to be solved, the benefits of success here seemed to be so enormous that no country could apparently afford to ignore them. Moreover there were by this time many areas in which research on nuclear weapons and research into nuclear power overlapped, a fact which the British Government was to find much to its advantage.

In October 1945 it was announced that Britain was to set up an Atomic Energy Research Establishment. The site was to be a disused airfield at Harwell, some twenty miles south of Oxford, the director John Cockcroft, now recalled from Canada. If it was not specifically spelt out that Harwell was to be the birthplace of Britain's nuclear power stations, this was the general impression given. There would also be research into the creation and utilization of radioactive isotopes, the new wonder tools that fission made possible.

This emphasis on peace-time uses was just as well. The belief that Britain would have to build her own bombs had been held before Trinity, before Hiroshima and Nagasaki, before the potentialities of the new weapons had been appreciated, and before their use had created worldwide turmoil. But in 1945 any announcement that the new Labour Government was planning to make nuclear bombs would have created a wave of alarm and despondency in a nation tired after a long war and unable to see how they could be meant for use against any force other than the Russians who had done so much to make the defeat of Germany possible.

However, there was no need to deceive the British public. Silence would be enough. British decisions on nuclear matters were being taken from August 1945 onwards by Gen 75, a committee of some half a dozen senior Ministers chaired by Clement Attlee, the Prime Minister, who informally called it the Atom Bomb Committee, and in mid-December the Committee decided that one nuclear pile should be built for the production of plutonium. A fortnight later, on New Year's Day 1946, Attlee received a memorandum from the Chiefs of Staff Committee which argued that Britain should build the greatest capacity for making nuclear weapons that her economics and her access to raw materials would allow. This memorandum, drawn up on the recommendation of Marshal of the Royal Air Force Lord Portal of Hungerford during his last few weeks as Chief of Air Staff, argued that Britain must have a considerable number of bombs, 'a stock to be measured in hundreds rather than scores. In the number of bombs will be our strength.'

In the view of the Government – or those of its members who were let into the secret – this was no more than a necessary precaution, since there were already signs that the new agreement for mutual nuclear co-operation signed between Truman, Attlee and the Canadian Prime Minister the previous month, was a distinctly fragile document.

The decision to build the first nuclear reactor in Britain, reinforced by the appeal from the Chiefs of Staff Committee, was quickly implemented and early in 1946 the organization to build it was announced. A public as yet innocent of nuclear complexities would have been justified in thinking that the first instalment of the fissile goodies which were to revolutionize civilian life would soon be on the way. Yet although the knowledge of plutonium to be gained during the next few years was eventually of considerable industrial use, this was as much by good management as by intention. 'It must be realized that the remit given to the new organization,' Sir Leonard Owen has stated, 'was the production of plutonium for military purposes. Beyond this remit was the thought that the plutonium route to fissile material for weapons, via reactors and chemical extraction plant would be easier than the alternative uranium 235 route via a diffusion plant. ... The target was, of course, the first weapon trial.'

But while work was continuing on the production of military plutonium, no final decision had been taken on how, and by whom, the fissile material would be fashioned into the weapons which, it was hoped, British bombers would be able to carry – the hope being necessary since specifications for the V-weapon carriers had not yet been drawn up.

Decision came only on 8 January 1947. On that date Attlee called a meeting with the Foreign Secretary, the Lord President of the Council, the Minister of Defence, the Dominions Secretary and the Minister of Supply. Before them was a note from Lord Portal, now Controller of Production of Atomic Energy in the Ministry of Supply. So secret did Portal consider the note that he had shown it to no one else in the Ministry which he now served.

'I submit,' it began, 'that a decision is required about the development of Atomic weapons in this country.' The Government could, he went on, decide not to develop atomic weapons at all. Secondly, they could develop them through the normal channels. However, there was a disadvantage about such a course which he then went on to spell out. It would, he said, be impossible to conceal that Britain was making nuclear weapons. 'Many interests are involved, and the need for constant consultation with my organization (which is the sole repository of the knowledge of atomic energy and atomic weapons derived from our wartime collaboration with the United States) would result in very many people, including scientists, knowing what was going on.' Who the 'very many people' were was spelt out in the next paragraph. 'Moreover,' this went, 'it would certainly not be long before the American authorities heard that we were developing the weapon "through the normal channels" and this might well seem to them another reason for reticence over technical matters, not only in the field of military uses of atomic energy but also in the general "know-how" of the production of fissile material.' However, there was an alternative to 'normal channels'.

'The Chief Superintendent of Armament Research (Dr Penney) has been intimately concerned in the recent American trials and knows more than any other British scientist about the secrets of the American bomb,' Portal continued. 'He has the facilities for the necessary research and development which could be "camouflaged" as "Basic High Explosive Research" (a subject for which he is actually responsible but on which no work is in fact being done). His responsibilities are at present to the Army side of the Ministry of Supply, but by special arrangements with the head of that Department he could be made responsible also to me for this particular work and I would arrange the necessary contacts with my organization in such a way as to secure the maximum secrecy. Only about five or six senior officials outside my own organization need know of this arrangement.

'I have already discussed this matter with the Chiefs of Staff who authorised me to say that they are in agreement with me in strongly recommending the special arrangements outlined ... above. If these were adopted, the Chiefs of Staff would see to it that security was not prejudiced by enquiries from the Service Departments. (The Chairman of the Defence Research Committee would of course be informed.)'

This elegant arrangement which was approved at the January meeting ensured that for the time being not only the rest of the Cabinet, Parliament, the public, the Americans, and the Russians, but also the Service Departments would be unaware of the new development. This was probably wise. In the United Nations Atomic Energy Commission the British representative had been strongly supporting the recommendation that all nuclear weapon research should be vested in an international authority. Any suggestion that Britain was preparing to go ahead with construction of her own bombs might have raised accusations of hypocrisy, so there were adequate reasons for Portal's proposed 'camouflage' quite apart from that of keeping the Americans in the dark. Moreover, the Americans had by this time reneged on the Tripartite Agreement of November 1945 by passing the McMahon Act.

The economic consequences of the decision do not appear to have been considered very carefully, and the percipient official historian of the Atomic Energy Authority has noted: 'Although the meeting was being held when Britain was almost at her darkest hour, with factories closing down for lack of coal, neither the Chancellor of the Exchequer nor the President of the Board of Trade was present.'

There appeared to be one very good reason for the Cabinet's decision: Britain's only prospect of survival lay in preventing the outbreak of another major war and it was hoped, on what seem to be rather slender grounds, that the threat of dropping a comparatively small number of nuclear bombs across the huge expanses of Russia would provide a sufficient deterrent. But Attlee's blunt admission in November 1945 that there was no defence against nuclear weapons applied far more damagingly to Britain, with her 50 million people packed into 90,000 square miles, than it did to Russia's 170 million spread across more than 8 million square miles. Deterrence, credible or not, did appear to be the only option and in the autumn of 1947 Portal's recommendation appeared to be justified when Tizard, now Scientific Adviser to the Minister of Defence, stated that 'no one can envisage an adequate defence of the civil population of this country in the event of a major war in the atomic-bomb age'. The only apparent answer, failing deterrence, was outlined in a Cabinet Defence Committee paper on civil defence policy. 'It may be,' this said, 'that the proper solution to our problem ... lies in a carefully planned large-scale dispersal of industry and population to the Dominions and Colonies overseas.' But the figures bruited around went as high as 15 million; their emigration was to be spread over twenty-five years, but the proposal soon sank under the weight of its own political problems. Deterrence appeared to be the only hope.

More than another year was to pass before it was found inexpedient to keep the secret any longer. On 12 May 1948, a Labour Member of Parliament, George Jeger, asked the Minister of Defence in a pre-arranged question 'whether he is satisfied that adequate progress is being made in the development of the most modern types of weapons'. Mr Alexander answered: 'Yes, sir. As was made clear in the Statement Relating to Defence 1948 (Command 7327), research and development continue to receive the highest priority in the defence field, and all types of modern weapons, including atomic weapons, are being developed.'

Meanwhile, the French had been following much the same path of surreptitious development although, with more leeway to make up, they were to be some years behind the British. The setting-up of the Commissariat à l'Énergie Atomique (C.E.A.) had officially been announced in October 1945, but according to the British Embassy in Paris it had, in fact, been formed seven months earlier. At the time of the Hiroshima attack a member of the Embassy staff had been shown 'a manuscript note in General de Gaulle's own hand dated early in March 1945, instructing the Secretary-General [Monsieur Joxe] to convene a secret Cabinet committee to study the possibility of the use of atomic energy and the uses to which the heavy water still in French hands might be put. The new Commissariat,' the Embassy reported to London, 'has in fact been secretly in existence since that date with Monsieur Dautry in charge, but the latter's appointment has not been officially announced in the *Journal Officiel.* In his note to Monsieur Joxe, General de Gaulle stated that he would preside over the secret Committee and instructed the former to arrange in particular for MM. Pleven (Finance), Dautry (Reconstruction), Giscobbi (Colonies) and Professor Joliot-Curie to be members of it. Monsieur Joxe was also told to ensure that no dossiers on the subject were passed from Ministry to Ministry. ...'

The French were lucky in discovering a few tons of uranium concentrate, overlooked both by the Germans and Allied scientific teams. The pre-war connection with Norway was renewed to ensure supplies of heavy water. War-worn machine tools were commandeered and makeshift factories and laboratories built with whatever materials could be found. Enthusiasm brought results.

Before the end of 1948 the Commissariat had commissioned a small experimental reactor – Z.O.E., for zero energy, uranium oxide, heavy water – and had found a small but valuable source of uranium in the centre of the country. Before long the first milligrams of plutonium were being extracted. But when, a few years later, the French began to build their own bomb, the work was begun in advance of any official decision to do so. In the words of a member of the French Atomic Energy Committee, important sums were spent on a programme of nuclear arms 'without the Government having taken the decision to make the weapons and also without a debate ... in Parliament to approve such a decision'.

There are other similarities between the decisions of Britain and of France to become the proud owners of nuclear weapons. In both cases the decision was governed partly by a demand for national prestige; partly by fear that the Americans might not be relied upon in the future – a recrudescence of Cherwell's being 'very much averse' to putting himself at the mercy of his neighbours; and, in the case of Britain, which planned production of the bomb before planning a plane to carry it in, without very much attention to detail. Both countries based their programmes on the assumption that Russia would be deterred from attack with conventional weapons – the huge land armies she had raised to defeat the Wehrmacht – by the threat of nuclear weapons flown from Western Europe. Given the vast spaces of Russia and the minimal number of nuclear bombs that Britain or France were likely to have available for such an assault, the chances of successful deterrence did not seem great, even before Russia acquired nuclear weapons of her own.

The first few years of the post-war world thus saw the genesis in Britain and France of the process that was later to be so harshly condemned: proliferation – meaning the acquisition by non-nuclear powers of the nuclear weapons one already has oneself. In the circumstances there appeared enough justifications for this. But it tended to pare down the remaining slim chance of bringing common sense to the new nuclear world. Russia, stalling in the United Nations until she, too, had her own nuclear weapon at the hip, was already making that a difficult enough task.

Eight years after Wells had written *The World Set Free* and three years after the end of what was still being called the Great War, he wrote a new preface to the book.

> It is the main thesis which is still of interest now, the thesis that because of the development of scientific knowledge, separate sovereign states and separate sovereign empires are no longer possible in the world, that to attempt to keep on with the old system is to heap disaster upon disaster for mankind and perhaps to destroy our race altogether. ... The dream of *The World Set Free*, a dream of highly educated and highly favoured and ruling men, voluntarily setting themselves to the task of reshaping the world, has thus far remained a dream.

That was in 1921. Now, in the aftermath of the bombs on Japan, the fictional chance had been transformed to factual opportunity. But it was an opportunity on a new scale and one which could be grasped successfully only by exceptional men. Niels Bohr, one of the few who combined scientific genius with political vision, was ruled out of the international scene both by his inexperience in the gentle art of practical diplomacy and, very probably, by the residual after-effects of Churchill's disastrous intervention with Roosevelt. In the United States, where leadership now rested, there was no replacement for the lost President, able and honest as Truman might be; and embryonic scientist-statesmen such as Bush and Conant were anxious to return to their home grounds. In Britain there was the stern figure of Sir John Anderson, now translated into Lord Waverley: a man of impeccable integrity and unqualified experience but suspect in American eyes through his connections with Imperial Chemical Industries. In Russia even scientists of Peter Kapitza's stature were kept subservient to the system, while in France any feeling for the concept of one world was qualified by the belief that France must lead it. Germany, where argument still continued among physicists as to whether their nuclear project had failed through inability or intent, had not even begun to work her passage back into decent society; even men like Otto Hahn were still treated with caution.

At one of the most critical moments in the history of the world, the world failed to provide the man or men for the job.

CHAPTER 13

Agreement to Disagree

Among the views on 'the bomb' held by laymen, scientists, military men and politicians in the immediate post-war years, one belief was almost universal: that international control of nuclear weapons was both the most important and the most intractable of problems to be solved if a sane peace was to follow a necessary war. But if the belief was strong, the follow-up was human; pursuit of what most men knew to be essential was often inhibited by attitudes and articles of faith that had little relevance in the nuclear age.

In the United States the genuine hopes for international control were distorted by America's industrial lead and by growing distrust of Russia's imperial aims. For some, the situation offered an ideal opportunity for forcing international agreement from the unwilling Soviets; men like Groves, with a long-standing record of anti-Soviet feeling, certainly expressed the situation crudely. Nevertheless, the general feeling was that of Byrnes who sincerely believed that, without threats, veiled or otherwise, the Russians would be more amenable to discussion if the West were known to be well armed with the new weapons.

In Britain, whose new Labour Government was constitutionally in favour of international solutions, feelings were modified by the fact that the Red Army was entrenched less than 400 miles away on the Elbe – the distance between New York and Cleveland. It is true that the enthusiasm of ordinary people for the Russians evaporated only slowly. But in Britain there existed, from 1945 onwards, a small but concentrated body of opinion which maintained, with as much force as its American counterpart, that the wasting asset of temporary nuclear superiority should be used to push the Russians to the conference table while there was still time.

Openly, this view was most vehemently advocated by Bertrand Russell, the philosopher, whose visit to Russia in 1920 had left him with an unqualified detestation for the Communist system. Russell, scientifically literate, knew better than most men that the 'secret' of the bomb was merely whether or not it would work; with no illusions about the ways of the world, he realized that from 6 August 1945 the Russians would have been trying with renewed vigour to wipe out the American lead. As a logician, he could be forced to only one conclusion. And less than a month after Hiroshima he was writing: 'There is one thing and one only which could save the world, and that is a thing which I should not dream of advocating. It is, that America should make war on Russia during the next two years, and establish a world empire by means of the atomic bomb. This will not be done.'

Maybe not; but before the end of October 1945, Russell was describing the dangers of giving the Soviets nuclear information, outlining a confederation which would monopolize nuclear weapons, and pointing out that Russia would be powerless to do anything as long as America retained her atomic lead. 'There might be a period of hesitation followed by acquiescence,' he wrote, 'but if the U.S.S.R. did not give way and join the confederation, after there had been time for mature consideration, the conditions for a justifiable war, which I enumerated a moment ago, would all be fulfilled. A *casus belli* would not be difficult to find.'

Russell's attitude, emphasized in numerous speeches and articles during the following months, continued to arouse little enthusiasm in a British public which had genuinely admired the fortitude of the Red Army during the bleakest months of the war and which now averted its eyes from the Continent rather than watch her wartime ally swallowing the 'poor little Poland' on whose behalf the war had begun, and then digesting eastern Europe piecemeal.

Yet Russell's line was much the same as that of Churchill, increasingly anxious as the Russians consolidated their position in Eastern Europe and looked about for more. Churchill revealed his attitude to the Canadian Prime Minister Mackenzie King who, visiting him in the autumn of 1947, during the London meeting of the Council of Foreign Ministers, asked how America could possibly mobilize forces for another war. 'He turned to me sharply,' King recorded in his diary,

> his eyes bulging out of his head, and said: they would, of course, begin the attack in Russia itself. You must know they have had plans all laid for this, for

over a year. What the Russians should be told at the present conference, if they are unwilling to co-operate, is that the nations that have fought the last war for freedom have had enough of this war of nerves and intimidation. We do not intend to have this sort of thing continue indefinitely. No progress could be made and life is not worth living. We fought for liberty and are determined to maintain it. We will give you what you want and is reasonable in the matter of boundaries. We will give you ports in the North. We will meet you in regard to conditions generally. What we will not allow you to do is to destroy Western Europe; to extend your regime further there. If you do not agree to that here and now, within so many days, we will attack Moscow and your other cities and destroy them with atomic bombs from the air. We will not allow tyranny to be continued.

It was, in fact, very unlikely that the Americans could have bludgeoned the Russians into a nuclear concordat either by threats or by war. As P. M. S. Blackett was to write in a secret report to the British Government in November 1945: 'That the U.S.S.R. would capitulate before a threat alone can be excluded as not remotely probable.' On the contrary, she would increase her nuclear research, her air defence, and her influence in semi-satellite countries. As for a war in which America tried to bomb Russia out of existence, the probable outcome 'would be to destroy much of the Soviet homeland, but also to extend Soviet power over nearly the whole of the European mainland, much of the Near and possibly also the Far East'. Even when the Victorian boast that: 'We have got the Maxim gun, and they have not', could be updated by altering Sir Hiram's gun to 'atom bomb', even then a preventive war to solve the problems raised by nuclear fission was no solution at all.

The situation was complicated by a number of factors. For as the knowledge of what nuclear fission really meant filtered down from the scientists to politicians, the military, and even to ordinary men and women, it became clear that 'the secret' was really non-existent; lack of technical expertise and know-how might delay certain processes for a while, or make them more costly in men or materials. But it gradually dawned on those worried about what was to happen next, that the Russians would eventually, whether in five years or ten, have their own nuclear weapons. Worse still, there was the spectre which everyone tried to ignore: the likely chance that within a few more years nuclear weapons would not be limited to the great industrial powers but might be available to almost any country with access to uranium and the will to put nuclear research at the top of the priority list.

If this did not present enough problems for the future there was another awkward fact that could no longer be denied: that production of nuclear power for peaceful purposes – overwhelmingly, it appeared, for the production of electricity – would have as a by-product the creation of plutonium, the element now known to be even more efficient than uranium 235 as a nuclear explosive. Control of a nation's armaments appeared to involve control of the way in which it solved its energy problems, a factor which inevitably complicated any prospect of international control.

Nevertheless, on 24 January 1946, the United Nations Atomic Energy Commission, which had been proposed at the tripartite meeting the previous November, came into being, composed of members of the Security Council plus Canada. Its terms of reference were that the Commission shall make specific proposals for control of atomic energy to the extent necessary to ensure its use only for peaceful purposes; for the elimination from national armaments of atomic weapons; for effective safeguards by way of inspection and other means to protect complying states against the hazards of violations and evasions.

The Americans were justifiably suspicious of Russian motives. The British were determined that they would no longer remain under the thumb of the Americans – 'No other Foreign Secretary should have to sit in front of the U.S. Foreign Secretary and be talked to as I have been talked to by Byrnes,' said Mr Bevin during one discussion of the nuclear future. 'It shouldn't happen and the only way to stop it is to have the plants in this country, with the Union Jack on top.' The Russians were determined to remove their susceptibility to nuclear blackmail. Thus the chances of success for the new U.N. body would have been slight even had the details of the Canadian spy case not become public knowledge less than a month later.

Igor Gouzenko, a cipher clerk in the Ottawa office of the Soviet Military Attaché, had defected in September 1945 and revealed to the Canadians that an extensive Russian network, one of whose aims was to gather nuclear information, had been operating in Canada for some while. The news, with the distinct hint of more to come, strengthened an undertow of feeling that it was unwise to trust the Russians. It increased, moreover, a feeling near to despair which was epitomized by a rather emotional entry which Mackenzie King made in his diary.

Churchill was right when he said it would not do to let the Russians have the secret of the atomic bomb. I thought Roosevelt was right when he said he felt an ally should know what we are doing in that regard. I can see that Churchill had the sounder judgement; had keener perceptions of what was at stake, what was going on. ... Indeed, this revelation gives one a new and more appalling outlook on the world than one has ever had before. Combined with the atomic bomb and what it may yet lead to, causes one to feel physically actually older in years than one has felt before. I cannot believe that this information has come to me as a matter of chance. I can only pray for God's guidance that I may be able to be an instrument in the control of powers beyond to help save a desperate situation, to maintain peace now that it has at least been nominally established.

As far as the Americans were concerned, God's guidance was also needed elsewhere, since the crucial question of how much it would be wise for them to tell the Russians, and under what conditions, soon became entangled with the question of how much they should tell the British. This in itself was almost inextricably linked with the manner in which nuclear fission was to be controlled domestically. Both Britain and the United States had to solve the problem of deciding in what way its possibilities for peace and for war could best be developed. In Britain, the wise words of the Scientific Advisory Committee in 1941 on the power project – that the 'matter should not be allowed to fall into the hands of private interests' – were by now taken for granted and the Government maintained control. In the United States the Atomic Energy Commission were to take charge, and the prospects of a private take-over of the nuclear industry were quenched, at least for the time being.

In Britain, control of nuclear research by the Ministry of Supply evoked, of itself, comparatively little public discussion or interest. One reason was that, as far as the public knew, nuclear weapons were not involved. Nuclear power was at best only a possibility in the distant future, and a nation still as sternly rationed for food and petrol as it had been during the war – and soon to be rationed for bread, which had not been necessary even at the height of the war – debated more pressing matters.

In the United States, where the production of nuclear weapons was taken for granted and nuclear power was expected to bring the millenium if not next week then perhaps next month, debate was more vigorous and more healthy. This was as well. As early as October 1945 the Army Department had sponsored legislation, the May–Johnson Bill, which would have brought all development of nuclear fission, civilian as well as military, under military control. The Bill, pushed forward with brisk determination, might have become law had not Congressman May rather overplayed his hand and tried to press it through without adequate hearings, thus causing what has been called 'the bitterest legislative battle since the slaughter in the Senate of the League of Nations'.

Among those aroused were Leo Szilard and many of the other physicists who earlier had failed in their efforts to influence either the use of the bomb or an approach to the Russians. Szilard has explained:

> For some four to six weeks after Hiroshima atomic scientists expressed no opinion on the political implications of the bomb, having been requested by the War Department to exercise the greatest possible reserve. Our response to this request does not mean that we were intimidated by the War Department. We kept silent because we all believed that Hiroshima was [to be] immediately followed by discussion between the United States, Great Britain and Russia, as indeed it should have been, and we did not want to embarrass the President or the Secretary of State.

But eventually it became clear that no such conversations were contemplated. Szilard and his colleagues now formed the first atomic lobby with the aim of blocking the May–Johnson Bill and bringing about the establishment of a Congressional committee on atomic energy which would give the subject the attention it deserved. The Federation of Atomic Scientists was founded, dedicated to informing the public of the nuclear facts of life, as well as ensuring that nuclear energy did not come under military control. The group was extremely successful at the art of catching the public ear and by the first months of 1946 the McMahon Bill was already being discussed as an alternative to the May–Johnson Bill. Eventually signed as the Atomic Energy Act by Truman on 30 July 1946, this provided for the setting up of an Atomic Energy Commission which would control the development of nuclear fission for either wartime or peace-time purposes and would thus make atomic energy 'an island of socialism in the midst of a free-enterprise economy'.

While the Bill kept total control from military hands, the Canadian spy case had created such a furore that an amendment was eventually added providing for a military board with power to review decisions made by the Commission and the right to take complaints to the President. More important were the clauses which enshrined the fallacy of America'a atomic monopoly and put an absolute ban on international co-operation, military or civilian, and on the transfer of fissionable material or information. These might of course be amended or removed by Congress in the light of agreements made at the United Nations. Nevertheless, they made it impossible for America to honour the agreements which Roosevelt had signed at Quebec and Hyde Park, or that which Truman had signed a mere eight months previously in Washington. As far as Britain and France were concerned the McMahon Bill justified their decision to go it alone and pushed the prospects of genuine international control even further into the future.

However, while settling domestic control of atomic energy, the Americans were at the same time trying to discover what proposals could best be put before the United Nations Atomic Energy Commission. The first attempt was made by a five-man committee headed by David Lilienthal, the head of the Tennessee Valley Authority since it had been set up by Roosevelt in 1933. Lilienthal had already created a sizeable band of enemies through his enthusiasm for what many Americans considered to be a wildly socialist enterprise. He tended to be constitutionally critical, and was under no illusions about the difficulties of his new assignment.

'Saw Acheson [Under Secretary of State] at 9.15 until 10.00,' he wrote in his diary for 16 January.

> He talked frankly and in detail. Those charged with foreign policy – the Secretary of State (Byrnes) and the President – did not have either the facts or an understanding of what was involved in the atomic energy issue, the most serious cloud hanging over the world. Commitments, on paper and in communiqués, have been made and are being made (Byrnes is now in London before the United Nations General Assembly) without a knowledge of what the hell it is all about – literally! The War Department, and really one man in the War Department, General Groves, has, by the power of veto on the ground of 'military security', really been determining and almost running foreign policy. He has entered into contracts involving other countries (Belgium and their Congo deposits of uranium, for example) without even the knowledge of the Department of State.

Lilienthal, Robert Oppenheimer, and the three industrialists who completed his team, reported in April. The nub of their proposals was that nuclear development should be divided into 'dangerous' activities, mainly the production of weapons, to be undertaken by an international authority, and 'safe' activities which could be carried out by individual countries under supervision. The Lilienthal panel agreed that there was at present no hard line dividing what was dangerous from what was safe, but rather ingenuously hoped that this could be provided by 'denaturing' plutonium so that it would not be suitable as an explosive.

The crux of the matter was that the status quo would remain until the international authority was operating. This might be some years ahead and until then the United States would retain control of her own stock of nuclear weapons. As Mark Oliphant put it when reviewing the report in *Nature*, 'The problem of reconciling a proper case for the security of the United States with essential handing over of information to the new body is solved in a manner which is eminently reasonable from the American point of view, but which may not seem so free from danger to the U.S.S.R.' But not only the Russians were involved. The British, already preparing their own fissile material, and with their Chief of Air Staff preparing the first requisition for an atomic bomb, to be presented to the Ministry of Supply in August, were hardly more enthusiastic.

From the Lilienthal Report, as it was to be known, there developed the Baruch Plan, America's formal proposal for international control of nuclear weapons, presented to the United Nations Atomic Energy Commission on 14 June 1946. Baruch was an unlikely man to launch the enterprise. A 75-year-old ex-Wall Street tycoon, he had little enthusiasm for the work. The slightly casual manner in which he took on the greatest problem of post-war diplomacy was revealed to Vannevar Bush who told him that he was the least qualified man in America for the difficult task. 'Doc,' said Baruch, 'you couldn't be more right. Put on your hat and let's go and tell the President, and that will let me out.' Bush warned him that if he 'got out' the enterprise would fail.

'If I stay in, I'm unqualified, and if I get out, it will blow up,' Baruch countered. 'What do you want me to do?'

'Oh, hell,' replied Bush, 'stay in.'

Baruch did have at least one qualification. He knew that the matter was, as he described it to one of his assistants, a matter of life and death. A few days later his assistant telephoned. 'I've got your opening line,' he said. 'It comes from the best possible source – the Bible.' So it was that on the 14th Baruch began his address to the first session of the U.N. Commission with the words. 'We are here to make a choice between the quick and the dead.' That, he went on, was their business. 'Behind the black portent of the new atomic age lies a hope which, seized upon with faith, can work our salvation. If we fail, then we have damned every man to be the slave of fear. Let us not deceive ourselves: we must elect world peace or world destruction.'

Behind these fine words the Baruch Plan appeared, at first glance, to be the essence of sweet reasonableness. An international body would control the use of all nuclear energy, whether developed for the bomb or 'the boiler'; the countries of the world would thus be frozen in their 1946 military and industrial positions, a situation which could only be changed by approval of the international body. The Baruch Plan was far-sighted in appreciating, even in 1946, that there was no firm line dividing military and non-military uses and that once a country had embarked on a programme for the industrial use of nuclear energy it became virtually impracticable to prevent the secret manufacture of nuclear weapons. Since those days the facilities for inspection, notably by satellite, have enormously increased; but so, with equal speed, have facilities for evasion.

However, the rocks on which the Baruch Plan was to founder consisted of the veto, or rather the lack of it, and the threat of 'condign punishment', which included war, which formed essentials of the Plan. Neither Lilienthal, nor most of his colleagues, believed that those provisions were necessarily wrong, objecting only to the ways in which they had been made. 'My colleagues and I agreed completely with the "no veto" principle,' Lilienthal was to write, 'but we felt that the injection of this issue would endanger consideration of the affirmative basis of the plan itself.' The question of condign punishment, moreover, 'perhaps wasn't fatal; that would depend upon the course of the negotiations'.

Yet the removal of the veto meant that no decision by the Commission, which in the nature of things would be dominated by American experience, could be blocked by the Russians. At the time, more than a third of a century ago, it was almost universally believed that the Russian intransigence which was to halt the Baruch Plan in its tracks sprang from the U.S.S.R.'s ambitions for nuclear rearmament. Today, when a great deal more is known about Russia's need to pull herself up by her bootstraps after the devastation of war, it is impossible to discount another factor. In the eyes of the Russians, agreement to the Baruch Plan would have given the Commission a stranglehold over the way in which they developed and used nuclear power for the re-industrialization of the country.

It is now known, following publication of General Eisenhower's papers, that any American surprise at Russian rejection of the Baruch proposals was entirely simulated. In April Baruch met Eisenhower and other military leaders to discuss the plan he was to put forward to the United Nations; those present included General Groves, General Spaatz and General Norstad of the Air Force, and ex-Secretary of State Cordell Hull. The editors of Eisenhower's papers say 'Baruch agreed with the military that no steps should be taken that would threaten U.S. security and that the problem should be approached "slowly and progressively".' Hull, it is added, had previously pointed out to Eisenhower that the Russian experts in both the State Department and the War Department were agreed that Soviet acceptance of the Baruch proposals was 'almost unthinkable'.

As far as the Russians were concerned they had not yet forgotten the Allied intervention after the Revolution and the reluctance with which the Western forces eventually withdrew. They no doubt knew that circumstances rather than caution had prevented the dispatch of an Anglo-French force to fight them in Finland in the autumn of 1939. And once Hiroshima and Nagasaki had been obliterated it was clear that the Western Allies had been as cautious about their secrets as the Russians had been about their plans to overrun central Europe. 'Russia,' as Alanbrooke noted in his diary, '... will be very loath to tie her hands to prevent her from taking advantage of a means of obtaining complete control of the Western Hemisphere, if not of the whole world. And who can blame her, judging matters on past international standards.'

If these factors made Russian acceptance of the Baruch Plan unlikely and Britain's plans for her own bombs made her acceptance even less so, there was also some doubt about the position of the United States herself. James Byrnes, leaving Washington for the United Nations meeting which was to establish the Atomic Energy Commission had commented:

> The language of the resolution makes clear that even as to the exchange of basic scientific information for peaceful purposes, the Commission has authority only to make recommendations. Therefore unless the United States concurs in the recommendation, it could not be adopted. If the United States concurred and the Security Council adopted the recommendation, it would still be for the Government of the United States by treaty or by Congressional action to determine to what extent the recommendation should be acted upon. If action is required by treaty it would take a two-thirds vote of the Senate to ratify the treaty.

This implication that any decision in the U.N. Commission might be annulled by Congress was no encouragement to the Russians. Neither were the British particularly impressed, even though they failed to put forward any proposals of their own. Their position was epitomized by the graphic and colloquially speaking Ernest Bevin, the British Foreign Minister. 'Let's forget about the Baroosh and get on with making the fissle,' he is reported to have said. And, as the official British historian comments: 'When, soon after this, Ministers asked a group of officials to produce proposals for international control, they too reported that, whatever arrangements might be made in the international field, Britain should undertake production of bombs on a large scale for her own defence as soon as possible; the Prime Minister should issue a directive to this effect.' On this, at least, the British and the Russians appeared to take the same view.

Andrei Gromyko put the Soviet proposals to the United Nations Commission on 19 June. These called for an international convention outlawing the production and use of atomic weapons and demanded the destruction of all those already in existence. The huge gap between this and the American proposal did no more than confirm what most of those involved had taken for granted: that in practice the Americans would be unlikely to yield the smallest fraction of their monopoly and that the Russians would make every effort to break it. While the Netherlands did not commit herself when the vote was taken, every other delegate voted as expected. Poland supported Russia and the Baruch Plan was supported by the rest – including Britain who could safely do so since there was little chance of the Plan's acceptance, an eventuality which might have gravely embarrassed her.

Before the U.N. vote, the start of a long series of wrangles which were to end without result in 1948, the Americans had announced a new series of nuclear tests planned to take place in the Pacific early in July. This could easily be represented as a tactful reminder to the Russians of existing American power. It was, however, the outcome of a suggestion made to Secretary of the Navy Forrestal by his special assistant, Lewis Strauss, only eight days after Hiroshima. The United States, he then suggested, should 'at once test the ability of ships of present design to withstand the forces generated by the atomic bomb'. The need for such a test was underlined by the lack of understanding still evident in more conservative Service circles. Thus the British Rear Admiral H. G. Thursfield, reviewing the lessons of the war in the authoritative *Brassey's Annual* for 1946, still insisted that a nuclear weapon was nothing but a 'bigger and better bomb' and added that while 'it may possibly prove to be capable of sinking any ship that may be within range of its destructive effect when it explodes', even that was 'mere speculation'.

The timing of the Bikini tests was unfortunate but, as was often the case with nuclear affairs, the military left hand often went on with its work without asking what the political right hand was planning to do. Thus in the U.S. Senate Scott Lucas was able to ask: 'If we are to outlaw the use of the atomic bomb for military purposes, why should we be making plans to display atomic power as an instrument of destruction?'

'Operation Crossroads', as the 1946 explosions were named, involved ninety Japanese and outdated U.S. warships, including battleships, aircraft carriers, submarines and landing craft, all anchored for the experiment in the 25-mile-long lagoon of Bikini atoll in the Marshall Islands. The native inhabitants were moved to Rongerik atoll, 130 miles away, and a large number of pigs, goats, and rats were brought to the atoll and the anchored vessels. A complex array of recording devices was placed on and around the test site.

The first test took place on 1 July when a bomb of similar power to that dropped on Hiroshima was exploded over the fleet. Five of the ships were sunk but the damage appears to have been less than expected. The second test, which took place on 25 July, produced rather more alarming results. A bomb of the same yield as the earlier one was exploded beneath one of the target ships in the world's first underwater nuclear explosion, and as results from a multitude of recording instruments were analysed the devastating effect of radioactive debris became evident.

'The second bomb,' said a report to Truman from his Civilian Evaluation Commission, 'caused a deluge of water loaded with deadly radioactive elements over an area that embraced ninety per cent of the target array. All but a few of the target ships were drenched with radioactive seawater, and all within the zone of evident damage are still unsafe to board. It is estimated that the radioactivity dispersed in the water was equivalent to that from many hundred tons of radium.' The effects were underlined in a second report, from the Army-Navy Joint Chiefs of Staff. 'These contaminated ships,' it pointed out, 'became radioactive stoves and would have burned all living things aboard them with invisible and painless but deadly radiation.' Admiral Thursfield's doubt about whether ships would be sunk by nuclear weapons was shown to be irrelevant. Whether sunk or not, all on board would be dead or dying. And the Federation of Atomic Scientists noted that the crossroads before mankind was not whether navies could survive but whether civilization itself could survive.

However, both navies and civilians now had a new problem, as was made clear in a report from the Bikini observers to the Chiefs of Staff: 'It is too soon to attempt an analysis of all the implications of the Bikini tests,' this stated. 'But it is not too soon to point to the necessity for immediate and intensive research into several unique problems posed by the atomic bomb. The poisoning of large volumes of water presents such a problem. Study must be given to procedures for protecting not only ships' crews but also the populations of cities against such radiological effects as were demonstrated in Bikini lagoon.'

However, there was one significant event on which no mention appears to have been made at the time. Three explosions had been planned for Bikini. The third one was cancelled.

The military reaction to Operation Crossroads is instructive. The report by the Joint Chiefs of Staff evaluating board included twenty-two conclusions and fifteen recommendations, the first recommendation being that the J.C.S. should work out 'an acceptable guaranty of international peace'. Failing this, the United States should stockpile atomic weapons and continue with research and development on better ones. More important, it was recommended that fresh legislation should require the President, after consulting the Cabinet, 'to order atomic bomb retaliation when such retaliation is necessary to prevent or frustrate an atomic weapon attack upon us'. Further recommendations were for selection of targets, establishment of a board to study the psychological aspects of atomic warfare and the development of passive defence, although it was admitted that this was impractical for 'urban and industrial areas'.

All this was a tribute to the impact of the first bombs at Bikini. But it was to mean that from the summer of 1946, as American diplomats at the United Nations argued for acceptance of some variation of the Baruch Plan, and experts evaluated the results of the Bikini explosions, the United States pushed ahead with two parallel policies. The first was that of international control. The second was that of building up her own stock of nuclear weapons and perfecting contingency plans for their use.

The McMahon Bill came into operation on 1 August and by the end of the year all was ready for the handover by the Army of the Manhattan Project's plant and personnel to the newly formed Atomic

Energy Commission, headed by David Lilienthal. Groves was pessimistic about the change, and not only because day-to-day nuclear control passed from the hands of the Services. 'Whatever the facts,' say the official historians, '[he] had come to think of the project as his own personal creation. On one occasion he frankly told a Commissioner that he was in the position of a mother hen watching strangers take all her chicks.' Other illusions remained. Shortly before the formal handing over, the new authority was telephoned by one newspaper; could they, it was asked, have a photograph of Groves handing over the secret to David Lilienthal.

During 1947 the world appeared to be quietly digesting the knowledge revealed at Hiroshima and Nagasaki. The quietness was illusory. In Britain the formal decision to build nuclear weapons was made and at Windscale in north-west England construction was started on the first piles which were to produce fissile material for the British weapons. In the United States the Air Force persuaded the Joint Chiefs of Staff to demand 'approximately 400 atomic bombs of destructive power equivalent to the Nagasaki-type bomb', and drew up details of 'Half-Moon', a plan which called for the nuclear bombing of twenty Russian cities. On 14 July the Joint Strategic Survey Committee proposed the faster stockpiling of nuclear weapons, development of a mobilization plan and supporting measures including the 'establishment of a top-secret project to exhaust every practicable means of gaining factual information concerning the progress of Soviet atomic research and the development of atomic and other possible weapons of mass destruction within the U.S.S.R. regardless of the cost or the means employed.' Ten days later Eisenhower, then Chief of Staff, described the proposals as 'a virtual – if less than complete – mobilization for war'.

Yet despite the advocates of preventive war, such plans were primarily devised as a deterrent centred on monopoly of the atomic bomb. Their efficacy was put to the test in 1948.

When the Allies were completing their plans for the occupation of Germany it had been agreed that the Russians should garrison an area bounded on the west by a line running south from the Bight of Lübeck on the Baltic. Berlin, eighty miles or more within the Russian Zone was to be occupied by Russia, the United States, Britain and France, each of whom had been allotted a sector of the capital. However, by an extraordinary oversight no watertight legal formula had been signed formally allowing the Western Allies to use the land routes which linked Berlin with the West across the Russian occupation zone. This provided the Russians with a weapon when, on 18 June, the three Western powers reformed the West German currency and vastly strengthened the West German powers of recovery. Six days later the Russians stopped all road and rail traffic between Berlin and the West.

The options open to the Western powers were four. They could argue a case at the United Nations where a Russian veto would no doubt be used to block any resolution satisfactory to the West. They could attempt to push an armed convoy across the Russian Zone to Berlin, a move which it was generally expected would start the Third World War. They could openly threaten the Russians with the bomb. Or they could mount an airlift to supply the German capital by transport planes.

The first move towards use of the weapon came less than a month after the start of the blockade. During a meeting at the White House on 21 July, the National Military Establishment presented a formal request that the President should turn over custody of the atomic bomb from the Atomic Energy Commission to the military establishment. Two days later Truman held back James Forrestal, Secretary of the Navy, after a Cabinet meeting. 'He told me,' Forrestal wrote in his diary, 'that he would make a negative decision on the question of the transfer of custody of atomic bombs and said that political considerations, at the immediate moment, had influenced his decision. He indicated that after election it would be possible to take another look at the picture.' Thus it appeared that the potentialities of Armageddon would be held back at least until the Presidential election had been settled in November. However, Truman may have had another motive: to stay the hands of the military until the effects of the Berlin airlift could be properly judged. Like Churchill, and despite the popular aggressive growlings of both men on occasion, Truman agreed with Churchill's belief that 'jaw-jaw is better than war-war'.

The difficulty of his task became evident throughout the late summer and autumn, especially after 25 October when the Russians vetoed a United Nations proposal to lift the blockade. At a meeting of nearly a score of newspaper publishers and editors in September, Forrestal, who on 28 July had been given the new post of Secretary of Defense, had found 'unanimous agreement that in the event of war the American people would not only have no question as to the propriety of the use of the atomic bomb, but would in fact expect it to be used'. Even in Britain, closer to Berlin, more tired after a longer war, more concerned about the dangers of another one, *The Observer* could comment: 'It is we who hold the overwhelming trump cards. It is our side, not Russia, which holds atomic and post-atomic weapons and could if sufficiently provoked, literally wipe Russia's power and threat to the world's peace from the face of the earth.'

During the next few months, entries in Forrestal's diaries reflect what he saw as current feeling about the use of the bomb. John Foster Dulles, then a U.S. delegate to the United Nations, was reported as having told General Marshall, 'the American people would execute you if you did not use the bomb in the event of war'. General Clay, who directed the Berlin airlift, had said that he 'would not hesitate to use the atomic bomb and would hit Moscow and Leningrad first'. Churchill had said that it was wrong to minimize the destructive effect of the bomb since this would only encourage the Russians. Just how objective Forrestal's reports are is a moot point – on 12 November 1948 he wrote in his diary that the British Prime Minister (Clement Attlee) had told him that there was 'no division in the British public mind about the use of the atomic bomb – they were for its use. Even the Church in recent days had publicly taken this position.'

But everyone held their hand. The airlift survived the difficulties of the winter and in May 1949, when it was plain that Berlin could be supplied by air indefinitely, the Russian blockade was lifted.

America, Britain, and the world in general breathed a sigh of relief. Although the threat of the bomb had not been openly used, the Russians knew the facts of the situation even without the help of their agent Donald Maclean. The great deterrent appeared to have deterred. If one persevered, perhaps even the Russians could be brought to the conference table.

Within a few months all was to be changed.

CHAPTER 14

The Whirlwind Arrives

For three years after the passage of the McMahon Bill, that is until the summer of 1949, it appeared that William Laurence's verdict after the Bikini test of 1946 was still correct: the public still continued to believe that nuclear energy was nothing for the average citizen to be worried about. To the military it presented the need for making decisions, but America's nuclear hegemony apparently prevented them from being very serious decisions. The blockade of Berlin seemed to reinforce the argument; the mere existence of the American bomb, it had appeared to many, had forced the Russians to back down.

Much the same complacent attitude prevailed elsewhere. In Britain, it is true, complacency was largely a result of the public being kept in ignorance, for so long, of what was being done in the public's name. Even when the cat was let out of the bag in May 1948, it was the tail rather than the whole animal which was presented for criticism or acclaim. On the Continent the overwhelming interest – with the solitary exception of France – was still in restoring normal life following the years of German occupation. There was a small, if growing, hope that the peaceful uses of nuclear fission might one day justify these extraordinary tamperings with nature; but even after the bombing of Japan, and the first instinctive horror at the news, mankind's obvious wish was to turn quietly over and to go to sleep once again.

The complacency ended in September 1949. The Russians, it was announced in Washington, had exploded their first nuclear weapon.

Surprise was so great that at first the announcement was not believed, particularly by some Service leaders in the United States, who had firmly refused to take Russian capabilities seriously. Yet as early as November 1947, there had been a hint of things to come. Speaking in the Bolshoi Theatre on the thirtieth anniversary of the Revolution, Molotov had commented: 'In the expansionist circles of the U.S.A. a new, peculiar sort of illusion is widespread – faith is placed in the secret of the atomic bomb, although this secret has long ceased to exist.' To those who noticed the words at all this was merely another example of Russian bluster or, at the most, a reference to what many scientists had always considered the main secret of the bomb: the secret of whether it would work.

A few months before Molotov's statement, General Eisenhower, then Chief of Staff, had given General Carl Spaatz the responsibility for 'detecting atomic explosions anywhere in the world'. The scheme was quickly in operation and from then onwards the frontiers of Russia were constantly patrolled by planes carrying contemporary equivalents of the Geiger counters used in the early days of nuclear research.

In September 1949 one of the aircraft sucked in a radioactive air sample. The cloud containing it was tracked from the North Pacific to Europe and its presence confirmed there by planes of the Royal Air Force. The results were checked and analysed, and on 23 September President Truman announced to his astounded countrymen that the Russians had, contrary to the expectations of all but a few awkward physicists, ended the American nuclear monopoly.

Surprise at the Russian success is at first a little difficult to understand, even granting the fact that Allied intelligence must have found Russia one of the more difficult countries from which to obtain information. The most likely explanation is that even in 1949 the West still tended to judge Russian progress by what had been learned of Kurchatov's efforts between the summer of 1942 and the end of the war in August 1945. This must have looked almost pathetically inadequate when compared with the Manhattan Project; and it was not realized that a change of *tempo* had taken place after Hiroshima.

Although Stalin's alleged remark at Potsdam – 'We'll have to have a talk with Kurchatov and get him to speed things up' – need not be taken too literally, it is now known that the Japanese bombings, plus Truman's reticence, had had their effect. One Russian account has described how the People's Commissar of Munitions, Boris L'vovich Vannikov and his deputies, were summoned to the Kremlin in mid-August, shortly after Stalin's return from Potsdam. They were apparently taken in to him by Kurchatov. 'A single demand of you, comrades,' said Stalin. 'Provide us with atomic weapons in the shortest possible time. You know that Hiroshima has shaken the whole world.

The equilibrium has been destroyed. Provide the bomb – it will remove a great danger from us.'

Kurchatov's biographer has described how he and the rest rose to the occasion. If the prose is in the best Party tradition, there is no reason to doubt the general line of the story it tells: the chain reaction – the first in Europe – on 26 December 1946, the first plutonium from the piles the following year; and, on 29 August 1949, the successful test of the bomb in Central Asia. The path followed was virtually identical to that of the Manhattan Project, which was hardly surprising; it was industrially the quickest route to the bomb, there had been at least helpful hints and probably much more from Fuchs, and there was, of course, the mass of useful information in the earlier, uncensored version of the Smythe Report to which the Russians had access.

First reaction to the Russian success was in some cases an embarrassed attempt to eat the words uttered only a year or so previously. General Eisenhower, by this time president of Columbia University, commented that he could see no reason why a development that was anticipated years ago should cause any revolutionary change in America's thinking or in her actions, a view which ignored the fact that the Second Coming in wrath had been repeated much earlier than forecast. The same platitudinous avoidance of the implications came from General Groves who shortly after the end of the war had told a Senate Committee that without help from another country Russia would need fifteen or twenty years to make a bomb – and five to seven years even with such help. The Joint (Intelligence) Committee, reporting on 13 October, was rather more frank, referring to 'Russia's ownership of the bomb, years ahead of the anticipated date'.

Only the scientists had been less optimistic. Irving Langmuir had forecast three years from 1946; Hans Bethe, three to six years. Their colleagues forecast a similar time-scale and James Franck, commenting on Truman's announcement, noted: 'There has probably been no other occasion when we have been so unhappy that our prediction turned out to be right.'

From some, the reaction was not unhappiness but belligerence. David Lilienthal, who as chairman of the Atomic Energy Commission was deeply involved in the repercussions of the Russian explosion, records spending two hours with Senator McMahon. 'Pretty discouraging,' he noted in his diary. 'What he is talking about is the inevitability of war with the Russians, and what he says adds up to one thing: blow them off the face of the earth, quick, before they do the same to us – and we haven't much time. He uses all sorts of words to justify this, and part of the time he is practising speeches on the floor of the Senate at us.'

There was also in some quarters a reaction which, if not exactly panic, yet showed more than worry. Thus on the main highway leading west from Cambridge, Massachusetts, there was erected an enormous sign: 'In case of enemy attack,' it read, 'this road will be closed to all but military vehicles.' Faith in the deterrent had evaporated overnight.

In Britain, where the Government had secretly appropriated large sums that it could ill afford for a weapon whose use in war had not even been considered, there had been as little expectation as in America that the Russians would become a nuclear power so quickly. The disillusion here was quite as great, and it was left to Lord Boyd-Orr, that ever-optimistic bastion of the U.N. Food and Agriculture Organization, to see some consolation from the development: 'The fact that both sides are armed with new weapons invented in the last few years will help prevent war,' he noted in the world's first suggestion of mutual nuclear deterrence.

Yet here, if anywhere, was the place on the slippery slope where men of any belief or of none might have been expected to dig in their heels and halt the downward slide. National annihilation in its contemporary meaning was for the United States not yet a feasible prospect; the horrors of fall-out were even less understood than they are today, and if they did form a cloud on the horizon it was one to which only alarmists drew attention. Yet despite the comparatively modest claims of nuclear weapons these did now enable either of the world's two superpowers to beat the other into insensibility with a few swift blows. This might, in a rational world, have been considered the time to catch an opportunity and halt the quickening arms race. Once again, the catch was missed.

It was not to be the case. After a brief hesitation, America decided to restore, for what common sense suggested could be only a short period, her superiority of annihilation. The means was to be 'the Super', the hydrogen bomb which had been discussed, almost behind raised hands, since serious work on nuclear weapons had got under way early in the war. The manner in which the United States decided to go ahead with what in the old-fashioned world of the early 1950s was called the ultimate deterrent is one of the most extraordinary in the story of nuclear fission. Yet it should be prefaced by an account which shows that the hydrogen bomb had a background far longer than most people were allowed to appreciate in the autumn of 1949.

For long it was popularly believed that the Super was a weapon thought up solely as a response to the Russians' first nuclear explosion. This is not so. The decision to transform a theoretical idea into lethal hardware was certainly a result of the news that America's monopoly no longer existed. Yet the idea had a long history and had been growing for some considerable while.

Ever since Aston's calculations of the early 1920s it had been realized that the fusion of light atoms would release an enormous amount of energy. The theory had been given experimental muscle as far back as 1934 when a paper by Rutherford, Oliphant and Harteck reported experiments of processes developed for 'the ultimate weapon' two decades later. The main problem in releasing the enormous amount of energy involved was that the constituent atoms would have to be raised to a temperature of many million degrees. During the first months of the war the Germans had speculated that there might be a way of creating such temperatures by using conventional explosives, and some experiments were actually made at the German Army's Kummersdorf research establishment. They failed to produce any useful results and it was scientists in Britain and the United States who had the idea of using a fission bomb to create a temperature sufficiently high to set off the fusion process.

Egon Bretscher, the Swiss who in 1940 had been drawn into the work of the M.A.U.D. Committee, considered the possibilities while working in Cambridge with Halban and Kowarski. When he moved with other members of the British team to the United States in 1942 he found discussion of the hydrogen bomb already in full swing. This had started earlier in the year in more than one U.S. university and Edward Teller has recorded how, when walking back to the laboratory one day after lunch with Fermi at Columbia, Fermi had said: 'Now that we have a good prospect of developing an atomic bomb, couldn't such an explosion be used to start something similar to the reactions in the sun?' Teller soon produced preliminary figures and was convinced that a fusion explosion could be produced. As he later wrote, 'We even thought we knew precisely how to do it.' And when Teller joined Oppenheimer's theoretical group at Berkeley in the summer, he found that the fusion weapon was constantly being discussed.

It was realized by this time that the best chance of success lay in using one of the two rarer isotopes of hydrogen as the raw materials in the fusion process. One was deuterium, with one proton and one neutron in its nucleus. The other was tritium, a radioactive isotope, with one proton and two neutrons, which can be made in a nuclear reactor. It was always known that manufacture of a fusion weapon would be an immensely costly business. Yet it would have two advantages over the fission bomb. One was that, gram for gram, it produced three times more energy. The other was that while the fission bomb was limited in size by the critical mass of the fissionable material used, there appeared to be no theoretical limit to the size of the fusion weapon.

The practicability of a bomb using one of the heavy isotopes of hydrogen rose and fell, as Teller later put it, almost daily, with the prospects of success changing as fresh barriers arose and were then removed.

> A spirit of spontaneity, adventure and surprise prevailed during those weeks in Berkeley, and each member of the group helped move the discussions toward a positive conclusion. The contributions of Konopinski and Bethe were especially remarkable. Konopinski suggested that, in addition to deuterium, we should investigate the reactions of the heaviest form of hydrogen, tritium. At the time, he was only making a conversational guess. It turned out to be an inspired guess. ... We were all convinced, by summer's end, that we could accomplish a thermonuclear explosion – and that it would not be too difficult. Oppenheimer was as interested by the prospect as any of us: He concluded: 'Now we really need another laboratory.'

Oppenheimer was in fact so taken with the possibility that he made a special journey to Arthur Holly Compton's summer retreat in Michigan to discuss the matter. He also arranged for basic nuclear studies of the very light elements to be made at Harvard and Minnesota, while a meeting to plan research on the subject was held in Chicago in September. Moreover when another laboratory was set up, that at Los Alamos, thermonuclear research was given considerable attention. One of the first buildings completed was designed for handling thermonuclear materials and one of the first tasks carried out was measurement of the properties of tritium. Several scientists, Teller was later to claim, were recruited to Los Alamos 'only because they were intrigued by the thermonuclear possibilities.' There was, moreover, no chance of turning a blind eye to the possibilities among the small circle working on the bomb. 'We resolved to keep the possibility of a fusion bomb a close secret,' Compton has written. 'But time and again throughout the war period men on the project would think of the idea independently and bring it to our attention. In certain circles at Los Alamos, including those frequented by Klaus Fuchs, the super bomb was a matter of occasional conversation.' But it was also something more. When Fermi moved from Chicago to Los Alamos in September 1944, four groups were organized under his general leadership. One, under Teller, studied the theoretical problems of the hydrogen bomb while another, under Egon Bretscher, was its experimental counterpart.

Discussion of 'the Super' thus ran parallel to the development of the fission bomb throughout the mid-1940s. 'I can still remember my shock and incredulity when I first heard about it,' wrote William Laurence, the *New York Times* correspondent. 'It seemed so fantastic to talk of a super atomic bomb even before the uranium, or the plutonium, bomb had been completed and tested – in fact even before anybody knew that it would work at all – that I was inclined at first to disbelieve it.'

Shortly afterwards he asked Hans Bethe if it could be as much as fifty times as powerful as the bomb on whose completion all efforts were now being concentrated. 'I shall never forget the impact on me of his quiet answer as he looked away toward the Sangre de Cristo (Blood of Christ) mountain range, their peaks blood-red in the New Mexico twilight. "Yes," he said, "it could be made to equal a million tons of T.N.T." Then, after a pause, "Even more than a million." '

Laurence was later shown by Oppenheimer a vial of clear liquid that looked like water. 'It was,' he has written, 'the first highly diluted minute sample of superheavy water, composed of tritium and oxygen, ever to exist in the world, or anywhere in the universe, for that matter. We both looked at it in silent, rapt admiration. Though we did not speak, each of us knew what the other was thinking. Here was something, our thoughts ran, that existed on earth in gaseous form some two billion years ago, long before there were any waters or any forms of life. Here was something with the power to return the earth to its lifeless state of two billion years ago.'

So much was 'the Super' a matter for general discussion that when Oppenheimer was, in late July 1945, discussing with Groves a final point about the combat use of bombs, he asked how Groves felt about the Super. '[He] was unclear whether his mandate and therefore mine extended to fiddling with this next project. I so reported to the people in the laboratory, who were thinking about it.'

This was still the situation when the bombs were dropped on Hiroshima and Nagasaki: the hydrogen bomb, a subject for extracurricular thought, was a hardly realistic possibility which should be kept under mental hatches until the war was won. But Groves considered its eventual development to be, as he put it, 'inevitable'.

The scientists' attitude changed quickly in the aftermath of Hiroshima and Nagasaki, and before the end of August 1945, Byrnes was being told that, according to Oppenheimer, 'the scientists prefer not to do that [super bomb] unless ordered or directed to do so by the Government on the grounds of national policy'. However, George L. Harrison, Stimson's link man with the Manhattan Project, thought that work could go on 'in improving present techniques without raising the question of "the Super" at least until after Congress has acted on our proposed Bill'.

Oppenheimer underlined his position early in 1945 when he was approached by Teller who had been asked to remain at Los Alamos but was unhappy that the work there was being run down. 'This has been your laboratory and its future depends upon you,' he said to Oppenheimer, who had recently left. 'I will stay if you will tell me that you will use your influence to help me accomplish either of my goals, if you will help enlist support for work toward a hydrogen bomb or further development of the atomic bomb.' According to Teller, Oppenheimer quickly replied: 'I neither can nor will do so.'

Ostensibly, the scientists' reaction was that a weapon of the Super's capability would be used primarily against the huge civilian targets of the great cities. Not everyone viewed it in this light. Lewis Strauss maintained to the President that the H-bomb could be used against a mobilized army. He pointed out that it could have destroyed the Allied fleets off Normandy in 1944, the Allied armies in the early days of the Normandy bridgehead, or the German forces in the Battle of the Bulge.

While all discussion of the Super remained secret in the United States, its possibilities were revealed in 1946 when Hans Thirring published *Die Geschichte der Atombombe* in Vienna, a book which used data to estimate the energy that would be released if the high temperature of a uranium bomb was used to trigger off a hydrogen-helium or lithium-helium reaction. Since lithium was a common enough element, Thirring pointed out, bombs with a much larger amount of fissile material would be more economical than those used over Japan. A weapon containing six tons of lithium appeared to be a practical possibility and might give as much as a thousand times the energy of a simple uranium bomb. There were complications, of activating mechanism and of weight, but to the few who troubled to read the treatise the implications for the future were obvious. Yet, as pointed out in the *Scientific American* when the bomb later became the subject of discussion, 'reviews of the book published in the U.S. deliberately refrained from mentioning that chapter or anything it contained'.

Thus the subject had not been publicly discussed – in the U.S.A., that is – when the members of the Atomic Energy Commission met on 5 October 1949, to debate the implications of the Russian explosion. Strauss had no doubts, as he made clear to the other members of the Commission. 'It seems to me,' he said, 'that the time has now come for a quantum jump in our planning (to borrow a metaphor from our scientist friends) – that is to say that we should now make an intensive effort to get ahead with the Super. By intensive effort, I am thinking of a commitment in talent and money comparable, if necessary, to that which produced the first atomic weapons. That is the way to stay ahead.' Other members were less confident, and the subject was referred to the Commission's main scientific body, the General Advisory Committee.

The opinion of most Committee members was typified by a letter which Oppenheimer wrote to Conant.

> What concerns me is really not the technical problem. I am not sure the miserable thing will work, nor that it can be gotten to a target except by ox cart. It seems likely to me even further to worsen the unbalance of our present war plans. What does worry me is that this thing appears to have caught the imagination, both of the congressional and of the military people, as the answer to the problem posed by the Russian advance. It would be folly to oppose the exploration of this weapon. We have always known it had to be done; and it does have to be done, though it appears to be singularly proof against any form of experimental approach. But that we become committed to it as the way to save the country and the peace appears to me full of dangers.

A number of very practical points supported a policy of caution. For one thing, the A.E.C. was already near to producing a hugely improved fission weapon which would yield the equivalent of 500,000 tons of T.N.T. and which would, it was confidently added, meet any conceivable military requirements of the United States. Such weapons, furthermore, would inevitably be affected by any crash programmes to produce the Super. The hydrogen bomb needed large amounts of tritium and the most feasible method of making this in quantity was to bombard lithium with neutrons; but neutrons used in the production of tritium could not be used to produce plutonium, so that fission and fusion weapons were almost inextricably linked – although with some confusion since at this stage no one knew how much tritium the Super would require.

The feasibility of the hydrogen bomb was, moreover, still unknown. It was at first considered that the tritium and deuterium involved would have to be kept in their liquefied state and at high pressure, thus involving heavy refrigerating equipment. In addition, one other point had to be considered – the position of the United States if and when the Russians were encouraged by American development to go ahead with their own hydrogen bombs. The American population was concentrated, to a far greater extent than the Russian, in a small number of exceptionally large cities and urban complexes. This made the country comparatively more vulnerable than Russia to the threat from weapons of mass destruction. Finally there was the ethical argument about developing weapons whose prime use would almost certainly be the killing of civilians; but that argument appeared to have been settled over Japan in August 1945.

The General Advisory Committee reported to the Commission on 30 October. Its recommendations, signed by Conant, Hartley Rowe, Cyril Stanley Smith, L. A. DuBridge, Oliver E. Buckley and Oppenheimer, were unqualified. 'We recommend strongly against' an all-out effort to develop the Super, they advised.

> We base our recommendations on our belief that the extreme dangers to mankind inherent in the proposal wholly outweigh any military advantage that could come from this development. Let it be clearly realized that this is a super weapon; it is in a totally different category from an atomic bomb. The reason for developing such super bombs would be to have the capacity to devastate a vast area with a single bomb. Its use would involve a decision to slaughter a vast number of civilians. We are alarmed as to the possible global effects of the radioactivity generated by the explosion of a few super bombs of conceivable magnitude. If super bombs will work at all, there is no inherent limit in the destructive power that may be attained with them. Therefore a super bomb might become a weapon of genocide.

While it might appear from such words that the moral issues had dominated the discussions, this was far from the case. In the report itself, signed by Oppenheimer, it was pointed out that a crash programme for the Super would involve a huge expenditure on something that might not work. Bearing in mind that plans were well advanced for the fission bomb equivalent to 500,000 tons of T.N.T., it appeared 'uncertain to us whether the super will be cheaper or more expensive than the fission bomb'. Finally, the signatories to the recommendation added: 'To the argument that the Russians may succeed in developing this weapon, we would reply that our undertaking it will not prove a deterrent to them. Should they use the weapon against us, reprisals by our large stock of atomic bombs would be comparatively effective to the use of a super.' Even if the ethics of killing civilians *à outrance* were ignored, both economic and military considerations seemed to weigh down the scales against the Super.

The Commission reported to Truman on 9 November and the following day he referred the issue to the National Security Council, whose Special Committee reported back to him at the end of January 1950. On the 31st, Truman made the great decision – or, more accurately, made a decision and left the great one to the future. His verdict was 'that an effort be made to determine the technical feasibility of a thermonuclear weapon; that the rate and scale of the effort be fixed jointly by the A.E.C. and the Department of Defense; and that the State and Defense Departments undertake concurrently to review the nation's foreign and military policies in light of the prospective nuclear capabilities of the Soviet Union.'

This was very different from the recommendation of the advisers appointed to advise. But it was still far less than the military had hoped for. Within a few days they were to be given unexpected support. On 2 February, Klaus Fuchs, the British-naturalized German who had been privy to many of the secrets of Los Alamos, was publicly revealed to have been a Russian spy, and for the second time since the end of the war the discovery of Russian espionage appears to have changed the course of history. Any chance of East-West compromise on the Baruch Plan had been greatly lessened by the uncovering of the Canadian ring in 1946. Now, with the news that Fuchs, well-informed as he was on the possibilities of the Super, had been passing information to the Russians for years, the U.S. Joint Chiefs of Staff asked for 'an all-out development of hydrogen bombs and means for their production and delivery'. Only to the British did Fuchs bring a chance of benefit. He had believed in the greatest dissemination of scientific knowledge, and it has never been denied that the British learned from him, before his conviction, more about the hydrogen bomb than they were ever to learn from the Americans.

On 1 March 1950 Truman called the Special Committee of the National Security Council together again. This time it recommended that research on the hydrogen bomb should be carried through to the test stage as quickly as possible; and as a further indication of urgency, preparations for quantity production of the necessary material should be started at once, even before the bomb was proved to be practical. Without delay, plans were made for the huge Savannah River plant for the production of tritium; cost for this alone would be $1·5 billion – roughly three-quarters that of the entire Manhattan Project.

The G.A.C. was 'right' in sensing that the development of the H-bomb would drive twentieth-century man deeper into the box that he [had] been building for himself with his military technology, doctrine, foreign policy, and cultural ethos. The G.A.C. was also 'right' in asserting that it was a time to stop, look and think. But the G.A.C. was not alone in seeing the dimensions of the box. It was every bit as apparent to most of the advocates of the Super program. The trouble was that no one had any good ideas of how to get out of the box. Nor are they apparent today.

Basically, the H-bomb decision is a story of international rather than domestic politics. It affords a classic example of the traditional security dilemma. Both the Soviet Union and the United States would no doubt have preferred a world in which neither had the H-bomb. Each, however, wished to avoid a world in which the other had the H-bomb and it did not. Both rushed to make it, and they ended in a worse position than that in which they had begun.

There is no reason to doubt that nuclear physicists in Russia had for long been as aware of the Super's potentialities as their counterparts elsewhere in the world. Andrei Sakharov, often described as 'the father of the Russian Super', has reported that in 1948 he was 'included in a research group working on the problem of a thermonuclear weapon'. It seems likely that until the autumn of 1949 his work was roughly equivalent to that being done at Los Alamos, a theoretical investigation of a weapon which no one had yet thought about building. But following the American decision on the Super, that position changed. 'In 1950,' Sakharov has written, 'our research group became part of a special institute. For the next eighteen years I found myself caught up in the rotation of a special world of military designers and inventors, special institutes, committees and learned councils, pilot plants and proving grounds. Every day I saw the huge material, intellectual, and nervous resources of thousands of people being poured into creating the means of total destruction, a force potentially capable of annihilating all human civilization.'

At the end of the war the Americans had the advantage in men, materials, money and expertise. Yet a careful chronology of events suggests that even by 1949 the Russians had narrowed the nuclear gap which had separated them from the U.S.A. in 1945; within the next few years it was to be still further reduced.

The first aim of the American programme was to show that a thermonuclear reaction could be set off by a fission bomb. What was required, Teller has written, was 'a significant test. Without such a test no one of us could have had the confidence to proceed further along speculations, inventions, and the difficult choice of the most promising possibility. This test was to play the role of a pilot plant in our development.' Code-named the George shot, it was successfully fired at Eniwetok in the Pacific on 8 May 1951. 'The mushroom cloud we saw rising beyond the lagoon,' Teller later wrote, 'showed only that we had been successful in asking a question. The answer had to come from the reports of the recording instruments.' The answer was encouraging. A very large fission explosion had set off an almost nominal amount of fusionable material.

But this was only a half-way house to the Super. Something very different was needed. It had been provided a few months earlier by Edward Teller and Stanislaw Ulam. This solution, 'the Teller–Ulam configuration', revolutionized the concept of the Super, making it possible for a theoretically limitless amount of fusionable material to be triggered off by a 'conventional' fission explosion. So much so that Oppenheimer was later to compare the pre- and post-Teller–Ulam situation in these words: 'The program we had in 1949 was a tortured thing that you could well argue did not make a great deal of technical sense. It was therefore possible to argue also that you did not want it even if you could have it. The program in 1951 was technically so sweet that you could not argue about that. The issues became purely the military, the political and the human problem(s) of what you were going to do about it once you had it.'

The 'technically so sweet' idea was apparently described by the Russian physicist L. I. Rudakov a quarter of a century later when he lectured in America to scientists in the summer of 1976. Although the meetings were attended by scores of physicists, and Rudakov's remarks were reported, the Americans still do their best to prevent the details from being known to Americans. Nor is it only they who adopt this attitude, presumably in the interests of non-proliferation. 'I agree,' the American physicist, Herbert York, writes of the Teller–Ulam configuration in his penetrating book *The Advisors: Oppenheimer, Teller and the Superbomb*, 'with the security officials of the five thermonuclear powers (U.S.A., U.S.S.R., U.K., France, and China) that this particular detail should remain as restricted as possible in its circulation.'

The Teller–Ulam configuration was tested in the 'Mike shot' of 1 November 1952, an outstanding success which produced an explosion equal to some 10 million tons of T.N.T., roughly the figure that had been predicted. However, it would be misleading to call this a test of the Super. That word took it for granted that the bomb would be deliverable. But the Mike shot's thermonuclear fuel consisted of liquid deuterium, which had to be cooled to a temperature below minus 250°C. Complicated refrigerating equipment was required, and the 'bomb' weighed some 65 tons.

There was, however, to be a compensation for the fact that the Mike shot could hardly be called a test of the Super. Two weeks later there came a test of the greatly boosted fission bomb that had been on the drawing-boards for some while. The yield, as planned, was 500,000 tons, a figure higher than the Russians were to produce for another three years.

Yet Kurchatov's groups were soon to have a success of their own. 'The U.S.A. has long since lost the monopoly in the matter of the production of atomic bombs,' said Georgy Malenkov, Chairman of the Council of Ministers, addressing the Supreme Soviet on 8 August 1953. 'The Government deems it necessary to report to the Supreme Soviet that the U.S. has no monopoly in the production of the hydrogen bomb either.' Many commentators in the West, having failed to learn the lesson of 1949, were sceptical. They were less so four days later when American patrol planes sucked in radioactive dust showing that the Russians had, in fact, exploded a thermonuclear device of their own.

The yield of the Russians' 12 August 1953 explosion has never been revealed but it was almost certainly less than that of the Americans' 500,000 ton fission bomb and thus far below that of the 10 megaton Mike shot. However, the Russians justify their claim that this was the world's first test of a genuine 'Super' on the grounds that their device, exploded on a tower-top like the Trinity explosion of 1945, was small enough to be carried in a plane. It appears to be true that it used as a thermonuclear fuel not the inconvenient liquid deuterium but the convenient lithium deuteride which demanded no refrigerating equipment. But the small yield suggested that the Russians had not yet discovered any equivalent to the Teller–Ulam configuration.

Nevertheless, it would be difficult to deny that both sides were now, in terms of potential annihilation, running neck and neck. As so often, Cherwell was one of the first to appreciate the position. 'People,' he wrote to James Tuck, 'do not seem yet to realize that the whole situation has changed now that the Russians appear to know pretty well as much as anyone else if not more.'

With the Mike shot ruled out by its weight and the Russians' August explosion ruled out by a yield less than that of the Americans' boosted-fission process, the way was now clear for an indisputable test of the Super. It came on 1 March 1954, the first of six tests spread over ten weeks at Bikini atoll. Code-named 'Bravo', it was to have repercussions far greater than the Americans expected. The yield appears to have been forecast as 8 million tons but turned out to be almost 15 million tons; the area over which it extinguished all life, some 200 square miles. If the competition was really that of proving that 'anything you can do I can do better', then the Americans had won a notable, if temporary, victory. But it was not only the size of the figures which was to give the first Bikini test its place in history.

As Lewis Strauss, chairman of the Atomic Energy Commission, was to explain before the month was out: 'Unfortunately the wind had failed to follow the prediction and had shifted southward so that the islands of Rongelap, Rongerik and Uterik had been in the path of the fall-out.' Luckily the natives of these islands lived in areas which received only the tail end of the fall-out. But the Atomic Energy Authority announced on 11 March that twenty-eight Americans and 236 natives had been unexpectedly exposed to some radiation and had been moved to Kwajalein. As Teller was later to admit, 'The situation was serious and we narrowly escaped some dreadful consequences ... On the atoll's northern tip only thirty miles from the native villages, crews measured the radioactive fall-out at 1,000 röntgens. Such a massive dose would have meant certain death in less than one month.'

But in the path of the fall-out there was also the small Japanese fishing-boat *Lucky Dragon*. Although 100 miles from the scene of the explosion, the ship was soon afterwards covered by a grey-white ash which fell from a cloudless sky. None of the twenty-three members of the crew knew its significance; even when one, then another, and finally all, fell ill with nausea, vomiting and diarrhoea, the cause was not evident. Only when the *Lucky Dragon* finally docked in Japan were doctors able to discover what had happened. Radioactive by-products of the H-bomb explosion had rained down on the ship, incapacitating all of the men, one of whom died, and revealing, for the first time, the significance of the warning given by Frisch and Peierls fourteen years earlier – that even a fission explosion would produce highly radioactive substances which 'would be fatal to living beings even a long time after the explosion'.

The danger zone was now trebled in size – making it as large as the British Isles plus Holland, Belgium, half of France and a considerable part of Germany. Special drugs for the treatment of the Japanese victims were hurriedly flown out from California; but the world's first hydrogen-bomb victim, Aikichi Kuboyama, died in Tokyo six months later.

It was found that fish in the fall-out area were radioactive; moreover, smaller fish found in the stomachs of bigger fish, and the plankton found in the stomachs of the small fish, were as radioactive as the big fish, between 1,000 and 50,000 times more so than the seawater in which they had been swimming. Thus it appeared that the radioactivity had been concentrated in the larger fish from the water they took in through their gills and from the plankton and smaller fish which they ate. Later in the year, after the Russians had tested a number of fission bombs, it was found in Japan – within the fall-out zone from Russia – that rain, vegetables, drinking water and even the dust on roofs and in houses had all become faintly radioactive.

But it was the fate of the *Lucky Dragon*, produced by a totally unexpected quirk of circumstances, which had set the alarm bells ringing, handed a formidable weapon to those calling for a nuclear moratorium, and justified those who maintained that neither scientists nor Servicemen really knew enough about the dangers of the weapons they were playing with.

At Westminster Lord Salisbury, a peer not notable for making extravagant statements, commented of the Super: 'We must, I am afraid, accept that fearful weapon not as a nightmare of the future but as a hideous reality of today.' And *The Times*, equally unnoted for wild statements, commented in a leading article on 26 March: 'Mankind, if it is to survive, can never give up trying. Some day the decisive date for good or ill will arrive. It is at least worth seeing whether that date was not 1 March.'

CHAPTER 15

A Small Cloud on the Horizon

The wind which showered radioactive ash on the deck of the *Lucky Dragon* marked the start of a new chapter in the story of nuclear fission. After Bikini, it became more difficult to explain away the dangers of radioactive fall-out. Suspicions that these had been, if not concealed, at least played down by all the authorities concerned, tended to increase. The situation was hardly simplified as it slowly became evident that neither the scientists and Governments on one hand nor those who became known as nuclear protesters on the other, had indisputable quantitative evidence of exactly how great the dangers of radioactivity really were. One result was that both sides tended to pitch their claims, mollifying or alarmist as they might be, rather higher than credibility demanded. However, the dangers of fall-out provided much material for the nuclear disarmament lobby which gained power for a decade or more after the mid-1950s.

A hint of the dangers revealed to an unsuspecting public by the Bikini H-bomb and God's unfortunate changing of the wind, had been given during the early days of the nuclear age. The perils of X-rays had become apparent soon after Röntgen's discovery at the end of the nineteenth century, and the first uses of radium that followed showed natural radioactivity to be a dangerous beast unless treated with proper care and attention. The physicists who successively built up a more detailed picture of the atom knew of the perils to be avoided and the Frisch–Peierls memorandum gave due warning of a 'cloud of radioactive material [which] will kill everybody within a strip estimated to be several miles long'.

Yet the problem of fall-out seems to have been given scant attention in disputes about the bomb and its use; and, as General Groves was to admit, his first knowledge about this complicating factor came only a few months before the Alamogordo test. The most obvious explanation of the comparative failure to face awkward facts is psychological: fall-out, to those who had an inkling of the truth, was a terror many of them preferred not to acknowledge. In addition, it was natural enough for both scientists and Servicemen to concentrate on the effects of nuclear weapons which they already understood. Heat and blast from a nuclear explosion would be so immense as to be the factors which really mattered. Radioactivity might well make the target area uninhabitable for a while, but this would surely be a small matter compared with the immense destruction which seemed to be so happily assured?

From the first there was great argument about how the data concerning radioactivity should be interpreted. There were genuine reasons for disagreement and while those who supported maximum rearmament in the West pointed to gaps in the evidence, those more favourable to détente claimed that these gaps made little difference to the main argument. A revealing example came after the Nobel Prize-winner, Linus Pauling, had broadcast on radioactivity's genetic dangers. Lord Cherwell, equally honest, equally sure of himself, protested to Sir Ian Jacob, Director-General of the B.B.C., that it was 'really monstrous to allow so many decent, honest people to be worried by these reckless scaremongers'. The argument 'that the British tests constitute a danger to the health of humanity', he later maintained 'is unmitigated nonsense. These tests could not in any circumstances cause any significant increase in the number of still births, or in leukaemia or cancer.'

Other men were not quite so sure. After spring and summer bomb tests on the American continent in 1955, radioactive iodine 131 was found in the thyroids of cattle and sheep between 1,200 and 1,500 miles from the explosions. 'On this evidence,' commented *The Lancet* on 14 July 1956, 'an appreciable lifetime radiation dose to the thyroid beginning early in embryonic life is to be expected if test explosions continue at their present rate. The ultimate effects of this type of irradiation in man are unknown.'

Some facts are beyond dispute. They are hardly comforting. Nuclear explosions can produce up to fifty different kinds of highly radioactive by-products. The period during which they remain dangerously radioactive varies considerably: thus, xenon 133 has a half-life of 5·3 days – in other words, its radioactivity decreases by half during the first 5·3 days after its creation, the remaining radioactivity again decreases by half within the next 5·3 days, and so on; but the half-life of technetium 99 is no less than 1 million years. Radiation can be corpuscular, electromagnetic, or a mixture of both. Corpuscular radiation can consist of protons, alpha particles, electrons or neutrons, while the electromagnetic can include gamma rays and other high-energy radiations which move at the speed of light.

While the harm done by radiation depends on many factors, including the type of radiation, its energy, and numerous physiological factors – a few millionths of a curie may be harmful in some circumstances but several curies almost harmless in others – both corpuscular and electromagnetic radiation have one thing in common: they produce ionization in the material through which they pass. In other words they break up into two parts the atoms forming the material, one part carrying a negative charge and the other a positive charge.

Once this fact is appreciated, the biological danger of radiation becomes clear. For the material affected includes human cells, the cells that normally replace themselves, as necessary and to a definite and pre-ordained pattern. That pattern is upset by radiation, and in a number of complicated ways that are not yet fully understood; but one change which can occur is that cells cease replacing themselves in the normal way and instead begin to multiply, a process known as the growth of a cancer. The extent of the cancer danger from radioactivity depends very largely on whether or not there is a threshold below which the danger is non-existent. One view is that below a certain 'intake' no danger exists; the other maintains that there is no 'safe' level and that a decrease in the intake merely reduces the numbers that will be affected.

The situation has rarely been described more clearly than by three members of the Atomic Scientists Association – Doctors Rotblat, Haddow, and Penrose – in discussing very small doses of strontium 90, a radioactive element which is easily incorporated in human bone due to its affinity to calcium, and which has a half-life of twenty-eight years.

There is here a fundamental difficulty in that the relationship between the damage produced and the amount of radiation is not known. If this relationship is such that there exists a threshold dose below which cancer cannot be induced, then it can reasonably be inferred that the small amount of strontium 90 which will accumulate in bone from current H-bomb tests would not result in any harm. If, however, the number of additional bone tumours resulting from radiation is directly proportional to the dose, then even a very small dose will give rise to a small but definite probability of bone cancer. This means that in a very large population a certain number of people would contract this disease as a result of having a small amount of strontium 90 in their bones.

The wisdom, or otherwise, of being dogmatic about the subject was suggested as late as 1978 by Sir John Hill, the chairman of the United Kingdom Atomic Energy Authority. 'Sir John,' states an official account of his Sylvanus Thompson Memorial Lecture that year, 'noted that radiation protection measures were based on the assumption that damage due to radiation was directly proportional to dose. So far, however, it had not been possible to establish whether or not this assumption was accurate, because the possible effects of low doses were completely swamped by statistical variations in the population and other facts which could cause or prevent cancer. We might nevertheless expect that low doses would be less damaging than a linear extrapolation would predict.'

It is true that the human race has always been subject to low doses from background radioactivity. Some comes in the bombardment of the earth by cosmic rays, the high-energy radiation from outer space whose strength varies with the altitude at which it is received; some consists of the natural radioactivity of the earth, which varies greatly in different parts of the world. And during the last few decades this 'background' has been augmented by the use of X-rays for medical and other purposes. All this has led to the supporters of nuclear weapons decrying the dangers of the last quarter-century, an attitude which has been described as 'like many arguments used in discussions about nuclear fission … a collection of half-truths held together by fallacies'.

Whatever the statistical chances, the initiation of cancer by radiation has been common knowledge since the early years of the century, even though minimal consideration appears to have been given to it in the natural excitement of making the world's first nuclear weapons. Yet another side effect was even more disregarded until after the bombs had been dropped on Japan: that of radiation on unborn generations.

It has been known for more than half a century that the characteristics of living things – human and the rest – are passed on from one generation to the next by means of particulate factors carried in certain cells. At times the chemical constitution of one or more of these factors is, for reasons that are not at all clear, changed by a process known as mutation. Such mutations have been compared to typing errors, which may make sense but usually fail to do so; in practice almost all mutations represent changes for the worse. Now as far back as 1928 it was known that ionizing radiations could produce mutations in insects and plants. Yet the possible genetic effects of nuclear explosions on human beings appear to have been almost entirely overlooked, although they were considered when it was believed that the Germans might release radioactive fission products on the Normandy invasion beaches. A member of Britain's Medical Research Council studied the possibility and pointed out that mutations might be produced even by cumulative weak doses of radioactivity, that such mutations 'are nearly always harmful ones' and that they 'may be expected to show up probably as a weakening of the race in the second and third generations'.

However, the subject still remained one of academic dispute. Only in 1946 or early 1947 did Dr Stafford Warren, formerly the Chief Medical Officer of the Manhattan Project, discussing nuclear test explosions, suggest to the American authorities that 'sufficient quantities of radioactive material might be widely distributed by such explosions throughout the atmosphere and later accumulated in proteins used as food by man to cause detrimental mutations in large masses of people within a limited number of generations.' Dr Conant and others consulted were sceptical.

More than thirty years later, the few incontrovertible figures available are still being used by both sides to support their individual attitudes, a reminder of Mark Twain's comment that there are three kinds of lies – 'lies, damned lies and statistics'. Yet there are a few guidelines through the swamp of uncertainty. One is provided by changes in the Maximum Permissible Dose of radiation which have taken place in the last half-century as more has been learned about the damage which it can do. In 1931 the dose was 1·4 röntgens per week. Five years later it was reduced to 0·7 röntgens and in 1951 it was reduced yet further. If it is true that radiation is, like alcohol or sugar, harmless below a certain level of intake, there is as yet no indication that that level has been discovered. It still looks as though, as the 'scaremongers' maintain, every dose may count. Even if this is not so, uncertainty remains.

Discussion of radioactivity and its release into the air by nuclear-weapon tests continued as during the decade which followed Bikini the major powers decided that, in the words of *The Times*, 'the decisive date for good or evil' had not yet arrived. The Russians finally exploded their own full-scale hydrogen bomb in 1955, and it became clear that only the long-range ballistic missile was required to give both superpowers all that was needed for mutual annihilation. The scientists soon obliged.

Britain and France, the two pre-war great powers least able to survive a nuclear war, followed as best they could, for reasons of national prestige even though what they could add to the West's nuclear arsenal was peripheral. Britain's position was particularly awkward. The building of her atomic bomb, tested in October 1952 at Monte Bello, had strained her industrial and scientific resources and an attempt to build the Super would strain them still further. As late as the end of 1952 Lord Cherwell, still the most determined advocate of Britain's nuclear armament, believed nevertheless that work on a hydrogen bomb was, for the moment, 'quite beyond [Britain's] means'. William Penney, with the tribulations of his work at the Ministry of Supply burned deep into his mind, was convinced that construction of such a bomb 'would be an intolerable strain on British resources'. However, rational thought often flies out of the window when nuclear discussion comes in at the door, and it was decided that whatever the cost to the country Britain must have her own H-bombs. In the House of Commons Julian Amery explained: 'It would seem that the hydrogen bomb, when we have it, will make us a world Power again.'

Of the reasons openly put forward for building the weapon, one of the least unrealistic was given by Churchill in the House of Commons. 'Unless we make a contribution of our own ...,' he said, 'we cannot be sure that in an emergency the resources of other powers would be planned exactly as we would wish, or that the targets which would threaten us most would be given what we consider the necessary priority, or the deserved priority, in the first few hours.

'These targets might be of such cardinal importance that it would really be a matter of life and death for us.' Churchill, no doubt remembering the acrimonious differences between British and American bombing policy in the later stages of the Second World War, and the problems of Anglo-U.S. collaboration on Tube Alloys, was tactfully implying that in a future conflict the United States might not be willing to sacrifice New York in the hope of saving London.

To Anthony Eden, the development of 'the Super' could almost assume the shape of a blessing in disguise. 'One consequence of the evolution from the atomic to the hydrogen bomb was to diminish the advantage of physically larger countries,' he was to write. 'All become equally vulnerable. I had been acutely conscious in the atomic age of our unenviable position in a small and crowded island, but if continents, and not merely small islands were doomed to destruction, all were equal in the grim reckoning.' But, as Orwell pointed out in a different connection, some are more equal than others; perhaps a relevant qualification when considering the 55 million inhabitants of Britain, closely packed into 90,000 square miles, much smaller than the danger area which the Americans had eventually declared for their Bikini tests.

Britain fired her Super at Christmas Island in the Pacific in May 1957. France followed up with her first A-bomb in February 1960 and her Super eight years later. China provided a mild surprise with her A-bomb in 1964 and a larger one with her hydrogen bomb of 1967. Information, possibly exaggerated, flowed from beneath the frontiers of Israel, India and Pakistan suggesting that they, too, wanted the weapon which would enable them to appear, fully armed, at the conference table. Despite the efforts of the greater powers, anxious to prevent their neighbours from acquiring the weapons they themselves already had, it seemed likely that the smaller countries would eventually get them.

In the wake of Bikini there also came the development of what were rather ingenuously called tactical nuclear weapons. The Americans had already produced a nuclear shell small enough to be fired from a cannon and this was soon being shipped to Europe for N.A.T.O. forces. The size was subsequently reduced so that the shoulder-held bazooka was able to fire a nuclear charge. Nuclear land-mines were successfully tested, as well as sea-mines and nuclear torpedoes. The package weight of the nuclear explosive was reduced to as little as 100 pounds and the yield became controllable to anything from ten kilotons down to a single ton of T.N.T.

This nuclear proliferation in the U.S.A., and no doubt Russia, matched to a much more limited extent in Britain and France, was accompanied by three developments whose interactions have dominated the nuclear scene for the last quarter century. There were the first tentative moves towards arms limitation; the rise of the nuclear disarmament movement; and the drive to exploit the peaceful uses of nuclear fission, a drive whose optimistic assumptions are being increasingly questioned.

The first faint indication that governments might be beginning, in spite of their overt actions, to appreciate the perils of mutual destruction, came in 1958 with an Anglo-U.S.-Russian moratorium on nuclear tests that lasted for three years. It was followed in 1959 with a treaty banning nuclear weapons from the Antarctic. Four years later came a partial test ban treaty; by this time, it was gloomily noted, there had already been 423 nuclear-weapon tests, including in 1962 alone more than forty in Russia and eighty in the United States. In January 1967 the United States, Russia and fifty-eight other countries signed a United Nations treaty banning the use in space of nuclear weapons of mass destruction, and the following month there was signed a similar plan banning nuclear weapons from Latin America. And in 1969 President Podgorny of Russia and President Nixon ratified a forty-seven-nation Nuclear Non-Proliferation Treaty concocted in the hope that nuclear materials, ostensibly procured for peaceful purposes, would not be diverted to the weapon industry.

Neither these treaties, nor those which were to follow, halted the nuclear arms race; indeed, it can be claimed that their main result was to put a premium on the technological ingenuity of evasion. Nevertheless, they did suggest that whatever bellicose words the politicians felt it necessary to utter, they intended the great deterrent to go on deterring.

Even if the stumble towards international agreements did reflect a developing awareness that 'the bomb' was something more than just another weapon of war, that awareness was insufficient to reassure many people. Moral distaste at the prospect of again killing more civilians than was absolutely necessary had been slowly reinforced after Hiroshima by both economic and military doubts about the effectiveness of nuclear arms. These had with good reason been strongest in Britain, a country defenceless against nuclear attack. This went back at least to 1952 when pacifists began to concentrate their efforts on a nuclear threat. They were given strong support in 1954 by 'Man's Peril', a pre-Christmas broadcast by Bertrand Russell, whose views on threatening the Russians had now changed and who was busy pointing out that in wars between nuclear-armed powers there could be no winners.

By 1957 there were signs that Britain would not indefinitely be able to carry the economic burden of being a 'world power'. These doubts were increased with the advent of the Russian sputnik; in the coming missile age, Britain would have to drop out of the race whatever face-saving statements had to be made for purposes of prestige. The daily horrors of the previous war were more than a decade away and many of those who had not been unduly worried by the civilian slaughter of area-bombing joined those who saw no benefit to Britain, moral or strategic, in retaining a deterrent whose plausibility rested on faith rather than fact.

The catalyst was provided in November 1957 when the English writer J. B. Priestley wrote an article in the *New Statesman* on 'Britain and the Nuclear Bombs'. 'Now that Britain has told the world she has the H-bomb,' he said, 'she should announce as early as possible that she has done with it, that she proposes to reject, in all circumstances, nuclear warfare.' The article quickly brought together a number of groups which believed the same, even though they disagreed about the best method to adopt. The outcome was the Campaign for Nuclear Disarmament, led by Bertrand Russell who felt, very logically, that since the Russians now deployed nuclear weapons, the idea of bombing them into submission was impracticable; mutual disarmament was thus the only hope.

With its annual Easter marches to the nuclear weapons research centre at Aldermaston in Berkshire, and with its rallies and demonstrations, C.N.D. made a considerable impact on the British public. By the early sixties it seemed likely that it might even accomplish its first aim, the renunciation of nuclear weapons by a Labour Government if and when one came to power. The campaign just failed. What it did do was to make an apathetic public a little less unaware of what nuclear war would mean. It educated a number of M.P.s, a few Ministers and even some civil servants, and impressed on them the need to keep an East-West conversation going. Perhaps more important, the British organization encouraged the birth of similar groups in the United States and in Europe. These were in themselves no more successful than C.N.D. in leading to tangible results that could be traced back to their work. Yet they reinforced the suspicions that when nuclear matters were under discussion 'the official view', the 'authoritative' statement, must be viewed with even more than normal caution. This was to be a factor in the next two decades as the emphasis changed and as dissent about nuclear weapons broadened into dissent about the exploitation of nuclear fission in peace as well as in war.

CHAPTER 16

Power or Peril

Creation of the Super, the fusion bomb triggered off by an old-fashioned Hiroshima-type explosion, brought to an end one line of development that had begun with the birth of nuclear fission only sixteen years earlier. Weapons were to grow still larger and more deadly, as well as smaller and more convenient for use on the battlefield. The intercontinental ballistic missile was to remove the security which the Atlantic and the Pacific had previously granted the Americans and to bring them, like the rest of men, beneath the shadow of the bomb. But once it had been demonstrated that the largest city in the world could be erased by a single explosion, progress became a demonstration of technological ingenuity. Physicists could refine the details of destruction, politicians could maintain that mutual destruction would be more mutual for some than for others and the world's defence establishments could play war games with computers that brought intellectual enjoyment to all concerned. But as far as war was concerned nuclear fission had done its job.

That the great opportunities offered by its discovery in 1938 had been subject to so much dispute and controversy had not been entirely due to man's inexperience in handling such extraordinary problems. The rise of the Third Reich in the 1930s had already begun to shatter the rebuilt internationalism of science which might otherwise have led to a totally different handling of the unique situation. At Potsdam seven years later distrust of Russia had blocked the road that might, just possibly, have brought results, and in the aftermath Russian distrust built the road-blocks higher.

The difficulties which have increasingly bedevilled the peaceful uses of nuclear fission are of a different order, spring from different causes, and are due only in part to enthusiasm for a new source of power.

These difficulties are almost exclusively concerned with the use of fission in nuclear power stations, but there are other civil uses which provide at least significant items on the credit side of the balance sheet. The most important is the use of artificially made radioactive isotopes, and their little-appreciated impact on the last thirty years should be recalled before the possibilities of Armageddon by accident are discussed.

The uses of man-made isotopes which continually betray their presence by radiation were glimpsed by the Curies and by Fermi in 1934. Yet it was only after the war, when reactors made it possible to produce 'tailor-made' radioactive isotopes, that it became possible to exploit their great potentialities. Much earlier, an illuminating if somewhat bizarre indication of how naturally occurring radioactive materials could be used was given by the Hungarian-Danish chemist, Georg von Hevesy. He was staying with a landlady whom he suspected of using up the remains of Sunday lunch in meals served later in the week, a suggestion which was firmly denied. 'The coming Sunday,' he later wrote, 'in an unguarded moment, I added some active deposit of thorium [i.e. the decay products of thorium emanation] to the freshly prepared pie, and on the following Wednesday, with the aid of an electroscope, I demonstrated to the landlady the presence of the active material in the soufflé.'

Some years later Hevesy laid the foundation for what is now the radio-isotope industry by watering plants with a radioactive isotope of lead which he had obtained from thorium breakdown products. By checking the points on the plants from which radiations were coming he was able to follow the absorption and distribution of the lead. This element was not a natural component of the plant, so the results had to be taken with caution. However, Hevesy realized that if it were possible to use radio-isotopes of elements which were normally present, then it would be possible accurately to trace how living tissue made use of such material.

This use of radioactive isotopes as tracers is one of the most important ones today. Green plants, for instance, take in carbon dioxide and water, transform these into sugars and starches and while doing so give out oxygen. But both carbon dioxide and water contain oxygen and until the coming of radio-isotopes it was impossible to know whether the oxygen produced by the plant came from the water or the carbon dioxide. The answer – that it comes from the water – was found by increasing the amount of the radioactive isotope oxygen 18 first in the water, then in the carbon dioxide. This simple use of radio-isotopes has been developed to show at what point in a plant's life-cycle the plant takes up fertilizers. The mechanism by which phosphorus in food is turned in man and animals into bone – itself largely a compound of calcium and phosphorus – has been revealed by the use of radioactive phosphorus isotopes, and in numerous other ways the by-products of nuclear fission have handed doctors and scientists a fresh tool for research in biology and agriculture, medical diagnosis and animal husbandry. Even the most dangerous feature of radiation, the way in which it produces mutations, has been conscripted to create, by irradiation, hardier plants, new species, rust-resistant varieties of wheat, and a radiation-induced mutant of rice which contains twice as much protein as earlier varieties.

Other uses for these new products of the nuclear revolution are more esoteric and more unexpected. Radioactive material incorporated into the outer layer of motor-car tyres has produced clearer information on tyre wear. Radio-activated stones have revealed the underwater movement of off-shore sea and river beds, while underground leaks from water mains can be pinpointed merely by passing a radioactive fluid through the pipes and then following its course with a detecting instrument.

A totally different kind of application is typified by the use of the potentially deadly cobalt 60 radio-isotope, not only in cancer-control work, where it has in some cases advantages over radium, but in insect irradiation. Thus in parts of Florida, where the screw-worm fly was a so-far irradicable menace, cobalt 60 was used to sterilize up to 50 million screw-worm flies a week. The irradiated flies were dropped from the air at a concentration of up to 3,000 per square mile; breeding declined; and the pest was wiped out from the area in less than two years.

This use of radioactivity, artificially produced to specification, is – like the analysis of radioactive carbon 14 to date prehistoric artefacts – a by-product of the nuclear revolution. It has been developed amicably throughout the world and protests have been limited to such local fears as that radioactive pebbles will not only reveal estuary currents but also endanger the estuary. Thus it is different from that other civil by-product of nuclear research, the use of nuclear explosions to stimulate gas production, produce storage cavities for oil or gas, or excavate harbour or canal sites, all of which have been limited by public fear. But what is almost an international isotope industry has developed without friction not only because its benefits are so obvious and its dangers so relatively slight. It is more than eight decades since doctors began experimenting with radium; experience of radioactivity has been accumulating over the years and the post-war multiplication of available materials has merely demanded an enlargement, if a considerable one, of that experience.

No such experience existed before the later 1940s in the harnessing of nuclear fission to power. It is true that as far back as the 1920s Aston had prophesied that a tumbler of water would one day be enough to drive a great liner across the Atlantic and back; and nuclear power for ship propulsion was seriously discussed from the first weeks of 1939 when U.S. naval research scientists noted its advantages for submarine propulsion. These were exploited after the war by the U.S.S. *Nautilus* during her voyage under the Arctic ice-cap, and by the British and Russian submarine fleets. The Russians and the Americans, with the nuclear-powered ice-breaker *Lenin* and the N.S. (Nuclear Ship) *Savannah*, demonstrated the possibilities for surface craft. Yet the reluctance of many ports to welcome nuclear-powered vessels soon tended to limit their usefulness, a foretaste of the nuclear arguments which were to grow during the 1960s and 1970s.

The French in the Collège de France had at first been as interested in nuclear power as in nuclear explosions. An appendix to Britain's M.A.U.D. Report had speculated on the new power which after the war might 'affect the distribution of industry over the world', and had warned that it was 'essential that Great Britain should take an active part in this research work so that the British Empire cannot be excluded by default from future developments'. Indeed, it was the potential machinations of that Empire that had cast a shadow over Anglo-U.S. relations during the war and done much to keep American suspicions alive during the immediate post-war years. But although prospects for what George Thomson had called 'the boiler', the nuclear reactor whose bucket of fuel would replace train-loads of coal or convoy-loads of oil in the production of electricity, had been at the back of men's minds for some while, practical experience of the industrial problems involved was nil.

In the aftermath of Hiroshima and Nagasaki the immediate prospects still looked comparatively dim except to the more optimistic politicians. Physicists, chemists, and engineers all considered nuclear power as a subject involving a series of appalling technical problems which could be solved, if they could be solved at all, only in the distant future. Fears about nuclear power were fears far less nagging than those about the stockpiles for nuclear destruction which in the late 1940s appeared to be the aim of every first-class nation's hope and endeavour. Nevertheless, nuclear power was sure to come, if not on the grounds of economics then on grounds of prestige. Moreover the technical problems had their own attraction. As Oppenheimer had said of the programme after the Teller–Ulam configuration had been worked out, the problems of successfully creating nuclear power were 'so sweet that you could not argue' about them.

The United States, Britain, France, and Russia, the four countries for whom this was a practicable goal during the immediate post-war years, each had their own individual approach to the subject, and each was circumscribed by individual economic or political factors. So, in much the same way, were the score or more of other countries who before long were to grapple with the trials, tribulations, and potentialities of power from nuclear fission.

In the United States, the great issue was whether what was to become a powerful new industry should be controlled by the Government or, in the American tradition, become the playground of private enterprise. In Britain, where the first moves towards a nuclear stockpile could be glossed over as research into peaceful power, success could bring its own problems: nuclear power which could make the jobs of 400,000 miners appear redundant would have been decidedly unwelcome to any Government, let alone to a Labour administration. This possibility was by no means ruled out in some of the over-euphoric forecasts of early 1946. Thus a paper on the potential economic results of nuclear energy prepared for Cabinet reading in February suggested that if the country adopted a comprehensive nuclear energy scheme, 'most of the labour now in coal-mines and gas/coke industries, together with a substantial fraction of other labour forces (railways, etc.) would be released for other work'. This was not the only idea which dissolved as the problems of nuclear power were tackled. The atomic power stations of the future, the Government was told in 1946, 'could quite safely be placed in large towns' although, despite the fact that power stations would be smaller, 'It did not seem likely ... that there would be more than one generating unit per town.'

In France the reasonable desire of any ambitious nation to see what could genuinely be made of nuclear power was augmented by another: the wish to regain for France the leading role in nuclear physics which she had held at the outbreak of war. In this she, too, had considerable advantages. She not only had Joliot-Curie but also Halban and Kowarski among others, men who had gleaned much from the

Americans and who, in the regretful words of the wartime British, could not be prevented from making use of their knowledge, although they should not be encouraged to do so.

There were other countries in Europe with special advantages. Norway had her heavy water and with the help of the Dutch was by 1951 using a heavy-water research reactor at Kjeller. Belgium, which had provided the initial loads of uranium for the Manhattan Project, was in a special position and after securing graphite from Britain was soon operating her own experimental reactor. Sweden, conveniently next door to Norway from whom she purchased heavy water, discovered uranium within her frontiers and was by 1954 operating her own heavy-water experimental reactor.

Despite the divergences of background, of access to the raw materials of nuclear power, and of economics and politics, all countries faced with the prospect of 'going nuclear' had the same range of options. Actual utilization of the heat produced by a chain reaction is the same in most nuclear power plants: the heat produced by fission is taken away by a coolant to produce steam, the steam drives a turbine and the turbine produces electricity. It is in the heart of the reactor that the options are numerous. The fissile material can be natural uranium, with or without plutonium, or enriched uranium in one of numerous forms. The moderator can be water, heavy water, beryllium metal, beryllium oxide, carbon or certain organic fluids. The coolant which takes away the heat can be one of various gases, water, heavy water, or one of a number of liquid metals or organic liquids. The permutations are considerable, and it has been estimated that almost a thousand types of reactor are theoretically possible although only about twenty are practicable.

They can be classified in a number of ways. One is into thermal reactors, in which the neutrons produced by fission are slowed down by a moderator, and fast reactors which contain little or no moderator. Thermal reactors can themselves be divided into heterogeneous, in which the fissile material and the moderator are separate and form a geometric pattern or lattice, and homogeneous reactors in which fissile fuel and moderator are mixed into a uniform medium. Reactors can also be classified according to their function: power reactors are designed primarily for the creation of electricity; propulsion reactors for use in ships; production reactors, on the other hand, may be breeder reactors which actually produce the same kind of fissile material that they burn; and converter reactors which change non-fissile material such as uranium 238 into fissile plutonium. There are also research reactors, primarily designed to investigate specific problems.

The reactions involved in the production reactor have been understood since the 1940s when it was appreciated that fissioning plutonium surrounded not by a moderator but by a 'blanket' of uranium 238 could not only maintain a chain reaction but at the same time convert the enveloping 238 into more plutonium. Practical investigation in the early 1950s, when Britain was desperately anxious to produce plutonium for her first nuclear weapons, led to construction of the breeder reactor at Dounreay in the far north of Scotland. Since then many countries have embarked on major research and development programmes and Russia and France are among the countries which already have breeder reactors and are planning larger ones.

Not unexpectedly, the research reactors came first. Outside the United States, Canada led the way with what was then still called a 'pile' going critical less than a month after Hiroshima. Britain followed with Gleep at the Harwell research station in August 1947 and its successor, B.E.P.O., the following year. Russia appears to have been operating more than one experimental reactor by this date, while France and Norway quickly followed suit. Russia certainly had reactors for the production of plutonium in action by 1948, and probably by 1947. Britain's counterparts at Windscale were operating from 1950 onwards. As for the record of reactors actually producing electricity there are 'firsts' for everyone according to definition. In April 1952 the Americans began operating a reactor at Arco, Idaho, which produced 100 kW of power. But this was all used by the reactor's auxiliaries while the reactor itself had been designed for the *Nautilus*, the world's first nuclear-powered submarine. In June 1954 a Russian reactor at Obninsk, outside Moscow, went critical and began pumping up to 5 MW of electricity into the grid. Two years later came Britain's Calder Hall, the first of a series of nuclear-powered generating stations whose design gave Britain a temporary lead in the field. But, as was barely mentioned at the time, Calder Hall was not purely an electricity producer; as it has been officially stated, 'it was built primarily to produce military plutonium'.

From the first, there were doubts about nuclear power, but they were mainly financial doubts, a questioning of whether this bounty, apparently the milk and honey about to flow from nuclear fission, would be economically competitive when compared with coal- or oil-fired generating stations. This was natural enough. The huge initial cost of building a nuclear station, the difficulty of accurately estimating subsequent costs, the number of variables and the fact that no one had yet operated such a station, let alone run it for a statistically significant period, all made honest comparison difficult. This situation, in which it was easy to pick a convenient set of figures to prop up any particular case, led more thoughtful observers to view the special pleadings of both pro-nuclear and anti-nuclear camps with some scepticism.

Another reason for caution was the link with 'the bomb' which cut both ways. To some, nuclear fission which had made it possible to kill some 100,000 men, women, and children with impunity, was the epitome of evil from which no good could come. To others, typified by David Lilienthal, the not entirely impartial first Chairman of the U.S. Atomic Energy Commission, there was the 'conviction, and one that I shared fully, and tried to inculcate in others, that somehow or other the discovery that had produced so terrible a weapon simply *had* to have an important peaceful use'. That '*had* to' could have been satisfied by the ever-expanding use of radio-isotopes, yet power was the nub of the matter, and it was nuclear power which was given an immense boost by the American Atoms for Peace programme launched by President Eisenhower at the United Nations in December 1953.

Proliferation of nuclear power plants would of course have followed in the 1950s even had Eisenhower not acted. Nevertheless the Atoms for Peace programme, followed by the passage through Congress of a new Atomic Energy Act and the setting up of the International Atomic Energy Agency in Vienna, dramatically changed the prospects for all countries other than Britain and France. Under the new Act, America was allowed to sell not only nuclear fuel but nuclear know-how to countries which would have been barred from both by the McMahon Act, and within a few years, more than twenty-five, including fifteen developing nations, had been given previously classified material under bilateral agreements. The results were quick and spectacular. When the first conference on the peaceful uses of fission was held in Vienna in 1955, the only nuclear reactor pumping power into a national grid was that in Russia with its input of 5 MW. By the time the second conference was held in 1958, world nuclear capacity had risen to 185 MW; another six years on it had reached 5,000 MW.

Reaction to the atoms for peace campaign was mixed. With its generous offers of know-how, fissile material, and the expertise of specialists, it enabled many small countries and emergent nations to acquire the means of producing electricity by nuclear power. But as David Lilienthal was quick to point out,

Much of this was an utterly meaningless and wasteful operation, for most of these countries had hardly a cadre of scientists or the necessary facilities to put this 'exchange' of atomic knowledge to any significant use. Even as a propaganda move it was self-defeating and naïve. A great many of these countries need and could use doctors and medicine, storage batteries, plows and fertilizers and seed – and good *elementary* instruction. Only the desire to prove somehow that Atoms were for peace could justify the absurdity of a separate program, not in the foreign aid part of the State Department but in the A.E.C.

However, the programme certainly prevented the countries from coming under the technical thumb of Soviet Russia, by this time quickly shortening the nuclear gap between itself and the United States and anxious to impress the Third World with its prowess.

So far, fears of nuclear power were still minimal. They grew during the second decade after Hiroshima, as though to fill a vacuum as nations increasingly began to 'live with the bomb'. Hahn and Strassmann, scientifically dedicated to their work in the heart of Berlin, Otto Frisch and his aunt, walking in the snow-covered Swedish woods, had revolutionized the familiar world known for so many centuries. Many ordinary men and women knew at the back of their minds that things would never be the same again; nevertheless, they had gradually taken that particular revolution for granted. The fears that slowly began to exercise them as nuclear power stations began to open across the world, usually with self-congratulatory plaudits from the authorities who had commissioned them, were at the lowest level based on confident ignorance. However many attempts were made to explain the contrasts between the uncontrolled chain reaction of a nuclear weapon and the controlled chain reaction of a power generator, the illusion remained: if something went wrong 'the thing would blow up'.

Yet if there was no possibility of a nuclear power station 'blowing up', there were other dangers to which even the most optimistic officials could not close their eyes. In Britain it had initially been planned to build the country's first reactors at a newly created centre near Mallaig in the far north-west of Scotland – not only because the site had access to the required large volumes of cooling water but because its isolation would limit the death-roll in the event of an accident. This policy had been followed in the United States where the wartime defence plants had a potential for serious accidents and it continued to be followed in Britain even though improvements in reactor design, plus quickly expanding experience, dramatically reduced the possibilities of accident.

Even the most optimistic of the world's nuclear authorities has never claimed 100 per cent protection against every possibility, and as *Nuclear Energy International* – hardly an alarmist journal – has stated in discussing the Harrisburg accident of 1979: 'One day there may be a disastrous accident at a nuclear power station, somewhere in the world.' A vital point is the statistical chance of such an accident, a chance which the supporters of nuclear power claim to be extremely remote and which its opponents claim to be much more likely. As reactors have continued to proliferate towards the present figure of some 700 throughout the world – with between 200 and 300 of them being power-station reactors – the supporters of nuclear power have frequently made the same point: that if complete safety was the criterion, then a great many everyday activities, ranging from driving to flying, and from smoking to drinking, would have to be abandoned completely. As for accidents it was – and is – regularly pointed out that the nuclear industry has a world-wide safety record far better than that of coal-mining or oil-extraction. These comforting truths are supported by the fact that – up to the time of writing – there have been no disasters; only a few close shaves.

Yet despite the statistics of safety rolled out by the nuclear industry, the anti-nuclear lobby has achieved considerable success during its relatively short life and certainly more than was achieved by 'ban the bomb' groups. In Austria, West Germany, Sweden and Spain, civil disobedience has been considerable and is increasing. The anti-nuclear lobby was largely responsible for unseating Sweden's Social Democratic Party in 1976, and in Germany has forced a nuclear moratorium on the North Rhine–Westphalia Government. Protests have been very different from the mainly peaceful operations of C.N.D. days, and in Germany pitched battles have been fought with police. Grappling hooks, iron bars, bayonets and Molotov cocktails have all been used. In Austria a national referendum has forced the Government to abandon plans for using its only, 250 million dollar, nuclear power station, which may eventually be transformed to use fossil fuel. In the United States, the Governor of New York State came to power in 1978 in support of a fifteen-year moratorium on nuclear power stations, while during the same year a Montana ballot in effect banned nuclear power from the state. The signs multiply. Only behind the Iron Curtain, where Russia is planning another eighty nuclear stations in her own and satellite states for the early 1980s, does the trend continue unopposed. But there, as *Nuclear Energy International* has commented, 'Questions of environmental protection appear to play no role as there is no public debate on this subject to act as a brake on the nuclear programme.'

The contrast between what appear to be the slightly casual operations east of the Iron Curtain and the complex network of controls within which the nuclear energy industry has to work elsewhere has led to suggestions that the protest movement in Europe and America is politically motivated. This seems unlikely. It is true that at first glance the strength of the movement, which seems likely to limit the future of nuclear energy in the West, appears strangely disproportionate to the threat. The explanation lies partly in the fear of the not-fully-understood, of the mysterious and insidious way in which radioactivity strikes; partly in a healthy distrust of authority and its spokesmen with their shining morning faces and built-in bias towards the nuclear industry; partly in the very enormity of a nuclear disaster which it is admitted may occur some day; and partly because the industry can still only hope that something will turn up to solve the growing problem of waste-disposal. On most of these questions the evidence is conflicting; and where statistics cannot be denied they are sometimes quoted by both sides in the argument to support diametrically opposed views.

The case for nuclear power rests not so much on the convenience argument of a matchbox of fuel for a month's electricity, as on the claim that nothing else can cope with the energy demands of the coming decades. It is a plausible argument, not seriously dented by the claims for solar or wind power or the proposal that the world should limit its energy demands to the energy available. On the debit side there are two main items. One is the possibility of a major nuclear power station accident. The other is the claim that the storage of radioactive nuclear wastes, accumulating in ever-increasing amounts as the number of nuclear reactors continues to grow, is not as safe or as foolproof as Governments maintain.

Release of radioactivity, the main danger in a reactor accident, is most likely to be brought about by a failure to take heat away from the reactor core, a melt-down of the core, a breaching of the protective shield, and a release into the atmosphere of lethal radioactivity. This, like any comparable nuclear disaster, would come about only after an unexpected failure of equipment and, following that, the failure of a series of fail-safe precautionary devices. At least, that is the argument of the industry, and it does seem unlikely that some totally unexpected nuclear phenomena will, after nearly a quarter-century of operations, occur to confound all the careful calculations. However, as *Hymns Ancient and Modern* once had it, 'a thousand ages in Thy sight are but as yesterday', and twenty-five years is not really very long.

The estimates of the casualties which such an accident would produce vary from the bland to the fearful, even allowing for the fact that much would depend on circumstances such as the particular site, the direction of the wind, and the speed and efficiency with which warning was given. The Rasmussen Report of 1974, produced by Professor Rasmussen of the Massachusetts Institute of Technology with the help of some sixty fellow scientists, estimated that an extremely serious accident would cause 3,300 immediate deaths, 45,000 latent cancer fatalities over thirty years, early illness of 45,000, genetic defects in 30,000 births during the subsequent 150 years and the creation of 240,000 thyroid nodules in the population during the thirty years after the accident. Other estimates put these figures as ridiculously high; but whatever they might be, it seems unlikely they would be acceptable.

When the chances of such an accident taking place are estimated the figures vary so considerably that faith in them rapidly disappears. According to the Rasmussen Report the chances are of one accident on this scale in about 200 million years of reactor operation. Even when operating reactors have risen in numbers to 1,000, this would still suggest only a single accident in some 200,000 years. However, the figures give no direct information on when the one accident would take place. And an equally well-informed research report, the Ford Foundation's *Nuclear Power: Issues and Choices*, reduces the time-scale by a factor of 10,000 – thus bringing the latest date for a serious accident no later than the turn of the present century.

Statistically, the dangers can easily be shown to be remote. This has been done by Fred Hoyle, the eminent astronomer who has claimed that the risk of living close to a nuclear power plant is that of driving a car for an additional 250 yards per day. 'If one prefers the estimates we attributed to the critics,' he has written, '[20 times larger], it does not pay to move away if doing so increases commuting distances by more than 1·5 miles per day. Even with the critics' estimates, living next to a nuclear power plant reduces life expectancy by only 0·03 years, which makes it 150 times safer than living in a city.'

From the tangle of conflicting figures one conclusion seems inescapable: that nuclear power presents a very remote chance of a very serious disaster. It can be argued – and often is – that natural disasters such as tidal waves, earthquakes, pestilence, and famine, produce equally horrific casualty rolls. The humble petrol engine logs up annually, in the air and on the road, deaths and mutilations comparable to those of a major nuclear disaster, and it is often grandly asked why the human race should pay this tribute without much ado but baulk at a so-far-unrealized penalty for power from fission. But this of course is a moral choice which should presumably be taken without too much regard for the way in which a growing industry tends to put its finger on the scales.

There have been a number of accidents, of which that at Windscale in October 1957 achieved international importance. The two Windscale reactors, each containing 2,000 tons of graphite and 70,000 fuel elements, had been built to provide plutonium for Britain's first nuclear weapons, and by 1957 they were also producing tritium for Britain's hydrogen bombs. The number one pile had been closed down a number of times for release of Wigner energy, stored energy in the graphite named after the Eugene Wigner who had worked with Fermi in 1939 and had been involved in Einstein's letter to Roosevelt. The operation had become increasingly difficult and in October 1957 it resulted in the release of radioactive iodine. The more dangerous strontium and caesium, it was revealed during the subsequent enquiry, had been trapped in the filters at the top of the release stacks. These had not been part of the original design but had been added as a safety measure after what has been called the determined interference of Sir John Cockcroft. Known derisively as 'Cockcroft's Folly', they had stopped the surrounding area from being deluged with fall-out and prevented an accident from turning into a disaster.

However, the radioactive iodine was known to contaminate pasture; the contamination was taken up by the cattle who ate the grass, and thereby presented a potential threat to children who would drink the milk from the cattle, a salutary indication of the ramifications of radioactivity. A ban on milk was therefore imposed by the authorities and eventually extended over 2,000 square miles. Both the Windscale reactors, which were nearing the ends of their operational life, were closed down but while the loss of plutonium and tritium was inconvenient it was quickly being balanced by increased production from other sources.

In the United States the first warning came in 1966 with an accident at the Enrico Fermi station some miles outside Detroit. According to one account, 'during the four-week period that followed, scientists, engineers, utility executives and members of the Atomic Energy Commission held their breath while viewing a situation raising the possibility that Detroit with more than 1·5 million inhabitants might have to be evacuated.' Contrariwise, *The Health Hazards of NOT Going Nuclear* – a book praised in the U.K.A.E.A.'s journal *Atom* as 'like a breath of fresh air amid the smoke screen of biased, inaccurate and sensational nonsense that is being poured out by the anti-nuclear movement' – claims that the accident never produced danger to anyone.

Perhaps the Harrisburg accident of early 1979 offers the best example of how nuclear danger, like beauty, tends to lie in the eye of the beholder. The opponents of nuclear power point out that the plant came to the brink of disaster, while the final report of President Carter's commission on the accident says: 'What is quite clear is that its impact, nationally and internationally, has raised serious concerns about the safety of nuclear power.' The other side points out that although almost every event at Harrisburg followed Murphy's law – that anything which can go wrong will go wrong – there were no deaths and no injuries. A multi-million-dollar power station that for many practical purposes no one can now enter, let alone operate, is a big entry on the debit side of the ledger but, in terms of nuclear disaster, not decisive.

The problems inherent in the disposal of nuclear waste are less dramatic but more widespread, the potential dangers resting, once again, very largely on the statistical chances of misfortune. As the fuel elements in a nuclear reactor produce heat by fission their efficiency gradually decreases and eventually they have to be replaced. The exhausted elements contain not only residual uranium but also a wide variety of fission products, the best known of these being plutonium. The uranium and plutonium can be recovered for future use but most of the other material must be kept for long periods under conditions where its radioactivity is safely confined. Storage as a liquid, in specially cooled tanks which are either buried underground or sunk deep in the sea, is the present method. Work is continuing to make the operation not so much less dangerous, as, in the words of the authorities, 'even more safe'. Vitrification, or containment in a glass solid which could be sunk into deep pits is one of the more favoured solutions. The size of the problem, like the chances of accident, appear considerable or minimal according to which figures are chosen. In the United States it was announced in 1967 that about 80 million gallons of highly radioactive material was already being stored; but as the United Kingdom Atomic Energy Authority says, the nuclear waste from the equivalent of ten tons of coal is only a quarter of an ounce.

The reasonable contention that the problem of waste disposal has been blown up into bogyman size by irrational fears and a failure to look at the facts objectively appears to have been given a severe shaking by additional details of the nuclear disaster in the Urals during the winter of 1957–8. The first detailed news of this came from the exiled Russian scientist, Dr Zhores Medvedev, who in 1976 maintained that an accident had laid waste large areas and caused many deaths. The story was immediately discounted with such phrases as 'pure science fiction', and 'figment of the imagination'. This was an almost inevitable reaction to Medvedev's claim that the disaster was caused by an accident involving nuclear waste; and Western scientists still find it difficult to believe that desolation of an area hundreds of square miles in extent was caused in such a way. However, Medvedev has now provided more, if still largely circumstantial, evidence. *Nuclear Safety*, the journal of the Oak Ridge National Laboratory, gives qualified confirmation of some major accident which the Russians have taken great pains to conceal. And if it is, as yet, impossible to lay the cause on nuclear waste disposal, the fresh evidence still tends to undermine the official line taken by most countries that there is little to worry about as the world goes increasingly nuclear.

The fission industry is likely to weather peace-time protests against power stations and nuclear waste disposal, at least until one or the other produces an accident of traumatic proportions. Less dramatic, more insidious and in the long run probably quite as important, are the changes inherent in 'the plutonium society', the boss phrase for a world dependent on plutonium rather than enriched uranium. The move towards this situation has now reached a point from which, it is claimed with some plausibility, it will be difficult to turn back. The increase in the world's stock of plutonium, an element which until 1940 existed only in minute and undetected quantities, will almost certainly lead to a more security-conscious world. Side-tracking 'peaceful' plutonium into the arms industry and hijacking by terrorists both savour of science fiction – as was true of nuclear power itself little more than half a century ago. To deal satisfactorily with these problems would almost certainly produce what Sir Brian Flowers has called in his estimate of the nuclear future 'long-term dangers to the fabric and freedom of our society' – police surveillance, for instance, that would have to be increased to a level accepted only in totalitarian states. This would be regrettable; but it is to be doubted whether a majority of the population would protest effectively since they would be battling against the kinetic energy which the concept of nuclear power by this time possesses. 'A new industry has suddenly appeared,' an Atomic Energy Commissioner once stated in America, 'of huge stature when first unveiled, growing in overall dimensions by the hour, already by conventional measurements the biggest single industry we have. It ramifies throughout the country and in no negligible sense throughout the economy.' And that was in 1948, when nuclear weapons were all, and nuclear power no more than a riddle at the end of the tunnel.

It is possible that the safety problems of nuclear power will before the end of the century be further minimized or almost removed completely by controlled nuclear fusion, the taming for power production of the energy released in the hydrogen bomb. Since the mid-1950s the United States, Britain, Russia, and possibly other countries, have been searching for a way of doing it. The prize would be big. There would be an absence of the radioactive by-products that fission produces. The basic fuel would be hydrogen – 'fuel from the sea' – while the burning of one gram of deuterium, one of the favoured isotopes of hydrogen, could theoretically produce 100,000 kilowatt hours of electricity.

Yet power from fusion still looks like a dream of the twenty-first century; the more immediate prospect is of the plutonium economy and the proliferation of weapons: unless, that is, disaster intervenes.

CHAPTER 17

Close of Play?

'Astronomers,' said Frédéric Joliot-Curie, speaking of nuclear transformations in 1935, 'sometimes observe how a star of modest brilliance suddenly grows in magnitude. A star invisible to the naked eye becomes very bright and visible without the use of an instrument. We say it is the appearance of a nova. This sudden flaring up of a star is perhaps caused by transmutations of an explosive character, processes which researchers will doubtless seek to stimulate, while taking, let us hope, all necessary precautions.' So far, the precautions have been taken; or, more accurately, it now appears that, great as man's potential achievements may be, they do not enable him to destroy, in a single blow, the planet on which he lives.

However, such total disaster is only one among the range of possibilities which have been discussed during the last third of a century. The cobalt bomb, whose feasibility as a lunatic's nightmare is regularly propounded and as regularly denied, could possibly leave the planet intact but devoid of all life, which would have been wiped out by the radioactive cobalt 60 dispersed throughout the world's atmosphere. Less unlikely than such calamities is the much-prophesied nuclear exchange, either between the two superpowers or at least on such a scale between less well equipped countries that the world would be left a diminishing archipelago of inhabitable islands surrounded by an encroaching ocean of radioactive debris. Of course no one in his right mind would allow it to happen; but accidents occur, misjudgements are made, and pulling down the pillars could still be an aberration of crazy rulers. The odds that Hitler, marooned in the bunker, would not have pressed the button had it been available, seem rather low. Since 1945 there has been more than one forecast that the holocaust would come before a named date that has now already passed. There should be little comfort in this. The more frequently the ball stops in a black, the greater the chance of a red; nuclear war tomorrow may be unlikely, but the odds shorten with time.

So, too, with the prospects of that occasion when, bad luck piled upon bad luck, the last valve jams, the final fail-safe system fails because someone has forgotten to oil it properly, and the panic starts. It is not, of course, inevitable. But, an accident, military or industrial, can eventually happen unless the world is shaken awake to a greater realization that the dangers, as well as the opportunities, of nuclear fission are of an order totally different from those with which *Homo sapiens* has yet had to deal. The imaginative leap necessary before this can happen may be impossible without the spur of a traumatic shock.

It would be tragic if a limited nuclear exchange, halted after each side had experienced the agony of one city put to the bomb, were necessary to bring men to their senses and to realize, at last, that there are no victors in nuclear war. It would be equally tragic if the melt-down of a power-producing reactor and the creation of a trail of disaster across Britain, the state of New York, or Western Europe, were needed to bring about a radical reassessment of how nuclear power should fit into the world's energy problem. But failing such events we may well go on to fumble the last catch of all.

BIBLIOGRAPHY

ANDERSON, Herbert L.: ' "All in our Time", Fermi, Szilard and Trinity', *Bulletin of the Atomic Scientists*, Vol. XXX, No. 8, Oct. 1974, pp. 40–7.

ANDRADE, E. N. da C.: The Rutherford Memorial Lecture, 1957, *Proceedings of the Royal Society*, A. Vol. 244, pp. 437–55.

ASTON, F. W.: 'The Structural Units of the Material Universe', 7th Earl Grey Memorial Lecture, at King's Hall, Armstrong College, Newcastle-on-Tyne, 5 Mar. 1925 (London, New York, etc., H. Milford, Oxford University Press, 1925).

BAINBRIDGE, Kenneth T.: ' "All in our Time" – A Foul and Awesome Display', *Bulletin of the Atomic Scientists*, Vol. XXXI, No. 5, May 1975, pp. 40–6.

BARUCH, Bernard M.: *The Public Years* (London: Odhams Press Ltd, 1961. New York: Holt, Rinehart and Winston, 1960).

BERNSTEIN, Barron J.: 'Roosevelt, Truman and the Atomic Bomb, 1941–1945', *Political Science Quarterly*, Vol. 90, No. 1, Spring 1975, pp. 23–69.

BORN, Max: *The Born-Einstein Letters, 1916–1955*, correspondence between Albert Einstein and Max and Hedwig Born from 1916 to 1955 with comments by Max Born (London: Macmillan 1971).

BORN, Max: 'Fifty Years of Physics', Lecture to the Royal Society of Edinburgh, 23 Oct. 1950, printed in J. L. Crammer (ed.), *Science News*, 19, pp. 46–60 (Harmondsworth, Middlesex: Penguin Books, 1951).

BRITISH COUNCIL OF CHURCHES: 'The Era of Atomic Power', Report of a Commission Appointed by the British Council of Churches (London: S.C.M. Press, 1946).

BRYANT, Arthur: *Triumph in the West, 1943–1946, Based on the Diaries and Autobiographical Notes of Field Marshal The Viscount Alanbrooke, K.G., O.M.* (London: Collins, 1959).

BUTLER, J. R. M.: *Grand Strategy*, Vol. II (London: H.M.S.O., 1957).

BYRNES, James F.: *All In One Lifetime* (London: Museum Press, 1960).

BYRNES, James F.: *Speaking Frankly* (London, Toronto: William Heinemann Ltd, n.d.).

CHADWICK, Sir James: 'Some Personal Notes on the Search for the Neutron', Proceedings of the Tenth International Congress of the History of Science, Ithaca, New York, 26 viii 1962–3 ix 1962, pp. 159–62 (Paris, Hermann, 1964).

CHURCHILL, The Rt. Hon. Winston S.: *Thoughts and Adventures* (London: Thornton, Butterworth Ltd, 1932).

CHURCHILL, Winston S.: *The Second World War*, Vol. I *The Gathering Storm*; Vol. IV

The Hinge of Fate, and Vol. VI *Triumph and Tragedy* (London: Cassell, 1948, 1951 and 1954).

CLARK, Ronald W.: *The Birth of the Bomb* (London: Phoenix House Ltd, 1961).

CLARK, Ronald W.: *The Rise of the Boffins* (London: Phoenix House Ltd, 1962).

CLARK, Ronald W.: *Tizard* (London: Methuen and Co. Ltd, 1965).

CLARK, Ronald W.: *The Man Who Broke Purple: The Life of Colonel William F. Friedman, Who Deciphered the Japanese Code in World War II* (London: Weidenfeld and Nicolson, 1977).

COCKCROFT, Sir John: *The Development and Future of Nuclear Energy*, The Romanes Lecture delivered in the Sheldonian Theatre, 2 June 1950 (Oxford, at the Clarendon Press, 1950).

COMPTON, Arthur Holly: *Atomic Quest: A Personal Narrative* (Oxford: Oxford University Press, 1956).

COMPTON, Arthur H.: 'The Birth of Atomic Power', *Bulletin of the Atomic Scientists*, Vol. IX, No. 1, Feb. 1953, pp. 10–12.

CONANT, James B.: *Anglo-American Relations in the Atomic Age*, address delivered at the London School of Economics and Political Science, 17 Mar. 1952 (London: Oxford University Press, 1952).

CURIE, Eve: *Madame Curie* (London: William Heinemann Ltd, 1938).

CURTIS, Richard, and HOGAN, Elizabeth: *Perils of the Peaceful Atom: The Myth of Safe Nuclear Power Plants* (London: Gollancz, 1970).

DALLEK, Robert: *Franklin D. Roosevelt and American Foreign Policy, 1932–1945* (New York: Oxford University Press, 1979).

DAVIS, Nuel Pharr: *Laurence and Oppenheimer* (London: Cape, 1968).

de BROGLIE, Louis: *New Perspectives in Physics*, translated by A. J. Pomerans (Edinburgh and London: Oliver and Boyd, 1962).

de GAULLE, Charles: *Mémoires de Guerre*, Vol. II *L'Unité* (Paris: Plon, 1956).

DOUGLAS, A. Vibert: *The Life of Arthur Stanley Eddington* (London: Edinburgh, Paris, Melbourne, Toronto and New York: Thomas Nelson and Sons Ltd, 1956).

EDEN, Anthony: *Memoirs: Full Circle* (London: Cassell, 1960).

EGGLESTON, Wilfred: *Canada's Nuclear Story* (Toronto, Vancouver: Clarke, Irwin and Company Ltd, 1965).

EINSTEIN, Albert: 'Atomic War or Peace', *Atlantic Monthly*, Nov. 1945.

EISENHOWER, Dwight D.: *The White House Years: Mandate for Change, 1953–1956* (London: Heinemann, 1963).

EVANS, Ifor: *Man of Power* (London: Stanley Paul and Co., 1939).

EVE, A. S.: *Rutherford, Being the Life and Letters of the Rt. Hon. Lord Rutherford, O.M.* (Cambridge, at the University Press, 1939).

FEIS, Herbert: *Japan Subdued: The Atomic Bomb and the End of the War in the Pacific* (Princeton: Princeton University Press, 1966).

FELD, Bernard T., and SZILARD, Gertrud Weiss (eds): *The Collected Works of Leo Szilard. Scientific Papers* (Cambridge, Massachusetts, and London, England: The M.I.T. Press, 1972).

FERMI, Enrico: *Collected Papers of Enrico Fermi*, Vol. II (Chicago: University of Chicago Press, 1965).

FERMI, Laura: 'Bombs or Reactors', *Bulletin of the Atomic Scientists*, Vol. XXVI, No. 6, June 1970, pp. 28–9.

FISHER, H. A. L.: *A History of Europe* (London: Edward Arnold and Co., 1936).

FRANK, Philipp: *Einstein: His Life and Times* (London: Jonathan Cape, 1948).

FREEDMAN, Max: *Roosevelt and Frankfurter: Their Correspondence 1928–1945*, annotated by Max Freedman (London: Bodley Head, 1967).

FRISCH, Otto R.: 'Somebody Turned the Sun on with a Switch', *Bulletin of the Atomic Scientists*, Vol. XXX, No. 4, April 1974, pp. 12–18.

FRISCH, Otto R., and WHEELER, John A.: 'The Discovery of Fission', *Physics Today*, Vol. 20, No. 11, Nov. 1967, pp. 43–8.

GALAMBOS, Louis (ed.): *The Papers of Dwight David Eisenhower*, Vol. VII *The Chief of Staff*, Vol. VIII *The Chief of Staff*, and Vol. IX *The Chief of Staff* (Baltimore and London: The Johns Hopkins University Press, 1978).

GALANTIÈRE, Lewis (ed.): *The Goncourt Journals, 1851–1870*, edited and translated from the *Journal* of Edmond and Jules de Goncourt with an Introduction, Notes and a Biographical Repertory by Lewis Galantière (London, Toronto, Melbourne and Sydney: Cassell, 1937).

GLASSTONE, Samuel: *Sourcebook on Atomic Energy*, third edition (Princeton, N.J., Toronto, London, Melbourne: D. Van Nostrand Co. Inc., 1950, 1958 and 1967).

GOLDSMITH, Maurice: *Frédéric Joliot-Curie* (London: Lawrence and Wishart, 1976).

GOLOVIN, I. N.: *I. V. Kurchatov: A Socialist-Realist Biography of the Soviet Nuclear Scientist*, translated from the Russian by William H. Dougherty (Bloomington, Indiana: The Selbstverlag Press, 1968).

GOWING, Margaret: *Britain and Atomic Energy, 1939–1945* (London: Macmillan and Co. Ltd.; New York: St. Martin's Press, 1964).

GOWING, Margaret, assisted by ARNOLD, Laura: *Independence and Deterrence: Britain and Atomic Energy, 1945–1952*, Vol. I *Policy Making*, Vol. 2 *Policy Execution* (London: Macmillan, 1974).

GOWING, Margaret: 'Reflections on atomic energy history', *Bulletin of the Atomic Scientists*, Vol. 35, No. 3, Mar. 1979, pp. 51–4.

GROVES, Leslie R.: *Now It Can Be Told: The Story of the Manhattan Project* (London: Andre Deutsch, 1963).

GROVES, Lt. Gen. Leslie R.: 'Some Recollections of July 16, 1945', *Bulletin of the Atomic Scientists*, Vol. XXVI, No. 6, June 1970, pp. 21–7.

HAHN, Otto: *Otto Hahn My Life*, translated by Ernst Kaiser and Eithne Wilkins (London: Macdonald, 1970).

HEWLETT, Richard G., and ANDERSON, Oscar E., Jr.: *The New World, 1939/1946*, Vol. I, *A History of the United States Atomic Energy Commission* (University Park: The Pennsylvania State University Press, 1962).

HOLLOWAY, David: 'Entering the Nuclear Arms Race: The Soviet Decision to Build the Atomic Bomb, 1939–45', Working Paper presented at International Security Studies Program colloquium at the Woodrow Wilson International Center for Scholars, Washington, 25 July 1979 (Washington, D. C.: The Wilson Center, 1979).

HOYLE, Fred: *Energy or Extinction: The Case for Nuclear Energy* (London: Heinemann, 1977).

HOWARTH, Muriel: *Pioneer Research on the Atom. The Life Story of Frederick Soddy* (London: New World Publications, 1958).

IRVING, David: *The Virus House* (London: William Kimber, 1967).

JONES, R. V.: *Most Secret War* (London: H. Hamilton, 1978).

JONES, R. V.: 'Winston Leonard Spencer Churchill', *Biographical Memoirs of Fellows of the Royal Society, 1966*, Vol. 12, pp. 35–105.

JUNGK, Robert: *Brighter than a Thousand Suns: The Moral and Political History of the Atomic Scientists*, translated into English by James Cleugh (London: Victor

Gollancz Ltd in association with Rupert Hart-Davis, 1958).
KRAMISH, Arnold: *Atomic Energy in the Soviet Union* (Stanford, California: Stanford University Press; London: Oxford University Press, 1960).
LASH, Joseph P.: *From the Diaries of Felix Frankfurter* (New York: W. W. Norton and Co., 1975).
LAURENCE, William: *The Hell Bomb* (London: Hollis and Carter, 1951).
LEAHY, Admiral William D.: *I Was There* (New York: Whittlesey House, 1950).
LILIENTHAL, David E.: *Change, Hope and the Bomb* (Princeton, N.J.: Princeton University Press, 1963).
LILIENTHAL, David E.: *The Journals of David E. Lilienthal*, Vol. II *The Atomic Energy Years, 1945–1950* (New York, Evanston and London: Harper and Row, 1964).
LODGE, Sir Oliver: *Atoms and Rays: An Introduction to Modern Views of Atomic Structure and Radiation* (London: Ernest Benn Ltd, 1924).
MCLAINE, Ian: *Home Front Morale and the Ministry of Information in World War Two* (London: Allen and Unwin, 1979).
MILLIS, Walter (ed.), with the collaboration of DUFFIELD, E. S.: *The Forrestal Diaries: The Inner History of the Cold War* (London: Cassell and Co. Ltd, 1952).
MOORE, Ruth: *Niels Bohr: The Man and the Scientist* (London: Hodder and Stoughton, 1967).
NERNST, Professor Walter: *Theoretical Chemistry from the standpoint of Avogadro's Rule and Thermodynamics*, revised in accordance with the fourth German edition (London: Macmillan and Co., 1904).
NICOLSON, Harold: *Public Faces* (London: Constable and Co. Ltd, 1932).
NIELSON, J. Rud: 'Memories of Niels Bohr', *Physics Today*, Vol. 16, No. 10, Oct. 1963, pp. 22–31.
OLIPHANT, Mark: *Rutherford: Recollections of the Cambridge Days* (Amsterdam, London, New York: Elsevier Publishing Company, 1972).
OWEN, Sir Leonard: 'Nuclear Engineering in the United Kingdom in the First Ten Years', *The Journal of the British Nuclear Energy Society*, Vol. 2, No. 1, Jan. 1963, pp. 23–32.
PICKERSGILL, J. W., and FOSTER, D. F.: *The Mackenzie King Record*, Vol. 1 1939–1944, Vol. 2 1944–1945, Vol. 3 1945–1946, and Vol. 4 1947–1948 (University of Toronto Press, 1960, 1968, 1970 and 1970).
PIRIE, A. (General Ed.): *Fall Out: Radiation Hazards from Nuclear Explosions* (London: Macgibbon and Kee, 1958).
READER, W. J.: *Imperial Chemical Industries: A History*, Vol. II *The First Quarter Century, 1926–1952* (London: Oxford University Press, 1975).
RICHARDS, Denis: *Portal of Hungerford* (London: Heinemann, 1978).
ROZENTHAL, S. (ed.): *Niels Bohr, His life and work as seen by his friends and colleagues* (Amsterdam: North-Holland Publishing Company, 1967).
RUTHERFORD, E.: *Radio-activity* (Cambridge, at the University Press, 1904).
RUTHERFORD, Lord: *The Newer Alchemy*, based on the Henry Sidgwick Memorial Lecture held at Newnham College, Cambridge, Nov. 1936 (Cambridge, at the University Press, 1937).
RUTHERFORD, Ernest: *The Collected Papers of Lord Rutherford of Nelson*, Vol. 3 (London: George Allen and Unwin Ltd, 1965).
RUTHERFORD, Ernest: 'The Transmutation of Energy', Watt Anniversary Lecture, Watt Hall, Greenock, Jan. 1936.
RUTHERFORD, E., and SODDY, F.: 'Radioactive Change', *Philosophical Magazine*, Ser. 6,

V, May 1903, pp. 576–91.

SAKHAROV, Andrei E.: 'How I Came to Dissent', *The New York Review of Books*, Vol. XXI, No. 4, 21 Mar. 1974, pp. 11–17.

SACHS, Alexander: 'Opening Testimony' in Hearings before the Special Committee on Atomic Energy, United States Senate, Seventy-ninth Congress, First Session, Pursuant to S. Res. 179, 27 Nov. 1945, 'Background and Early History Atomic Bomb Project in Relation to President Roosevelt', Revised Transcript, pp. 553–73 (Washington: United States Government Printing Office, 1946).

SALAMAN, Esther: 'A Talk With Einstein', *The Listener*, 8 Sep. 1955.

SCHILLING, Werner R.: 'The H-Bomb Decision: How to Decide without actually choosing', *Political Science Quarterly*, Vol. LXXVI, 1961, pp. 24–46.

SCHILPP, Arthur (ed.): *Albert Einstein: Philosopher-Scientist* (Evanston, Ill.: The Library of Living Philosophers Inc., 1949).

SCHRIFTGIESSER, Karl: *The Lobbyists: The Art and Business of Influencing Lawmakers* (Boston: Little, Brown, 1951).

SEELIG, Carl: *Albert Einstein: A Documentary Biography* (London: 1956).

SHEINMAN, Lawrence: *Atomic Energy Policy in France under the Fourth Republic* (Princeton, N.J.: Princeton University Press, 1965).

SHERWIN, Martin J.: *A World Destroyed: The Atomic Bomb and the Grand Alliance* (New York: Alfred A. Knopf, 1975).

SHERWOOD, Robert E.: *The White House Papers of Harry L. Hopkins: An Intimate History* (London: Eyre and Spottiswoode, 1948, 1949).

SHTEMENKO, S. M.: *The General Staff at War, 1941–45* (Moscow: Progress Publishers, 1970).

SMITH, Alice Kimball: *A Peril and a Hope: The Scientists' Movement in America: 1945–47* (Chicago and London: University of Chicago Press, 1965).

SMITH, Alice Kimball: 'Los Alamos: Focus of an Age', *Bulletin of the Atomic Scientists*, Vol. XXVI, No. 6, June 1970, pp. 15–20.

SMITH, Alice Kimball, and WEINER, Charles (eds): *Robert Oppenheimer, Letters and Recollections* (Cambridge, Massachusetts, and London, England: Harvard University Press, 1980).

SNOW, C. P.: 'A New Means of Destruction', *Discovery*, New Series Vol. 2, No. 18, Sept. 1939, pp. 443–4.

SODDY, Frederick: *The Interpretation of Radium*, Being the substance of six free popular experimental lectures delivered at the University of Glasgow, 1908 (London: John Murray, 1909).

STEINER, Arthur: 'Scientists, Statesmen, and Politicians: The Competing Influences on American Atomic Energy Policy, 1945–64', *Minerva. A Review of Science, Learning and Policy*, Vol. XII, No. 4, Oct. 1974, pp. 469–509.

STIMSON, Henry L., and BUNDY, McGeorge: *On Active Service in Peace and War* (London, New York, Melbourne, Sydney, Cape Town: Hutchinson and Co. Ltd, n.d.).

STRAUSS, Lewis L.: *Men and Decisions* (London: Macmillan and Co. Ltd, 1963).

SZILARD, Gertrud Weiss, and WINDSOR, Kathleeen R. (eds): 'Reminiscences' by Leo Szilard, *Perspectives in American History*, Vol. II (Cambridge, Massachusetts: Harvard University, 1968).

SZILARD, Leo: 'We Turned the Switch', *Nation*, Vol CLXI, 22 Dec. 1945, pp. 718–19.

TELLER, Edward, with BROWN, Allen: *The Legacy of Hiroshima* (London: Macmillan and Co. Ltd, 1962).

THIRRING, J. H.: *The Ideas of Einstein's Theory* (London: Methuen and Co., 1921).

TRUMAN, Harry S.: *The Memoirs of Harry S. Truman*, Vol. 1 *Year of Decisions, 1945* (London: Hodder and Stoughton, 1955).

UNITED STATES ATOMIC ENERGY COMMISSION: 'In the Matter of J. Robert Oppenheimer: Transcript of Hearing before Personnel Security Board, Washington, D.C., April 12, 1954, through May 6, 1954' (Washington, D.C.: Government Printing Office, 1954).

UNITED STATES DEPARTMENT OF STATE: Foreign Relations of the United States, Diplomatic Papers, 'The Conferences at Washington and Quebec, 1943' (Washington, D.C., 1970); 'The Conference of Berlin (The Potsdam Conference), 1945', Vols I and II (Washington, D.C., 1960); and 'Diplomatic Papers, 1945', Vol. II 'General: Political and Economic Matters', Vol. V 'Europe' (Washington, D.C., 1967).

van der POEL, Jean (ed.): *Selections from the Smuts Papers*, Vol. VI (Cambridge, at the University Press, 1973).

WEART, Spencer R.: *Scientists in Power* (Cambridge, Massachusetts, and London, England: Harvard University Press, 1979).

WEART, Spencer R., and SZILARD, Gertrud Weiss (eds): *Leo Szilard: His Version of the Facts* (Cambridge, Massachusetts, and London, England: The M.I.T. Press, 1978).

WEBSTER, Sir Charles, and FRANKLAND, Noble: *The Strategic Air Offensive Against Germany, 1939–1945*, Vol. IV (London: H.M.S.O. 1961).

WELLS, H. G.: *The World Set Free: A Story of Mankind* (London: Macmillan and Co. Ltd, 1914).

WHARTON, Michael (ed.): *A Nation's Security: The Case of Dr J. Robert Oppenheimer*. Edited from the official transcript of evidence given before the Personnel Security Board of the United States Atomic Energy Commission (London: Secker and Warburg, 1955).

WHEELER-BENNETT, John W.: *John Anderson, Viscount Waverley* (London, New York: Macmillan and Co. Ltd, 1962).

YORK, Herbert F.: *The Advisors, Oppenheimer, Teller and the Superbomb* (San Francisco: W. H. Freeman and Company, 1976).

ZHUKOV, G. K.: *The Memoirs of Marshal Zhukov* (London: Jonathan Cape, 1971).

Notes and References

Page

3. 'by a gigantic system': H. A. L. Fisher, *A History of Europe*, p. 1215.
4. 'One thing is certain': quoted A. Vibert Douglas, *The Life of Arthur Stanley Eddington*, p. 43.
4. the revolutionary paper: Albert Einstein, 'Zur Elektrodynamik bewegter Körper', *Annalen der Physik*, Ser. 4, Vol. 17, 1905, pp. 891–921.
4. a brief paper: Albert Einstein, 'Ist die Trägheit eines Körpers von seinem Energieinhalt abhängig?', *Annalen der Physik*, Ser. 4, Vol. 18, 1905, pp. 639–41.
4. 'This thought is amusing': Einstein – Conrad Habicht, Summer 1905, quoted Carl Seelig, *Albert Einstein: A Documentary Biography*, p. 76.
5. 'Atoms or systems': Alexander Pope, *An Essay on Man*, I, 87.
5. 'the discovery of an outlying island': J. J. Thomson, President of the Royal Society, Address to Fellows of the Royal and the Royal Astronomical Societies, Burlington House, London, 6 Nov. 1919, quoted *The Times*, 8 Nov. 1919.
6. 'how God created this world': quoted Esther Salaman, 'A Talk With Einstein', *The Listener*, Vol. LIV, No. 1384 8 Sep. 1955, pp. 370–371, here p. 371.
6. 'All these things being consider'd': Sir Isaac Newton, *Opticks*, Book 3, Part 1, Query 31, p. 400. Based on the Fourth Edition, London, 1730.
7. 'more as a visualizing symbol': Paul Arthur Schilpp (ed.), *Albert Einstein: Philosopher–Scientist*, p. 19.
7. 'know of what the atom': quoted Edmund and Jules de Goncourt, Entry for 7 April 1869, in Lewis Galantière (ed.), *The Goncourt Journals, 1851–1870*, edited and translated from the *Journal* of Edmond and Jules de Goncourt with an Introduction, Notes and a Biographical Repertory by Lewis Galantière, p. 276, (afterwards referred to as 'Goncourt').
8. 'man would be' and 'To all this we raised': quoted Goncourt p. 276.
8. 'I was brought up': quoted E. N. da C. Andrade, The Rutherford Memorial Lecture, 1957, *Proceedings of the Royal Society*, A. Vol. 244, pp. 437–55 (1958), here p. 439 (afterwards referred to as 'Andrade').
8. 'Are you God': quoted Mark Oliphant, *Rutherford: Recollections of the Cambridge Days*, p. 145 (afterwards referred to as 'Oliphant').
8. 'We are all much obliged'; and 'You know, Bohr': quoted Oliphant, p. 28.
9. '... science is international': Ernest Rutherford, Radio address from Broadcasting House, London, to Fifth Pacific International Science Congress,

Vancouver, on 11 May, 1933, quoted A. S. Eve, *Rutherford, Being the Life and Letters of the Rt. Hon. Lord Rutherford, O.M.*, p. 373 (afterwards referred to as 'Eve').

9. 'the first strong breeze': H. T. Tizard, 'Rutherford' in *Dictionary of National Biography, 1931–1940*, p. 766.
9. 'enormous quantities of energy': Max Born, 'Fifty Years of Physics', Lecture to the Royal Society of Edinburgh, 23 Oct. 1950, printed in J. L. Crammer (ed.), *Science News*, 19, pp. 46–60, here p. 51.
10. 'One may also imagine': Pierre Curie, Nobel Lecture, 6 June 1905, quoted Eve Curie, *Madame Curie*, p. 239.
11. 'Don't call it transmutation': quoted Muriel Howarth, *Pioneer Research on the Atom. The Life of Frederick Soddy*, p. 84.
11. 'The difference between the energy': E. Rutherford, *Radio-activity*, pp. 337 and 338.
11. a paper on radioactive change: E. Rutherford and F. Soddy, 'Radioactive Change', *Philosophical Magazine*, Ser. 6, V, May 1903, pp. 576–591.
12. 'playful suggestion that': Sir W. C. D. Whetham–Rutherford, 26 July 1903, quoted Eve, p. 102.
12. 'Today I have made a discovery': quoted Max Born, 'Max Karl Ernst Ludwig Planck 1858–1947', *Obituary Notices of Fellows of the Royal Society, 1948–49*, Vol. VI, pp. 161–88, here p. 170 (afterwards referred to as 'Born, Planck Obit.').
13. 'though the actual production': Planck, 'Bemerkungen zum Prinzip der Aktion und Reaktion in der allgemeinen Dynamik', Verh. Ges. dtsch. Naturf. Arzte, Köln; Verh. dtsh. Phys. Ges. 10, 728–732; Phys. Z. 9, 828–830, quoted Born, Planck Obit., p. 174.
13. 'as if you had fired': quoted Andrade, p. 444.
15. 'dreamers of dreams': Arthur O'Shaughnessy, 'Ode'.
15. 'a porridge composed of Mr. Wells's vivid imagination': Review of *The World Set Free*, *The Times Literary Supplement*, 14 May 1914, p. 238.
15. 'long passages to the eleventh chapter': H. G. Wells, *The World Set Free: A Story of Mankind*, Dedication (afterwards referred to as 'Wells').
16. 'The nightmare will not come true': 'Musings without Method. "The World Set Free" – Mr. Wells' Revolution', *Blackwood's Magazine*, Vol. CXCV, No. MCLXXXIV, June 1914, pp. 859–864, here p. 864.
16. 'to be borne in mind': Rutherford, 'Radiations from Radium', Address at New Islington Public Hall, 7 Feb. 1916, quoted Eve, p. 253.
16. 'he could scarcely believe': R. V. Jones, 'The Rt. Hon. Viscount Cherwell, P.C., C.H., F.R.S.', Obituary, *Nature*, Vol. 180, 21 Sep. 1957, pp. 579–581, here p. 581 (afterwards referred to as 'Jones, Cherwell Obit.').
16. 'Do you really think': H. T. Tizard, quoted R. V. Jones to author.
17. 'which if it could be utilized': quoted *The Times*, 18 Sep. 1919.
18. 'And if ever the human race': Sir Oliver Lodge, *Atoms and Rays: An Introduction to Modern Views of Atomic Structure and Radiation*, p. 54, (afterwards referred to as 'Lodge').
18. 'He felt that we were on the brink': Sir Oliver Lodge, 'Sources of Power, Known and Unknown', Trueman Wood Lecture to Royal Society of Arts, London, 10 Dec. 1919, quoted *The Times*, 11 Dec. 1919.
19. 'an audacious speculation'; and 'I conceive': quoted Lodge, p. 46.
20. 'Aston's laboratory was a dingy darkened room': Oliphant, p. 45.
20. 'by its peculiar sequence': F. W. Aston, 'Atomic Weight of Caesium: Use of the

word "Mass-spectrograph"', *Nature*, Vol. CXXVII, 30 May 1931, p. 813.

22. 'I have little doubt myself': Aston, 'The Structural Units of the Material Universe', 7th Earl Grey Memorial Lecture, at King's Hall, Armstrong College, Newcastle-on-Tyne, 5 Mar. 1925, p. 23.
23. 'We sometimes dream that man': Sir Arthur Eddington, Presidential address to Section A of the British Association, Cardiff, 24 Aug. 1920, printed 'The Internal Constitution of the Stars', *The Observatory, A Monthly Review of Astronomy*, Vol. XLIII, No. 557, Oct. 1928, pp. 341–358, here p. 353.
23. 'it is hoped will one day be available': Professor Lindemann, 'Discussion on Isotopes', 3 March 1921, *Proceedings of the Royal Society of London, Series A*, Vol. XCIX, Sept. 1921, pp. 87–104, here p. 104.
23. 'to think of what might happen': J. H. Thirring, *The Ideas of Einstein's Theory*, p. 92.
23. 'You haven't lost anything'; quoted Philipp Frank, *Einstein: His Life and Times*, p. 211.
24. 'Should the research worker': F. W. Aston, 'Atomic Weights and Isotopes', Lecture to the Franklin Institute, Philadelphia, 10 Mar. 1922, *Journal of the Franklin Institute*, Vol. 193, No. 5, May 1922, pp. 581–608, here p. 606.
24. 'May there not be methods': Winston Churchill, 'Shall we all commit suicide?', *Thoughts and Adventures*, p. 250.
25. 'outlook for gaining useful energy': Lord Rutherford, *The Newer Alchemy*, based on the Henry Sidgwick Memorial Lecture held at Newnham College, Cambridge, Nov. 1936, p. 65.
25. 'The fact is, at this time': E. O. Lawrence, 'Atoms: New and Old', Sigma Xi lecture, 1938, quoted Samuel Glasstone, 'Sourcebook on Atomic Energy', third edition, p. 507 (afterwards referred to as 'Glasstone').
25. 'the Italian navigator has just landed': Arthur Compton to James B. Conant, 2 Dec. 1942, Arthur H. Compton, 'The Birth of Atomic Power', *Bulletin of the Atomic Scientists*, Vol. IX, No. 1, Feb. 1953, pp. 10–12, here p. 12 (afterwards referred to as 'Compton, At. Power').
26. 'I had a sudden inspiration': Louis de Broglie, *New Perspectives in Physics*, p. 138 (afterwards referred to as 'de Broglie').
26. 'lifted a corner of the great veil': Einstein–Paul Langevin, 1923, quoted de Broglie, p. 139.
27. 'I, at any rate, am convinced': Einstein–Max Born, 12 Dec. 1926, Max Born, *Born–Einstein Letters, 1916–1955*, p. 91.
27. 'It appeared that the mountainous barriers': Sir John Cockcroft, *The Development and Future of Nuclear Energy*, The Romanes Lecture delivered in the Sheldonian Theatre, 2 June 1950, and printed, p. 7 (afterwards referred to as 'Cockcroft, Romanes Lecture').
28. 'There were frequent vacuum troubles': Oliphant, p. 84.
28. 'When the voltage and the current of protons': quoted Oliphant, p. 85.
29. 'It seems not unlikely': J. D. Cockcroft and E. T. S. Walton letter to Editor, 'Disintegration of Lithium by Swift Protons', *Nature*, Vol. 129, No. 3261, 30 April 1932, p. 649.
29. 'The social unsettlement of the age': quoted Oliphant, p. 88.
29. 'an electrically neutral massless molecule': Professor Walter Nernst, *Theoretical Chemistry from the standpoint of Avogadro's Rule and Thermodynamics*, p. 392.
30. 'Under some conditions': Rutherford, 'Nuclear Constitution of Atoms',

Bakerian Lecture delivered to Royal Society, 3 June 1920, printed *Collected Papers of Lord Rutherford of Nelson*, Vol. 3, pp. 14–38, here p. 34.

30. 'like an invisible man': quoted Ifor Evans, *Man of Power*, p. 195.
30. 'wanted someone to talk to': Sir James Chadwick, 'Some Personal Notes on the Search for the Neutron', 'Proceedings of Tenth International Congress of the History of Science', Ithaca, New York, 26 VIII 1962–3 IX 1962, pp. 159–162, here p. 159 (afterwards referred to as 'Chadwick').
30. 'so far-fetched as to belong': Chadwick, p. 159.
31. 'Not many minutes afterwards': Chadwick, p. 161.
33. 'knew only that the experts': Harold Nicolson, *Public Faces*, p. 17.
33. 'The destiny of mankind': Desmond MacCarthy, 'Science and Politics', *The New Statesman and Nation*, Vol. III, No. 63 (New Series), 7 May 1932, pp. 584–585, here p. 585.
33. 'These transformations of the atom': quoted Eve, p. 374.
33. *The Times* quoted him: *The Times*, 12 Sep. 1933.
34. 'I believe that he was fearful': Oliphant, p. 141.
34. 'searching for a new source of power': quoted Cockcroft, Romanes Lecture, p. 8.
34. 'the recent discovery of the neutron': Rutherford, 'The Transmutation of Energy', Watt Anniversary Lecture, Watt Hall, Greenock, Jan. 1936.
35. 'not *this* version of the facts': Preface, Spencer R. Weart and Gertrud Weiss Szilard (eds), *Leo Szilard: His Version of the Facts*, p. xvii (afterwards referred to as 'Weart and Szilard').
35. 'Of course, all this is moonshine': Leo Szilard–Sir Hugo Hirst, 17 Mar. 1934, Weart and Szilard, p. 38.
35. 'It suddenly occurred to me': Weart and Szilard, p. 17.
35. 'Surely I have explained often enough': quoted Oliphant, p. 141.
36. 'I had one candidate for an element': Weart and Szilard, p. 17.
36. 'This invention has for its object': Szilard, 'Improvements in or relating to the Transmutation of Chemical Elements', British Patent Specification No. 630,726, 28 Sep. 1949, in Bernard T. Feld and Gertrud Weiss Szilard (eds), *The Collected Works of Leo Szilard. Scientific Papers*, p. 639 (afterwards referred to as 'Szilard, Coll. Works').
37. 'The object of this Patent': Szilard–Dr C. S. Wright, 26 Feb. 1936, Szilard, Coll. Works, p. 733.
37. 'I am naturally somewhat less optimistic': Probably from Prof. F. A. Lindemann–Dr C. S. Wright, 26 Feb. 1936, Szilard, Coll. Works, p. 733.
38. 'Even if I am grossly exaggerating': Szilard–Prof. Lindemann, 3 June 1935, Weart and Szilard, p. 41.
38. letters the following year: Szilard–Lord Rutherford, 27 May 1936, and Szilard–Dr Cockcroft, 27 May 1936, Weart and Szilard, pp. 45–6.
38. 'as generous with his ideas': quoted Szilard, Coll. Works, p. xvi.
38. 'I feel that I must not consider': Szilard–Enrico Fermi, 13 Mar. 1936, Szilard, Coll. Works, p. 730.
39. 'If surveying the past': Frédéric Joliot-Curie, Nobel Prize Speech, Stockholm, 12 Dec. 1935.
39. 'We were working very hard': *Collected Papers of Enrico Fermi*, Vol. II, 'United States 1939–1954', p. 926.
40. 'given a small bottle of element 93': Laura Fermi, 'Bombs or Reactors', *Bulletin of the Atomic Scientists*, Vol. XXVI, No. 6, June 1970, pp. 28–9, here p. 28.

40. 'One could assume equally well': Ida Noddack, *Zeitschrift für Angewandte Chemie*, Vol. 47, 1934, p. 653.
42. 'They felt protactinium was their own baby': Otto R. Frisch, 'How It All Began', in Otto R. Frisch and John A. Wheeler, 'The Discovery of Fission', *Physics Today*, Vol. 20, No. 11, Nov. 1967, pp. 43–8, here p. 46 (afterwards referred to as 'Frisch').
42. 'Being Jewish she had to wear a yellow badge': quoted Lewis L. Strauss, *Men and Decisions*, p. 170 (afterwards referred to as 'Strauss').
43. 'Exciting experiment with mesothorium': Otto Hahn, Notebook entry, 17 Dec. 1938, quoted Otto Hahn, *My Life*, p. 164 (afterwards referred to as 'Hahn').
44. 'It is now practically eleven o'clock at night': Hahn–Lise Meitner, 19 Dec. 1938, Hahn, p. 150.
44. 'However, as "nuclear chemists" ': Hahn and Strassmann, 'Uber den Nachweis und das Verhalten der bei der Bestrahlung des Urans mittels Neutronen Entstehenden Erdalkalimetalle', *Die Naturwissenschaften*, Jahrgang 27, Heft 1, 6 Jan. 1939, p. 11–15, here p. 15.
44. 'felt like an imbecile': Wells, p. 34.
45. 'In a small hotel in Kungälv': Frisch, p. 47.
45. 'It took her a little while': Frisch to author, quoted Ronald W. Clark, *The Birth of the Bomb*, p. 13, (afterwards referred to as 'Clark (1961)').
45. 'The picture was that of two fairly large nuclei': Frisch to author, quoted Clark (1961), p. 15.
46. 'I had hardly begun to tell him': quoted Ruth Moore, *Niels Bohr: The Man and the Scientist*, p. 226 (afterwards referred to as 'Moore').
46. 'At first Placzek did not believe the story': Frisch, p. 47.
47. 'it was computed that the energy': Frisch to author, quoted Clark (1961), p. 15.
48. 'Bohr has just come in': quoted, Edward Teller with Allen Brown, *The Legacy of Hiroshima*, p. 8 (afterwards referred to as 'Teller').
49. 'New hope for releasing the enormous stores of energy': Robert Potter, 'Science Service' Report, 27 Jan. 1939.
49. 'Do you know what this new discovery means?': quoted Herbert F. York, *The Advisors: Oppenheimer, Teller and the Superbomb*, p. 29 (afterwards referred to as 'York').
50. 'When I heard [of Hahn's discovery]': Leo Szilard, 'Reminiscences', edited by Gertrud Weiss Szilard and Kathleen R. Windsor, in *Perspectives in American History*, Vol. II, 1968, pp. 94–151, here p. 106 (afterwards referred to as 'Szilard, Reminiscences').
50. 'This is entirely unexpected': Szilard–Lewis L. Strauss, 25 Jan. 1939, quoted Strauss, p. 172.
51. 'Nuts!': quoted Szilard, Reminiscences, p. 107.
51. 'Ten per cent is not a remote possibility': quoted Szilard, Reminiscences, p. 107.
52. 'The idea of such destructive power': Jones, Cherwell Obit., p. 581.
52. 'Performed today proposed experiment': Szilard–Strauss, telegram, 6 Mar. 1939, quoted Strauss, p. 174.
52. the idea of a 'lattice-reactor', and the rights for nothing: quoted Albert Wohlstetter, 'Technology, Prediction and Disorder', 'Bulletin of the Atomic Scientists', Oct. 1964, pp. 11–20, here p. 18.
53. 'This is a tremendous military advantage': Ross Gunn–Director of Naval Research Laboratory, 1 June 1939, quoted Strauss, p. 436.

54. 'News from Joliot': Victor Weisskopf and Szilard–Halban, cable, 31 Mar. 1939, Weart and Szilard, p. 71.
54. 'It was the morning of 1 April'; and 'Then, I began to realise': Hans Halban to author, quoted Clark (1961), p. 20.
55. 'From the time of these experiments': Halban to author, quoted Clark (1961), p. 21.
55. 'Learned last week': Joliot, Halban, Kowarski–Szilard, cable, 5 Apr. 1939, Weart and Szilard, p. 73.
55. 'Question studied': Joliot–Szilard, cable, 6 Apr. 1939, Weart and Szilard, p. 74.
56. 'Then he walked back through the evening lights': Wells, p. 36.
58. 'It is sobering to realize': Leslie R. Groves, *Now It Can Be Told: The Story of the Manhattan Project*, p. 33 (afterwards referred to as 'Groves').
58. 'I began to consider': G. P. Thomson to author, quoted Clark (1961), p. 35.
59. 'a scientific but remote possibility': quoted John W. Wheeler-Bennett, *John Anderson, Viscount Waverley*, p. 289.
59. 'He arranged for me to meet': quoted P. B. Moon, 'George Paget Thomson', 3 May 1892–10 Sept. 1975 *Biographical Memoirs of Fellows of the Royal Society*, Vol. 23, 1977, pp. 529–56, here p. 541.
59. 'as a result of a recent discovery': A. M. Tyndall, Paper for Committee of Imperial Defence, 3 May 1939, File AB 1/9, Public Record Office, Kew, London (afterwards referred to as 'P.R.O.').
59. 'Ismay ... is anxious': Sir Henry Tizard–David Pye, 9 May 1939, AB 1/219, P.R.O.
60. 'Sir Henry asked me to grant' and 'Be careful': Sir Edgar Sengier to author, quoted Ronald W. Clark, *Tizard*, p. 184 (afterwards referred to as 'Clark (1965)').
60. 'Uranium Ore, Product of the Belgian Congo': Strauss, p. 181.
61. 'If it is true that the Germans': quoted Clark (1965), p. 184.
61. 'been an immense success': quoted Clark (1965), pp. 185 and 186.
61. 'horrified at the thought': quoted Clark (1965), p. 186.
61. 'Our main work consisted': G. P. Thomson to author, quoted Clark (1961), p. 36.
62. 'There are indications': Winston S. Churchill, *The Second World War*, Vol. I *The Gathering Storm*, p. 301.
62. 'In the summer of 1939': quoted Robert Jungk, *Brighter than a Thousand Suns: The Moral and Political History of the Atomic Scientists* p. 87.
63. 'It must be made, if it really is': C. P. Snow, 'A New Means of Destruction', *Discovery*, New Series Vol. 2, No. 18, Sep. 1939, pp. 443–4, here p. 444.
67. 'The show was going before that letter': *Boston Globe*, 2 Dec. 1962.
67. 'had not been aware': Szilard–Carl Seelig, 19 Aug. 1955, Die Einstein-Sammlung der Eidgenossische Technische Hochschule Bibliothek, Zurich.
68. 'So we began to think through': Szilard, Reminiscences, p. 111.
68. 'It occurred to me then that Einstein': Szilard, Reminiscences, p. 111.
69. 'I had, however, an uneasy feeling': Szilard, Reminiscences, p. 112.
70. 'Sir, Some recent work by E. Fermi and L. Szilard': Albert Einstein–President Roosevelt, 2 Aug. 1939, Franklin D. Roosevelt Library, Hyde Park, New York.
71. 'I did not, in fact, foresee': Albert Einstein as told to Raymond Swing, 'Einstein on the Atomic Bomb', *The Atlantic Monthly*, Vol. 176, No. 5, Nov. 1945, pp. 43–45, here p. 44.
71. 'with a view to highlighting that': Alexander Sachs, 'Opening Testimony' in

Hearings before the Special Committee on Atomic Energy, United States Senate, Seventy-ninth Congress, First Session, Pursuant to S. Res. 179, 27 Nov. 1945, 'Background and Early History Atomic Bomb Project in Relation to President Roosevelt', Revised Transcript, pp. 553–73, here p. 558 (afterwards referred to as 'Sachs').

71. 'There are those about us who say': quoted Sachs, p. 558 as F. W. Aston, *Forty Years of Atomic Theory*.
71. 'Alex, what you are after' and 'Precisely': quoted Sachs, p. 558.
72. 'This requires action': quoted Sachs, p. 558.
72. 'No one can be sure that it will go off': Lord Cherwell, quoted Clark (1961), p. 194.
72. 'It sounds like a professor's dream to me': Admiral William D. Leahy, 'I was There', p. 502–3 (afterwards referred to as 'Leahy').
72. 'I knew of no explosive that would develop': Leahy, p. 502–3.
73. 'Many scientists were there': Sachs, p. 559.
73. 'dissatisfied with the scope': Sachs, p. 562.
73. 'Since the outbreak of the war': Einstein–Sachs, 7 Mar. 1940, Weart and Szilard, p. 120.
74. 'a time convenient to you and Dr Einstein': Roosevelt–Sachs, 5 Apr. 1940, Weart and Szilard, p. 122.
74. 'we shall prepare a polite letter of regret': Szilard–Einstein, 19 Apr. 1940, Szilard Papers.
75. 'I am convinced as to the wisdom': Einstein – Dr Briggs, 25 April 1940, quoted Sachs, p. 565.
76. 'We are approaching a tremendous revolution': quoted Arnold Kramish, *Atomic Energy in the Soviet Union*, p. 6 (afterwards referred to as 'Kramish').
76. 'We were worried because it was the middle of the year': A. F. Ioffe, 'Can Science be Planned?', *Moscow News*, 16 June 1945, quoted Kramish, p. 7.
77. 'Physics stands at the threshold': *Izvestia*, 31 Dec. 1940, quoted Kramish, p. 59.
77. 'It was too early to speak': V. G. Khlopin, quoted David Holloway, 'Entering the Nuclear Arms Race: The Soviet Decision to Build the Atomic Bomb, 1939–45', Working Paper presented at International Security Studies Program colloquium at the Woodrow Wilson International Center for Scholars, Washington, 25 July 1979, and printed p. 15 (afterwards referred to as 'Holloway').
78. 'I'm glad we didn't have it': quoted Groves, p. 334.
78. 'I believe the reason': quoted Groves, p. 335.
78. 'I don't believe that': quoted Groves, p. 335.
79. 'I think it absurd': quoted David Irving, *The Virus House*, p. 16 (afterwards referred to as 'Irving').
79. 'We would not have had the moral courage': quoted Irving, p. 110.
79. 'We were delighted [with the nuclear project]': quoted Irving, p. 46.
79. von Weizsäcker has been quoted: Professor von Laue, quoted Irving, p. 46.
80. 'We take the liberty of calling': Professor Paul Harteck and Dr Wilhelm Groth–German War Office, 24 Apr. 1939, quoted Irving, p. 34.
80. 'that one cubic metre of uranium oxide': quoted Irving, p. 37.
81. 'He complained of Hitler's bad grammar': R. V. Jones to author, quoted Ronald W. Clark, *The Rise of the Boffins*, p. 100.
81. 'They proposed to effect the fission': Sir Edgar Sengier to author, quoted Clark (1961), p. 28.

82. 'of producing, in a medium containing uranium': Sealed document No. 11620 at Académie des Sciences, quoted Maurice Goldsmith, *Frédéric Joliot-Curie*, p. 75.
83. 'Dr Aubert was very glad': quoted Clark (1961), p. 70.
85. 'was of the order of tons': Professor Rudolf Peierls to author, quoted Clark (1961), p. 43.
85. 'Fortunately our progressing knowledge': O. R. Frisch, 'Nuclear Fission', *Annual Reports on the Progress of Chemistry for 1939*, pp. 8–19, here p. 16.
86. 'I think one can say': Professor Chadwick–Dr Appleton, 5 Dec. 1939, AB 1/219, P.R.O.
86. 'I gather that we may sleep': quoted Margaret Gowing, *Britain and Atomic Energy, 1939–1945*, p. 39 (afterwards referred to as 'Gowing (1964)').
86. 'For example, if I throw a ball': Peierls to author, quoted Clark (1961), p. 44.
87. 'the building of Canada's first': Wilfred Eggleston, *Canada's Nuclear Story*, p. 21 (afterwards referred to as 'Eggleston').
87. a gift of $5,000: Eggleston, p. 24.
87. 'Any competent nuclear physicist': Peierls to author, quoted Clark (1961), p. 51.
88. 'We sat down to work out' and 'In fact, our first calculation': Peierls to author, quoted Clark (1961), p. 51.
88. 'It was the problem': Peierls to author, quoted Clark (1961), p. 52.
88. 'I worked out the results': Peierls to author, quoted Clark (1961), p. 53.
89. 'Write to Tizard': quoted Clark (1961), p. 55.
90. 'Most of it will probably be blown into the air': O. R. Frisch and R. Peierls, 'On the Construction of a "Super-bomb"; based on a Nuclear Chain Reaction in Uranium' [The Frisch-Peierls Memorandum, Part I], p. 3, AB 1/210, P.R.O.
90. 'Owing to the spreading of radioactive substances': O. R. Frisch and R. Peierls, 'Memorandum on the Properties of a Radioactive "Super-bomb" ' [The Frisch-Peierls Memorandum, Part II], AB 1/210, P.R.O.
90. 'What I should like would be': quoted Clark (1965), p. 218.
91. 'You must realise that any project': quoted Clark (1961), p. 58.
91. 'to examine the whole problem': quoted Clark (1965), p. 218.
92. 'raised the question': Thomson–Cockcroft, 5 Apr. 1940, AB 1/210, P.R.O.
92. 'entered the project with more scepticism': 'Report by M.A.U.D. Committee on the Use of Uranium for a Bomb' Part I, July 1941, AB 1/37, P.R.O.
92. 'M. Allier made a statement': Thomson's hand-written Minutes of Meeting held at the Royal Society, 10 Apr. 1940, AB 1/8, P.R.O.
93. 'M. Allier seems very excited': Tizard–Wing Commander Elliot, 11 Apr. 1940, Tizard Papers.
94. 'We have taken all necessary action': Tizard–Brigadier Charles Lindemann, 15 Apr. 1940, quoted Clark (1965), p. 220.
94. 'desirable that those engaged': Minutes of M.A.U.D. Committee Meeting, 8 Jan. 1941, AB 1/8, P.R.O.
94. 'The lab boy will be paid £1 a week': Halban–B. G. Dickins, 17 Mar. 1941, AB 1/20, P.R.O.
94. 'It is not inconceivable that practical': A. V. Hill, Report from Washington, 16 May 1940, AB 1/219, P.R.O.
95. 'It was decided to make up' and 'Dr Frisch produced some notes': Minutes of third meeting of C.S.S.A.W. Sub-Committee on U-Bomb, 17 May 1940, AB 1/8, P.R.O.

95. 'Met Niels and Margarethe recently': quoted O. W. R. Richardson–Cockcroft, 15 May 1940, AB 1/210, P.R.O.
96. 'You will see that the last three words': Cockcroft–Chadwick, 20 May 1940 (copy to Thomson), AB 1/210, P.R.O.
96. 'Since last three words do not appear': Military Intelligence note, AB 1/210, P.R.O.
96. 'all very wild but just sufficiently reasonable': R. H. Fowler–Tizard, 28 May 1940, quoted Clark (1965), p. 221.
96. 'The cryptic message from Bohr': Oliphant–Cockcroft, 21 May 1940, AB 1/210, P.R.O.
96. 'In order to avoid': Thomson–Pye, 20 June 1940, AB 1/219, P.R.O.
96. 'M.A.U.D. means': Edward Appleton–author, 25 Mar. 1960.
96. 'The M.A.U.D. Committee' and 'What meaning?': Churchill–Cherwell, Diary of T. A. Events, PREM 3/139/9, P.R.O.
96. 'M.A.U.D. was a code name': Cherwell–Churchill, 26 July 1945, PREM 3/139/9, P.R.O.
97. 'I believe that a sufficient quantity': Chadwick–Lindemann, 20 June 1940, Cherwell Papers.
97. 'My instructions were to take': Halban to author, quoted Clark (1961), p. 95.
98. 'Luckily all that was really vital': Halban to author, quoted Clark (1961), p. 96.
98. 'I was told to make all possible efforts': Halban–Dr Ben Lockspeiser, 2 Sep. 1940, AB 1/229, P.R.O.
98. 'Suffolk had got the crew too drunk': *Radio Times*, 20 Sep. 1973.
100. 'a substance known as "Heavy Water" ': Minute, 14 Sept. 1940, AB 1/229, P.R.O.
100. 'Dr Halban is not a person carrying on business': Minute, 20 Feb. 1941, AB 1/229, P.R.O.
100. 'I remember one member of the Committee': Cockcroft, Romanes Lecture, p. 12.
100. 'to carry out Halban's experiments': Minutes of M.A.U.D. Committee Meeting, 10 July 1940, AB 1/8, P.R.O.
101. 'Dr Halban described the events': Minutes of M.A.U.D. Committee Meeting, 10 July, 1940, AB 1/8, P.R.O.
101. 'Joliot had already some personal contact': Hans Halban, Report on Relations with Union Minière to Dr B. G. Dickins, Ministry of Aircraft Production, 11 Dec. 1940, AB 1/19, P.R.O.
102. 'Tizard is sure to try and pump you' and 'I duly whispered': James Tuck to author, quoted Clark (1965), p. 212.
102. 'think that any useful purpose': Peierls–Tizard, AB 1/219, P.R.O.
103. 'By train this takes two days': Halban–Dickins, 19 Sep. 1940, AB 1/18, P.R.O.
103. 'seemed to think that because of': Peierls–Dr Thewlis, 16 Sep. 1941, AB 1/220, P.R.O.
104. 'In the summer of 1940 American scientists': Richard G. Hewlett and Oscar E. Anderson, Jr., *The New World, 1939/1946*, p. 27, (afterwards referred to as 'Hewlett and Anderson').
105. 'For example are our Prime Minister': C. G. Darwin–Lord Hankey, 2 Aug. 1941, AB 1/37, P.R.O.
105. 'to break down the enemy's means of resistance': quoted, Sir Charles Webster and Noble Frankland, *The Strategic Air Offensive against Germany, 1939–1945*, Vol. IV, p. 72 (afterwards referred to as 'Webster and Frankland').

105. 'the impression produced': Webster and Frankland, p. 77.
105. 'civilian life [would] be endangered': Webster and Frankland, p. 83.
105. 'direct the main effort of the bomber force': Webster and Frankland, p. 136.
106. 'aim of the Combined Bomber Offensive': Harris–Air Ministry, 25 Oct. 1943, quoted Ian McLaine, *Home Front Morale and the Ministry of Information in World War Two*, p. 161.
106. 'I cannot say that at that time' and 'We knew what it was': Max Born to author, quoted Clark (1961), pp. 82 and 83.
107. 'I had to explain': Peierls to author, quoted Clark (1961), p. 84.
107. 'He left me to travel back home': Peierls to author, quoted Clark (1961), p. 85.
107. 'a riddle wrapped in a mystery': Churchill, Broadcast, 1 Oct. 1939.
108. 'It was like getting a doctor': Frisch to author, quoted Clark (1961), p. 87.
108. 'the place in which to bring up': quoted Clark (1961), p. 88.
109. 'Simon tells me he has seen you': Peierls–Lindemann, 2 June 1940, AB 1/106, P.R.O.
109. 'I do not know him sufficiently well': quoted Gowing (1964), p. 47.
109. 'Dr Mayneord said that a few micrograms of dust': Report of meeting of Professor Cockcroft and Dr Vick with Dr Mayneord, 24 May 1940, AB 1/210, P.R.O.
110. 'What we want is something like this': Dr Nicolas Kurti to author, quoted Clark (1961), p. 89.
110. 'The first thing we used was "Dutch cloth" ' and 'it was necessary to use something': Dr Nicolas Kurti to author, quoted Clark (1961), p. 90.
111. 'An experimental investigation at the Liverpool University': 'Contract for an experimental investigation regarding the possibilities of developing a new explosive at Liverpool University', 12 Sep. 1940, AB 1/227, P.R.O.
111. 'still premature to think': Thomson–Pye, 22 Oct. 1940, AB 1/227, P.R.O.
114. 'In one experiment in which': Halban–M.A.U.D. Committee, 8 Jan. 1941, Alteration to the Minutes of the Meeting, AB 1/222, P.R.O.
115. 'I do not believe that the problem': Chadwick–Cockcroft, 18 July 1940, AB 1/210, P.R.O.
115. 'Two distinct lines of research': Halban, 'Report of Work carried out at Cambridge, Sept.–Dec. 1940', AB 1/233, P.R.O.
115. 'This was the first time': Thomson to author, quoted Clark (1961), p. 121.
116. 'So much is [the Halban report]': Fowler–Cockcroft, 15 Mar. 1941, AB 1/210, P.R.O.
117. 'critical study by Chadwick': Fowler–Thomson, cable read at M.A.U.D. Committee meeting, 9 Apr. 1941, AB 1/8, P.R.O.
117. 'This being purely academic scientific work': Ministry of Aircraft Production minute, AB 1/233, P.R.O.
117. 'the present meeting' and 'The Chairman stated': Minutes of M.A.U.D. Committee meeting, 9 Apr. 1941, AB 1/8, P.R.O.
118. 'An extremely important new possibility': E. O. Lawrence report, July 1941, quoted Glasstone, p. 525.
118. written to Lindemann: Simon–Lindemann, 19 Nov. 1940, Cherwell Papers.
118. 'with the purpose of co-ordinating research': M. W. Perrin, note dated 4 Mar. 1941, AB 1/175, P.R.O.
119. 'I was coming to work one day': S. S. Smith to author, quoted Clark (1961), p. 128.
120. 'Halban's work on nuclear energy' and 'would agree to let the Government':

'M.A.U.D. Halban Scheme', n.d., AB 1/175, P.R.O.

120. 'I am very unhappy about the question': Oliphant–Dickins, 30 June 1941, AB 1/220, P.R.O.

122. 'I was working on various committees' and paper circulated: Sir Geoffrey Taylor to author, quoted Clark (1961), p. 92.

122. 'Between June and October 1941': W. J. Reader, *Imperial Chemical Industries: A History*, Vol. II *The First Quarter Century, 1926–1952*, p. 290 (afterwards referred to as 'Reader').

122. 'The practicability of the preparation': Summary, enclosed Lord Melchett–Thomson, 2 July 1941, AB 1/10, P.R.O.

123. 'Dear Professor Lindemann, I would particularly like': Melchett–Lindemann, 2 July 1941, AB 1/106, P.R.O.

123. '8. This explosive effect': Summary, enclosed Melchett–Lindemann, 2 July 1941, AB 1/106, P.R.O.

123. 'worth serious consideration': 'Report by M.A.U.D. Committee on the use of Uranium as a Source of Power' Part I, 15 July 1941, AB 1/37, P.R.O.

124. 'On 23 July 1941, the Secret War Committee': Reader, p. 290.

124. 'The committee considers that the scheme': 'Report by M.A.U.D. Committee on the Use of Uranium for a Bomb', Part I, July 1941, AB 1/37, P.R.O.

125. 'I hear that you have been discussing': Tizard–Lord Hankey, 5 Aug. 1941, Tizard Papers.

126. 'We have worked in close collaboration': Cherwell minute to Churchill on use of atomic energy for bomb, 27 Aug. 1941, AB 1/170, P.R.O.

126. 'General Ismay for Chiefs of Staffs Committee': Churchill–Chiefs of Staffs Committee, 30 Aug. 1941, PREM 3/139.8A, P.R.O.

126. 'we are to have additional support': quoted J. R. M. Butler, *Grand Strategy*, Vol. II, p. 354.

127. 'I am a little worried': Colonel Moore-Brabazon–Hankey, 12 Sep. 1941, AB 1/220, P.R.O.

128. 'the development of the uranium bomb' and 'The matter should not be allowed': Scientific Advisory Committee: Defence Services Panel, Report, 21 Sept. 1941, p. 9, CAB 90/8, P.R.O.

128. 'that I.C.I. Ltd is peculiarly fitted': Pye, minute dated 29 Sep. 1941, AB 1/220, P.R.O.

128. 'Dear Brab., In view of your discussion': Lord McGowan–Colonel Moore-Brabazon, 10 Oct. 1941, quoted Reader, p. 290.

129. 'So, ended I.C.I's bid': Reader, p. 291.

129. 'rather nervous phrases': Reader, p. 291.

130. 'The Medical Research Council should be invited': Scientific Advisory Committee: Defence Services Panel, Report, 21 Sept. 1941, p. 10, CAB 90/8, P.R.O.

131. 'This verges on the Gilbertian!': M.A.P. Memorandum to Moore-Brabazon, 16 Oct. 1941, AB 1/37, P.R.O.

133. 'but that Russia was our enemy': quoted Michael Wharton (ed.), *A Nation's Security: The Case of Dr J. Robert Oppenheimer*, p. 80 (afterwards referred to as 'Wharton').

135. 'The British haughtily replied': Margaret Gowing, 'Reflections on atomic energy history', *Bulletin of the Atomic Scientists*, Vol. 35, No. 3, Mar. 1979, pp. 51–4, here p. 52 (afterwards referred to as 'Gowing, Reflections').

135. 'This report gave Bush and Conant': Hewlett and Anderson, p. 43.

136. 'at the top': Roosevelt–Bush, 9 Oct. 1941, quoted Hewlett and Anderson, p. 259.
136. 'It appears desirable': Roosevelt–Churchill, 11 Oct. 1941, PREM 3/139, P.R.O.
136. 'thought it desirable that this work': Statement by Hovde, quoted AB 1/207, P.R.O.
137. 'My dear Mr President, Thank you so much': Churchill–Roosevelt, December 1941, PREM 3/139.8A, P.R.O.
137. 'There is, however, some doubt': Cherwell–Churchill, n.d., PREM 3/139.9, P.R.O.
137. 'We have checked our file': Franklin D. Roosevelt Library, Hyde Park–author, 12 Sep. 1979.
137. 'We can proceed at once': Norman Brook–W. A. Akers, 1 Dec. 1941, AB 1/207, P.R.O.
138. 'I must say that I like [Groves]': Akers–Halban, 1 Jan. 1943, AB 1/128, P.R.O.
138. 'the same complete freedom': W. A. Akers and H. H. Halban, 'Note on Relations between the British, Anglo-Canadian and American Groups. History of Collaboration between British and American Groups', 22 Jan. 1943, AB 1/128, P.R.O.
139. 'the pilot plant on either side' and 'highly desirable that future action': Bush–Anderson, 20 Apr. 1942, AB 1/207, P.R.O.
139. 'We will continue, on this matter': Bush–Anderson, 19 June 1942, AB 1/207, P.R.O.
139. 'In fact, he states that his intention': Akers–Perrin, 16 Nov. 1942, AB 1/128, P.R.O.
140. 'As I told you over the telephone': Dr Frank Aydelotte–Bush, 19 Dec. 1941, United States Atomic Energy Commission (afterwards referred to as 'U.S.A.E.C.').
140. 'I wish very much that I could place': Bush–Aydelotte, 30 Dec. 1941, U.S.A.E.C.
140. 'After full consideration of the position': Anderson–Bush, 5 Aug. 1942, AB 1/207, P.R.O.
141. 'I fear that you have a misconception': Bush–Anderson, 1 Sept. 1942, AB 1/207, P.R.O.
141. 'a most extraordinary interview' and 'on an after-the-war military basis': V. Bush, 'Memorandum of Conference with the President [Roosevelt], 24 June 1943', U.S. Department of State, 'Foreign Relations of the United States. Diplomatic Papers. The Conferences at Washington and Quebec, 1943', pp. 631–2, here p. 631, (afterwards referred to as 'F.R.U.S. (1943)').
141. 'Doubts about British motives': Hewlett and Anderson, p. 271.
142. 'It was clear to me that Mr Akers': Groves, footnote, H[arvey] H. B[undy], 'Memorandum of Meeting at 10 Downing Street on 22 July, 1943', F.R.U.S. (1943), pp. 634–6, here p. 635.
142. 'If Halban goes to the U.S.A.': Internal Minute of Ministry of Aircraft Production, 5 June 1941, AB 1/220, P.R.O.
142. 'no direct application to the war effort': Pye–Sir Charles Darwin, 17 June 1941, AB 1/220, P.R.O.
142. '(notably a group of French refugee scientists)': United Kingdom Atomic Energy Authority, 'Atomic Energy in Britain, 1939–1975', p. 2.
142. 'it is likely to prove the quickest method': Anderson–Malcolm Macdonald, 6 Aug. 1942, AB 1/207, P.R.O.

143. 'furnish to each other on request': Anglo–Russian Agreement, 29 Sep. 1942, Tizard Papers.
144. 'on this whole subject by the application of the principle': James B. Conant–Dean Mackenzie, 2 Jan. 1943, quoted Eggleston, pp. 64–6.
145. 'The army and the higher military police': quoted Eggleston, p. 71.
145. 'It appears that this principle': Anderson–Churchill, 11 Jan. 1943, PREM 3/139.8A, P.R.O.
146. 'I should be very grateful for some news': Churchill–Harry Hopkins, 15 Feb. 1943, PREM 3/139.8A, P.R.O.
146. 'entirely destroys': Churchill–Hopkins, 27 Feb. 1943, PREM 3/139.8A, P.R.O.
146. 'Time is passing': Churchill–Hopkins, proposed cable, March 1943, PREM 3/139.8A, P.R.O.
146. 'collaboration appears to be at a standstill': Churchill–Hopkins, cable, March 1943, PREM 3/139.8A, P.R.O.
146. 'The adopted policy is that information': quoted Robert E. Sherwood, *The White House Papers of Harry L. Hopkins: An Intimate History*, p. 701.
147. 'the authorities feel': quoted Eggleston, p. 76.
147. 'For a time the Canadian government': Eggleston, p. 79.
147. 'That we should each work separately': Churchill–Hopkins, 1 Apr. 1943, PREM 3/139.8A, P.R.O.
147. 'and the necessary staff': Churchill–Anderson, 15 Apr. 1943, PREM 3/139.8A, P.R.O.
147. 'The President agreed that the exchange': Churchill–Anderson, cable, 26 May 1943, PREM 3/139.8A, P.R.O.
148. 'The matter finally came down to the point': V. Bush, Memorandum, 25 May 1943, F.R.U.S. (1943), pp. 209–211, here p. 210.
148. 'most extraordinary interview with Cherwell': V. Bush, 'Memorandum of Conference with the President [Roosevelt], 24 June 1943', F.R.U.S. (1943), p. 631.
148. 'The matter of Tube Alloys is in hand': Hopkins–Churchill, cable, 17 June 1943, F.R.U.S. (1943), p. 630.
148. 'When I recounted that Cherwell': V. Bush, 'Memorandum of Conference with the President [Roosevelt], 24 June 1943', F.R.U.S. (1943), p. 631.
149. 'My experts are standing by': Churchill–Roosevelt, cable, 9 July 1943, F.R.U.S. (1943), p. 632.
149. 'I wish, therefore, that you would renew': Roosevelt–Bush, 20 July 1943, F.R.U.S. (1943), p. 633.
149. 'I thought I was instructed to review': Bush, Memorandum, 'Sequence of Events Concerning Interchange with the British on the Subject of S-1, 4 Aug. 1943', F.R.U.S. (1943), pp. 642–645, here p. 645.
149. 'We cannot afford after the war': Anderson minute to Churchill, 21 July 1943, PREM 3/139.8A, P.R.O.
149. 'was disturbed to know': Churchill note to Anderson, 18 July 1943, PREM 3/139.8A, P.R.O.
150. 'suggested that he would be in favour': H[arvey] H. B[undy], 'Memorandum of Meeting at 10 Downing Street on 22 July 1943', F.R.U.S. (1943), pp. 634–6, here p. 636.
150. '... any post-war advantages': Churchill, n.d., PREM 3/139.8A, P.R.O.
150. 'I do not, of course': Anderson–Churchill, 29 July 1943, PREM 3/139.8A, P.R.O.

150. 'if all available'; 'this agency' and 'communicate any information about Tube Alloys': 'Articles of Agreement Governing Collaboration between the Authorities of the U.S.A. and the U.K. in the matter of Tube Alloys' [The Quebec Agreement], PREM 3/139.8A, P.R.O.
151. 'subject to the control'; 'between those in the two countries' and 'in view of the heavy burden': 'Articles of Agreement Governing Collaboration between the Authorities of the U.S.A. and the U.K. in the matter of Tube Alloys' [The Quebec Agreement], PREM 3/139.8A, P.R.O.
151. 'It seemed to me that we had signed away': R. V. Jones, *Most Secret War*, p. 474.
151. 'They are trying to avoid': Harvey Bundy–Stimson, Memorandum, 6 Aug. 1943, F.R.U.S. (1943), p. 648.
152. 'I am absolutely sure': Churchill–Cherwell, 27 May 1944, PREM 3/139.11A, P.R.O.
152. 'Explain to me shortly': Churchill–Cherwell and Anderson, 13 Apr. 1944, PREM 3/139.2, P.R.O.
152. 'claim a [post-war] industrial monopoly': Cherwell–Churchill, n.d., PREM 3/139.2, P.R.O.
152. 'When the inside history': James B. Conant, 'Anglo-American Relations in the Atomic Age', address delivered at the London School of Economics and Political Science, 17 Mar. 1952, and printed, p. 32 (afterwards referred to as 'Conant').
154. 'I cannot escape the feeling': Groves, p. 408.
155. 'Our point was that scientific work': quoted Kramish, p. 49.
156. 'no time must be lost': quoted I. N. Golovin, 'I. V. Kurchatov: A Socialist–Realist Biography of the Soviet Nuclear Scientist', p. 39, (afterwards referred to as 'Golovin').
156. 'By then the Soviet Government' and 'Three days later, having been appointed': Golovin, p. 40.
157. 'No one could guess whether the slugs': Hewlett and Anderson, p. 210.
158. 'To build a pile the engineer had to know': Hewlett and Anderson, p. 206.
159. 'The only reason for doubt': Arthur Holly Compton, *Atomic Quest. A Personal Narrative*, p. 137 (afterwards referred to as 'Compton (1956)').
160. 'So I assumed the responsibility myself,': Compton (1956), p. 138.
160. 'The frame supporting [it] was made': quoted Herbert L. Anderson, ' "All in our Time", Fermi, Szilard and Trinity', *Bulletin of the Atomic Scientists*, Vol. XXX, No. 8, Oct. 1974, pp. 40–7, here p. 43 (afterwards referred to as 'Anderson').
161. 'Thus we could tell': Anderson, p. 44.
162. 'At first you could hear the sound': Anderson, p. 44.
162. 'The pile has gone critical' and 'Zip in!': quoted Anderson, p. 44.
162. 'No cheer went up': Anderson, p. 45.
162. 'The Italian navigator': Compton, At. Power, p. 12.
164. 'I have been calculating the temperature': Thomson–Simon, 13 May 1941, AB 3/15, P.R.O.
165. 'More than any other man': H. A. Bethe, 'J. Robert Oppenheimer', *Biographical Memoirs of the Fellows of the Royal Society*, Vol. 14 (1968), pp. 391–416, here p. 391.
165. 'They were not in sympathy': quoted Wharton, p. 74.
165. 'What made Los Alamos different' and 'The isolation, the sun-drenched

mesas': Alice Kimball Smith, 'Los Alamos: Focus of an Age', *Bulletin of the Atomic Scientists*, Vol. XXVI, No. 6, June 1970, pp. 15–20, here p. 17 (afterwards referred to as 'Smith, Los Alamos').

166. 'changed everything': Smith, Los Alamos, p. 18.
166. 'scientists should be on tap': The phrase is credited to the British junior minister who said: 'We must keep the scientists on tap but never let them get on top'; Sir Thomas Merton, 'Science and Invention', *The New Scientist*, Vol. 25, p. 377.
166. 'On one occasion I was making': Otto R. Frisch, 'Somebody Turned the Sun on with a Switch', *Bulletin of the Atomic Scientists*, Vol. XXX, No. 4, April 1974, pp. 12–18, here p. 17.
169. 'But their very success': Anderson, draft cable, 27 April 1944, PREM 3/139.2, P.R.O.
169. 'I do not think any such telegram': Churchill, minute on Anderson's draft cable, PREM 3/139.2, P.R.O.
169. 'As soon as we get into discussion' and 'This minute has been prepared': Anderson–Churchill, 21 Mar. 1944, PREM 3/139.2, P.R.O.
170. 'Something will have to be done': Field Marshal Smuts–Churchill, 15 June 1944, PREM 3/139.11A, P.R.O.
170. 'thought the time had come to tell': Roosevelt–Mackenzie King, quoted J. W. Pickersgill and D. F. Foster, *The Mackenzie King Record* Vol. 2 1944–1945, p. 326 (afterwards referred to as 'Pickersgill and Foster').
170. 'thought he agreed' and 'troubled by the possible effect': Stimson diary entry 30 and 31 Dec. 1944, quoted Barron J. Bernstein, 'Roosevelt, Truman and the Atomic Bomb, 1941–1945', *Political Science Quarterly*, Vol. 90, No. 1, Spring 1975, pp. 23–69, here p. 30 (afterwards referred to as 'Bernstein').
171. 'Russian interest in the project': Anderson–Churchill, 7 Mar. 1945, PREM 3/139.11A, P.R.O.
171. 'There is no scientist in the world': quoted Aage Bohr, 'The War Years and the Prospects Raised by the Atomic Weapons', pp. 191–204, here p. 193 (afterwards referred to as 'Aage Bohr'), in S. Rozenthal (ed.), *Niels Bohr, His life and work as seen by his friends and colleagues*.
171. 'to assist in the protection' and 'Above all, I have': quoted Aage Bohr, p. 194.
172. 'They didn't need my help': quoted J. Rud Nielson, 'Memories of Niels Bohr', *Physics Today*, Vol. 16, No. 10, Oct. 1963, pp. 22–31, here p. 29.
173. 'a man weighed down with a conscience': Frankfurter, quoted Gowing (1964), p. 349.
173. 'worried him to death': Roosevelt–Frankfurter, quoted Gowing (1964), p. 350.
173. 'My father was entrusted': Aage Bohr, p. 203.
173. 'in a most encouraging manner': Bohr Memorandum to Roosevelt, quoted *Roosevelt and Frankfurter: Their Correspondence 1928–1945*, annotated by Max Freedman, p. 731 (afterwards referred to as 'Freedman').
174. 'I do not know how to thank you': and 'Notwithstanding my urgent desire': Niels Bohr–Peter Kapitza, 29 Apr. 1944, quoted Moore, pp. 336 and 337.
174. 'Mackenzie King spoke to me on Sunday': Cherwell–Churchill, 10 May 1944, PREM 3/139.11A, P.R.O.
175. 'I cannot avoid the conviction': Sir Henry Dale–Churchill, 11 May 1944, PREM 3/139.11A, P.R.O.
175. 'It was terrible': quoted R. V. Jones, 'Winston Leonard Spencer Churchill 1874–1965', *Biographical Memoirs of Fellows of the Royal Society 1966*,

Vol. 12, pp. 35–105, here p. 88 (afterwards referred to as 'Jones, Churchill Obit.').

175. 'Churchill was in a bad mood': Jones, Churchill Obit., p. 88.

176. 'able but quaint English': Felix Frankfurter–Roosevelt, 10 July 1944, Freedman, p. 728.

176. 'Continued secrecy would poison': Freedman, p. 726.

177. '1. The suggestion that the world': Aide-mémoire of Conversation between the President and the Prime Minister at Hyde Park, 19 Sep. 1944, PREM 3/139.9, P.R.O.

177. 'The last paragraph': Field Marshal Wilson–Stimson, 20 June 1945, PREM 3/139.9, P.R.O.

177. 'The President and I are much worried': Churchill–Halifax for Cherwell, 20 Sep. 1944, PREM 3/139.8A, P.R.O.

178. 'a German bomb which emitted some liquid': Churchill–Cherwell, 10 Jan. 1944, PREM 3/139.11A, P.R.O.

178. 'following his frequent practice': Bernstein, p. 29.

178. 'Aware through Bohr himself': Robert Dallek, *Franklin D. Roosevelt and American Foreign Policy, 1932–1945*, p. 471.

178. '[He had] scented danger' and 'Bohr was declared': Freedman, p. 725.

179. 'to bring about an internationalization': Einstein–Bohr, 12 Dec. 1944, diplomatic source.

179. 'quite realized the situation': Bohr memorandum, Dec. 1944, diplomatic source.

179. a comparable argument about the French: In *Scientists in Power* Spencer R. Weart gives bibliographical details of the different, and sometimes conflicting, French, American and British versions of this incident.

180. 'with [his] authorization': Charles de Gaulle, *Mémoires de guerre*, Vol. II *L'Unité*, p. 242.

180. 'only a promise which he might disregard': Cherwell–Anderson, 14 Oct. 1944, PREM 3/139.11A, P.R.O.

180. 'Joliot, who has great influence': Cherwell–Anderson, 19 Oct. 1944, PREM 3/139.11A, P.R.O.

181. '(a) I wanted to try to clinch': Anderson–Churchill, 2 Nov. 1945, PREM 3/139.5, P.R.O.

181. 'we should all have it in mind': Anderson–Churchill, 26 Jan. 1945, PREM 3/139.5, P.R.O.

181. 'If there is real danger': Churchill marginal note, Anderson–Churchill, 26 Jan. 1945, PREM 3/139.5, P.R.O.

182. 'It was Guéron, the only one of us': Bertrand Goldschmidt, *Rivalités atomiques*, pp. 88–9, quoted Spencer R. Weart, *Scientists in Power*, p. 203.

182. 'that the French had approached': Cherwell–Churchill, 27 Mar. 1945, PREM 3/139.6, P.R.O.

182. 'On board the "Quincy" yesterday': Churchill–Anderson and Cherwell, 16 Feb. 1945, PREM 3/139.11A, P.R.O.

182. 'I think the best way of putting it': Cherwell–Churchill, 26 Jan. 1945, PREM 3/139.11A, P.R.O.

182. 'made no objection of any kind': Churchill–Anderson and Cherwell, 16 Feb. 1945, PREM 3/139.11A, P.R.O.

183. 'fully into the picture': Anderson–Churchill, 19 Jan. 1945, PREM 3/139.11A, P.R.O.

183. 'The Chiefs of Staff have already been informed': Churchill–Anderson, 21 Jan. 1945, PREM 3/139.11A, P.R.O.
183. 'Personally, he [Anderson] would see no objection': General Ismay–Tizard, 6 Apr. 1945, Tizard Papers.
183. 'Pray see my previous minute': Churchill–Ismay, 19 Apr. 1945, PREM 3/139.11A, P.R.O.
185. 'I remember vividly the succinct explanation': Alice Kimball Smith, *A Peril and a Hope: The Scientists' Movement in America: 1945–47*, p. 5 (afterwards referred to as 'Smith (1965)').
185. 'Almost everyone knew': United States Atomic Energy Commission, *In the Matter of J. Robert Oppenheimer*, p. 13.
186. 'to the last day of the European war': 'Before Hiroshima' and 'A Report to The Secretary of War' [11], June 1945, *Bulletin of The Atomic Scientists*, Vol. 1, No. 10, 1 May 1946, pp. 2–4 and 16, here p. 2 (afterwards referred to as 'Franck Report').
186. 'What is the purpose of continuing': Weart and Szilard, p. 181.
187. 'raised the question of whether' and 'in connection with': Bush–Conant, 23 Sep. 1944, quoted Bernstein, p. 33.
187. 'there should be arranged': quoted Nat. S. Finney, 'How F.D.R. Planned to Use the A-Bomb', *Look*, Vol. XIV (14 Mar. 1950), p. 24, quoted Smith, p. 26.
188. 'the real warmongering': Sachs, p. 554.
188. 'You may be cleverer than I' and 'Sir, I agree with one half': 'James Franck, 1882–1964; Leo Szilard, 1898–1964', 'Bulletin of the Atomic Scientists', Oct. 1964, pp. 16–20, here p. 16.
189. 'I hope that you will get the President': quoted Szilard, Reminiscences, p. 124.
189. 'As the Einstein letter had indicated': James F. Byrnes, *All In One Lifetime*, p. 284 (afterwards referred to as 'Byrnes (1960)').
189. 'When I spoke of my concern': Szilard, Reminiscences, p. 124.
190. 'In Russia, too': 'Franck Report', p. 2.
190. '[They are] known to be mining radium': Franck Report, p. 2.
190. 'such a quantitative advantage': Franck Report, p. 2.
190. 'the race for nuclear armaments': Franck Report, p. 3.
190. 'It may be very difficult': Franck Report, p. 3.
191. 'This may sound fantastic': Franck Report, p. 3.
192. 'to study and report on': George L. Harrison–Stimson, 1 May 1945, quoted Martin J. Sherwin, *A World Destroyed: The Atomic Bomb and the Grand Alliance*, p. 295 (afterwards referred to as 'Sherwin').
193. 'As I heard these scientists': Byrnes (1960), p. 283.
193. 'At the luncheon following the morning meeting': Compton (1956), p. 238.
193. 'The reputation of the United States': Stimson Diary, 16 May 1945, quoted Arthur Steiner, 'Scientists, Statesmen, and Politicians: The Competing Influences on American Atomic Energy Policy, 1945–64', *Minerva. A Review of Science, Learning and Policy*, Vol. XII, No. 4, Oct. 1974, pp. 469–509, here p. 473 (afterwards referred to as 'Steiner').
194. 'the Secretary expressed the conclusion': R. Gordon Arneson, 'Notes of the Interim Committee Meeting, 31 May, 1945', quoted Sherwin, p. 302.
194. 'to be focused on the morale': Webster and Frankland, Vol. IV, p. 144.
194. 'to better our political relations with Russia': Arneson, 'Notes of the Interim Committee Meeting, 31 May 1945', quoted Sherwin, p. 301.
194. 'we can propose no technical demonstration' and 'that before the weapons':

'Science Panel: Recommendations on the Immediate Use of Nuclear Weapons', 16 June 1945, quoted Sherwin, p. 305 and p. 304.

194. 'there would be considerable advantage': Arneson, 'Notes of the Interim Committee Meeting, 21 June 1945', quoted Sherwin, p. 215.

195. 'I believe that these attempts': Irving Langmuir, U.S. Congress Hearings on Senate Resolution No. 179, quoted Kramish, p. 84.

195. 'I shall certainly continue' and 'Even six months will make': Churchill–Anthony Eden, 25 Mar. 1945, PREM 3/139.6, P.R.O.

197. 'The hazard that I feared the most': Lt. Gen. Leslie R. Groves, 'Some Recollections of July 16, 1945', *Bulletin of the Atomic Scientists*, Vol. XXVI, No. 6, June 1970, pp. 21–7, here p. 24.

198. 'It was extremely important': Groves, p. 291.

198. 'As the time intervals grew smaller': Groves, Memorandum for the Secretary of War, 18 July 1945, quoted Groves, p. 436.

199. 'a lighting effect ...': quoted Groves, p. 433.

199. 'It was golden, purple violet': quoted Groves, p. 437.

199. 'a great blinding light': quoted Gowing 'Reflections', p. 52.

199. 'There floated through my mind': quoted Nuel Pharr Davis, *Lawrence and Oppenheimer*, p. 240, (afterwards referred to as 'Davis').

199. 'Finally I could remove the goggles': Kenneth T. Bainbridge, ' "All in our Time" – A Foul and Awesome Display', *Bulletin of the Atomic Scientists*, Vol. XXXI, No. 5, May 1975, pp. 40–6, here p. 46.

199. 'It was just like the sun': Mrs H. E. Weiselman, quoted Teller, p. 5.

199. Groves remembered: Groves, Memorandum for the Secretary of War, 18 July 1945, quoted Groves, p. 439.

199. 'My God, it worked': quoted Hewlett and Anderson, p. 379.

199. 'Oppie, you owe me ten dollars': quoted Hewlett and Anderson, p. 379.

200. 'This is the end of traditional warfare': quoted Davis, p. 241.

200. 'The effects ... indicate that': Groves, Memorandum for the Secretary of War, 18 July 1945, quoted Groves, p. 434.

201. 'the President advise the Russians': Arneson, Notes of the Interim Committee meeting, 21 June 1945, quoted Sherwin, p. 215.

201. 'changed their view': Anderson–Churchill, 29 June 1945, PREM 3/139.8A, P.R.O.

202. '(1) Not to say anything': Cherwell–Churchill, 12 July 1945, PREM 3/139.9, P.R.O.

202. 'if he found that he thought': Stimson Diary, 3 July 1945, quoted Steiner, p. 489.

203. 'Operated on this morning,': Harrison–Stimson, 16 July 1945, quoted Hewlett and Anderson, p. 383.

203. 'Doctor has just returned': Harrison–Stimson, 18 July 1945, quoted Hewlett and Anderson, p. 386.

203. 'As he [Churchill] walked down to the gate': Stimson's Diary entry, 17 July 1945, quoted U.S. Department of State, 'Foreign Relations of the United States. Diplomatic Papers. The Conference of Berlin (The Potsdam Conference), 1945', Vol. II, p. 47, (afterwards referred to as 'F.R.U.S. (Potsdam)').

203. 'The following subjects were touched upon': Summarized note of Prime Minister's conversation with President Truman at luncheon, 18 July 1945, PREM 3/139.11A, P.R.O.

204. 'At twelve o'clock Lord Cherwell called': Stimson's diary entry, 19 July 1945,

quoted F.R.U.S. (Potsdam), Vol. II, p. 111.

204. 'Stimson, what was gunpowder?': quoted Strauss, p. 186.

204. 'The responsible heads at Los Alamos': Groves–Editor, *Science Magazine*, Dec. 1959, quoted Herbert Feis, *Japan Subdued: The Atomic Bomb and the End of the War in the Pacific*, p. 28 (afterwards referred to as 'Feis').

205. '[Churchill] now not only was not worried': Stimson's diary entry, 22 July 1945, quoted F.R.U.S. (Potsdam), Vol. II, p. 225.

205. 'He was already seeing himself capable': Alanbrooke, quoted Arthur Bryant, *Triumph in the West, 1943–1946*, based on the Diaries and Autobiographical Notes of Field Marshal The Viscount Alanbrooke, K.G., O.M., p. 478 (afterwards referred to as 'Bryant').

205. 'We had a brief discussion': Stimson's diary entry, 23 July 1945, quoted F.R.U.S. (Potsdam), Vol. II, p. 260.

205. 'I casually mentioned to Stalin': Harry S. Truman, *The Memoirs of Harry S. Truman*, Vol. I *Year of Decisions, 1945*, p. 346.

206. 'How did it go?': Winston S. Churchill, *The Second World War*, Vol. VI *Triumph and Tragedy*, p. 580 (afterwards 'Churchill, "Triumph and Tragedy" ').

206. 'seemed to be delighted': Churchill, 'Triumph and Tragedy', p. 579.

206. 'On returning to his quarters': G. K. Zhukov, *The Memoirs of Marshal Zhukov*, p. 675.

207. 'Antonov [Stalin's interpreter] told me later': S. M. Shtemenko, *The General Staff at War, 1941–45*, p. 347.

207. 'Stalin was – or pretended to be': Memorandum of Stalin–Harriman talk 8 Aug. 1945, quoted Feis, p. 115.

208. 'Then, plump, came the atomic bomb': Sir Archibald Clark-Kerr–Ernest Bevin, 3 Dec. 1945, U.S. Department of State, 'Foreign Relations of the United States. Diplomatic Papers, 1945. Vol. II General: Political and Economic Matters', pp. 82–84, here p. 83.

208. 'The U.S.S.R. is out to get the atomic bomb': Thomas P. Whitney report to Secretary of State, 24 Dec. 1945, U.S. Department of State, 'Foreign Relations of the United States. Diplomatic Papers, 1945. Vol. V Europe', pp. 933–936, here p. 934.

209. '[Attlee] said he thought it was a mistake': Pickersgill and Foster, Vol. 3, p. 59.

209. 'The wartime policy on atomic energy': Bernstein, p. 68.

210. 'We have discovered the most terrible bomb': Entry in Truman Journal, 25 July 1945, quoted *The Times*, 3 June 1980.

212. 'the bitterness ... caused': Stimson Diary, 24 July 1945, quoted Bernstein, p. 49.

213. 'The gist of his final message': James Forrestal, Diary, 15 July 1945, Walter Millis (ed.), *The Forrestal Diaries: The Inner History of the Cold War*, p. 88 (afterwards referred to as 'Millis').

214. 'important and might be just the thing': and 'hoped that the President would watch': Stimson Diary, 24 July 1945, quoted Steiner, p. 475.

214. 'The Japanese were given fair warning': Truman–editor, Dec. 16, 1946, *The Atlantic Monthly*, Feb. 1947, p. 27.

214. 'I feel we are approaching': Pickersgill and Foster, Vol. 2, p. 447.

214. 'There are only a few days left': quoted Strauss, p. 189.

215. 'If only I had had a channel': quoted Ronald W. Clark, *The Man Who Broke Purple: The Life of Colonel William F. Friedman, Who Deciphered the Japanese*

Code in World War II, p. 156.

216. 'I can testify personally': Philip Morrison, 'Blackett's Analysis of the Issues', *Bulletin of the Atomic Scientists*, Vol. 5, No. 2, Feb. 1949, pp. 37–40, here p. 40.

216. 'The request presented us': James F. Byrnes, *Speaking Frankly*, p. 207.

216 'We are sending this as a personal message': Luis Alvarez, Robert Serber and Philip Morrison–Dr Ryokichi Sagane, 9 Aug. 1945, quoted Compton (1956), p. 259.

217. 'believe that history might find': Henry L. Stimson and McGeorge Bundy, *On Active Service in Peace and War*, p. 372 (afterwards referred to as 'Stimson and Bundy').

217. 'The growing feeling of apprehension': Forrestal Diary, 10 Aug. 1945, Millis, p. 95.

218. 'Mr Byrnes did not argue': Szilard, 'A Personal History of the Atomic Bomb', University of Chicago *Round Table*, No. 601, 25 Sep. 1949, pp. 14–15.

218. 'As we understood it in July': Stimson and Bundy, p. 365.

218. 'we would have kept on fighting': quoted Karl Compton, 'If the Atomic Bomb had not been used', *The Atlantic Monthly*, Vol. 178, No. 6, Dec. 1946, pp. 54–56, here p. 54.

219. 'It is apparent that the effect' and 'Based on a detailed investigation': U.S. Strategic Bombing Survey, Washington, 1946.

219. 'When we first began to develop': Groves, p. 265.

220. 'there was never a moment's discussion': Churchill, Triumph and Tragedy, p. 553.

220. 'that he did not think it ever': quoted Pickersgill and Foster, Vol. 3, p. 133.

220. 'grave misgivings [about the use of the bomb]': Dwight D. Eisenhower *The White House Years: Mandate for Change, 1953–1956*, p. 483.

220. 'he was surprised that the second attack': quoted Pickersgill and Foster, Vol. 3, p. 236.

221. 'Could we have avoided the tragedy': Teller, pp. 19–20.

221. 'turn out to have been a sound decision': Forrestal Diary, 29 May 1947, Millis, p. 270.

221. 'I daresay this sudden collapse': Field Marshal Smuts–M. C. Gillett, 16 Aug. 1945, quoted Jean van der Poel (ed.), *Selections from the Smuts Papers*, Vol. VI, p. 550.

221. 'Of course that is not what Bohr thinks': Felix Frankfurter, Diary, 6 Nov. 1946, quoted Joseph P. Lash, *From the Diaries of Felix Frankfurter*, p. 289.

222. 'If atomic bombs are to be added': Oppenheimer, 16 Oct. 1945, Los Alamos, accepting Secretary of War's Certificate of Appreciation from General Groves; Alice Kimball Smith and Charles Weiner (eds) *Robert Oppenheimer: Letters and Recollections*, p. 310.

225. 'an entirely new situation': quoted 'The Era of Atomic Power', Report of a Commission Appointed by the British Council of Churches, p. 8 (afterwards referred to as 'B.C.C.').

225. 'I think it very undesirable': Cherwell note, 26 July 1945, PREM 3/139.9, P.R.O.

225. 'reached a secret agreement at Quebec': President Truman at Press Conference, 31 Oct. 1945, quoted Churchill, House of Commons, *Hansard*, Vol. 415 col. 1299, 7 Nov. 1945.

226. 'Subject to anything': Churchill, House of Commons, *Hansard*, Vol. 415, col. 1299, 7 Nov. 1945.

227. 'greatly stimulated': Allan F. Matthews, 'Minor Metals', U.S. Department of the Interior, Bureau of Mines, *Minerals Yearbook, 1943*, p. 825.
227. 'I do not know whether': Dr Haden Guest, House of Commons, *Hansard*, Vol. 381, col. 459, 1 July 1942.
228. 'Then I believe': H. H. Dale–Editor, 7 Aug. 1945, *The Times*, 8 Aug. 1945.
228. 'We are determined to return': Samuel K. Allison at Shoreland Hotel, Chicago, 1 Sep. 1945, quoted *Chicago Tribune*, 2 Sep. 1945, p. 5, quoted Smith (1965), p. 89.
229. 'a great extension of the practice': B.C.C., p. 50.
229. 'This counsel was not taken': Monsignor Fulton J. Sheen, sermon, 7 Apr. 1946, quoted *Bulletin of the Atomic Scientists*, Vol. 1, No. 10, 1 May 1946, p. 12.
229. 'It seems logical to me': Father Siemes, S. J., 'Hiroshima – August 6, 1945' *Bulletin of the Atomic Scientists*, Vol. 1, No. 11, 15 May 1946, pp. 2–6, here p. 6.
230. 'Responsible statesmen of the great powers': Attlee–Truman, 25 Sep. 1945, quoted Margaret Gowing, assisted by Laura Arnold, *Independence and Deterrence. Britain and Atomic Energy 1945–1952*, Vol. 1. *Policy Making*, Vol. 2. *Policy Execution*, here Vol. 1, p. 80 (afterwards referred to as 'Gowing (1974)').
230. 'Britain [was] peculiarly vulnerable': Memorandum by Prime Minister (Clement Attlee), 5 Nov. 1945, 'International Control of Atomic Energy', CAB 129/4, P.R.O.
231. 'no single nation can in fact' and 'We are not convinced': 'Washington Declaration', 15 Nov. 1945.
231. 'extending between all nations': 'Washington Declaration', 15 Nov. 1945.
231. 'full and effective co-operation': Groves–Anderson Memorandum, 16 Nov. 1945, Gowing (1974), Vol. 1. pp. 85–6.
232. 'would shorten as much as possible': quoted Gowing (1964), p. 324.
233 'a stock to be measured': quoted Denis Richards, *Portal of Hungerford*, p. 350 (afterwards referred to as 'Richards').
233. 'It must be realized': Sir Leonard Owen, 'Nuclear Engineering in the United Kingdom in the First Ten Years', *The Journal of the British Nuclear Energy Society*, Vol. 2, No. 1, Jan. 1963, pp. 23–32, here p. 23.
234. 'I submit that a decision is required': Note by the Controller of Production of Atomic Energy, 31 Dec. 1946, covered by Memorandum by the Minister of Supply, 3 Jan. 1947, CAB 130/16, P.R.O.
235. This elegant arrangement which was approved: Meeting of Ministers, Gen. 163/1st Meeting 8 Jan. 1947, Confidential Annex. Minute 1, CAB 130/16, P.R.O.
235. 'Although the meeting was being held': Gowing (1974), Vol. I, p. 182.
236. 'no one can envisage': Tizard, note to Minister of Defence, 25 Nov. 1947, PREM 8/750, P.R.O.
236. 'It may be that the proper solution': Cabinet Defence Committee paper on Civil Defence Policy (D.O.(47)66), 7 Oct. 1947, PREM 8/750, P.R.O.
236. 'whether he is satisfied' and 'Yes, sir. As was made clear': House of Commons, *Hansard*, Vol. 450, col. 2117, 12 May 1948.
236. 'a manuscript note in General de Gaulle's own hand': Adrian Holman–Bevin, 9 Nov. 1945, CAB 130/8, P.R.O.
237. 'without the Government having taken the decision': François de Rose, 'Aspects Politiques des Problèmes Posés Par L'Armament Nucléaire Française'. Speech given before the Institut des Hautes Etudes de la Défense Nationale, 18

Nov. 1958, Paris, quoted Lawrence Scheinman, *Atomic Energy Policy in France under the Fourth Republic*, p. 94.

238. 'It is the main thesis': H. G. Wells, Preface signed Easton Glebe, Dunmow, 1921, to new edition *The World Set Free*, Collins, n.d.

241. 'There is one thing': Bertrand Russell–Gamel Brenan, 1 Sep. 1945, The Henry W. and Albert A. Berg Collection, New York Public Library, Astor Lenox and Tilden Foundations.

241. 'There might be a period of hesitation': Betrand Russell, 'Humanity's Last Chance', *Cavalcade*, 20 Oct. 1945.

241. 'He turned to me sharply'; Pickersgill and Foster, Vol. 4, p. 112.

242. 'That the U.S.S.R. would capitulate': 'Atomic Energy: An Immediate Policy for Great Britain', Memorandum by P. M. S. Blackett, Gowing (1974), Vol. I, pp. 194–206, here p. 198.

242. 'We have got the Maxim Gun': H[ilaire]. B[elloc]. and B.T.B., *The Modern Traveller*, p. 41.

243. 'No other Foreign Secretary': Ernest Bevin, quoted Richards, p. 368.

243. 'Churchill was right': Pickersgill and Foster, Vol. 3, p. 19.

245. 'the bitterest legislative battle': Karl Schriftgiesser, *The Lobbyists: The Art and Business of Influencing Lawmakers*, p. 242 (afterwards referred to as 'Schriftgiesser').

245. 'For some four to six weeks after Hiroshima': Szilard, 'We Turned the Switch', *Nation*, Vol. CLXI, 22 Dec. 1945, pp. 718–19.

245. 'an island of socialism': Schriftgiesser, p. 238.

246. 'Saw Acheson [Under Secretary of State] at 9.15': David E. Lilienthal Diary, 16 Jan. 1946, *The Journals of David E. Lilienthal*, Vol. II *The Atomic Energy Years, 1945–1950*, p. 10 (afterwards referred to as 'Lilienthal Journals').

247. 'The problem of reconciling a proper case': M. L. Oliphant 'Control of Atomic Energy', *Nature*, Vol. 157, No. 3995, 25 May 1946, pp. 679–80, here p. 679.

247. 'Doc, you couldn't be more right' and 'I've got your opening line': Bernard M. Baruch, *The Public Years*, p. 334 (afterwards referred to as 'Baruch').

247. 'We are here to make a choice': Baruch, p. 339.

248. 'My colleagues and I agreed completely': Lilienthal Journals, Vol. II, p. 60.

249. 'Baruch agreed with the military': Louis Galambos (ed.), *The Papers of Dwight David Eisenhower*, Vol. VII *The Chief of Staff*, p. 1078 (afterwards referred to as 'Eisenhower Papers').

249. 'Russia ... will be very loath'; Alanbrooke Diary, 20 Sep. 1945, quoted Bryant, p. 488.

249. 'The language of the resolution makes clear': Byrnes (1960), p. 346.

250. 'Let's forget about the Baroosh': quoted Gowing (1974), Vol. I, p. 164.

250. 'When, soon after this': Gowing (1974), Vol I, p. 164.

250. 'at once test the ability of ships': Strauss–Forrestal, 14 Aug. 1945, quoted Strauss, p. 209.

250. 'bigger and better bomb': Rear-Admiral H. G. Thursfield, 'The Lessons of the War', *Brassey's Naval Annual, 1946*, pp. 1–12, here p. 10.

251. 'If we are to outlaw the use of the atomic bomb': quoted Hewlett and Anderson, p. 581.

251. 'The second bomb caused a deluge' and 'these contaminated ships': quoted 'The Bikini Tests – Radiological Effects', *Bulletin of the Atomic Scientists*, Vol. 2; Nos. 3 and 4, 1 Aug. 1946, p. 26.

252. 'It is too soon to attempt an analysis': quoted Smith (1965), p. 479.

252. 'an acceptable guaranty of international peace': Joint Chiefs of Staff Board, 'An Evaluation of the Atomic Bomb as a Military Weapon', quoted Eisenhower Papers, Vol. IX, p. 2047.
253. 'Whatever the facts [he] had come to think': Hewlett and Anderson, p. 644.
253. 'approximately 400 atomic bombs': *The Journal of American History*, June 1979.
253. 'the establishment of a top-secret project': Joint Strategic Survey Report, 14 July 1947, quoted Eisenhower Papers, Vol. VIII, p. 1857.
253. 'a virtual – if less than complete – mobilization': Eisenhower–Joint Chiefs of Staff, 25 July 1947, Eisenhower Papers, Vol. VIII, p. 1855.
254. 'He told me that he would make': Forrestal Diary, 23 July 1948, Millis, p. 433.
254. 'unanimous agreement that in the event of war': Roy Roberts, *Kansas City Star*, quoted Forrestal Diary, 14 Sep. 1948, Millis, p. 456–7.
255. 'It is we who hold': *The Observer*, 27 June 1948.
255. 'the American people would execute you': quoted Forrestal Diary, 12 Nov. 1948, Millis, p. 457.
255. 'would not hesitate to use the atomic bomb': quoted Forrestal Diary, 13 Nov. 1948, Millis, p. 457.
255. 'no division in the British public mind': Forrestal Diary, 12 Nov. 1948, Millis, p. 488.
257. 'detecting atomic explosions': Eisenhower–Carl Spaatz, 16 Sept. 1947, Eisenhower Papers, Vol. IX *The Chief of Staff*, p. 1918.
257. 'A single demand of you': quoted Holloway, p. 41.
258. 'Russia's ownership of the bomb': quoted Strauss, p. 439.
258. 'There has probably been no other occasion': James Franck, quoted 'Did The Soviet Bomb Come Sooner Than Expected?' *Bulletin of the Atomic Scientists*, Vol. V, No. 10, Oct. 1949, p. 262.
258. 'Pretty discouraging. What he is talking about': Lilienthal Diary, 1 Nov. 1949, quoted David E. Lilienthal, *Change, Hope and the Bomb*, p. 143 (afterwards referred to as 'Lilienthal').
259. 'In case of enemy attack': quoted Conant, p. 9.
259. 'The fact that both sides': Sir John Boyd Orr, *Bulletin of the Atomic Scientists*, Vol. V, No. 11, Nov. 1949, p. 323.
260. 'Now that we have a good prospect': quoted Teller, p. 37.
260. 'We even thought we knew precisely': Teller, p. 38.
261. 'A spirit of spontaneity': Teller, p. 39.
261. 'only because they were intrigued': Teller, p. 39.
261. 'We resolved to keep the possibility': Compton (1956), p. 128.
262. 'I can still remember my shock': William Laurence, *The Hell Bomb*, p. 3 (afterwards referred to as 'Laurence').
262. 'I shall never forget the impact on me': Laurence, p. 4.
262. 'It was the first highly diluted minute sample': Laurence, p. 17.
262. '[He] was unclear whether his mandate': quoted Wharton, p. 13.
263. 'inevitable': Groves, p. 387.
263. 'the scientists prefer not to do that' and 'in improving the present techniques': George L. Harrison Memorandum for the Record, 18 Aug. 1945, quoted Sherwin, p. 315.
263. 'This has been your laboratory': Teller, p. 23.
264. 'reviews of the book published in the U.S.': *The Scientific American*, Vol. 182, No. 3, March 1950, pp. 11–15, here p. 14.

264. 'It seems to me that the time': Strauss, p. 216.
264. 'What concerns me is really not': Oppenheimer–Conant, 21 Oct. 1949, quoted York, p. 55.
265. 'We recommend strongly against': quoted York, p. 156.
266. 'uncertain to us whether the Super': quoted York, p. 155.
266. 'To the argument that the Russians': quoted York p. 157.
266. 'that an effort be made to determine': quoted Werner R. Schilling, 'The H-Bomb Decision: How to Decide without actually choosing', *Political Science Quarterly*, Vol. LXXVI, 1961, pp. 24–46, here p. 36 (afterwards referred to as 'Schilling').
266. 'an all-out development of hydrogen bombs': quoted Schilling, p. 44.
267. 'The G.A.C. was "right" in sensing': Schilling, p. 46.
267. 'included in a research group' and 'In 1950 our research group became': Andrei D. Sakharov, 'How I Came to Dissent', *The New York Review of Books*, Vol. XXI, No. 4, 21 Mar. 1974, pp. 11–17, here p. 11.
268. 'a significant test': Edward Teller, 'The Work of Many People', *Science*, Vol. 121, 25 Feb. 1955, pp. 267–75, here p. 272.
268. 'The mushroom cloud': Teller, p. 51.
268. 'The program we had in 1949': U.S.A.E.C., *In the Matter of J. Robert Oppenheimer*, p. 251, quoted York, p. 81.
269. 'I agree with the security officials': York, p. 8.
269. 'The U.S.A. has long since lost the monopoly': quoted Kramish, p. 124.
270. 'People do not seem yet to realize': Cherwell–James Tuck, 22 Dec. 1953, Cherwell Papers.
270. 'Unfortunately the wind had failed to follow': Strauss at President Eisenhower's Press Conference, 31 Mar. 1954, quoted *The Times*, 1 Apr. 1954, p. 6.
270. 'The situation was serious': Teller, p. 172.
271. 'We must, I am afraid, accept that fearful weapon': The Marquess of Salisbury, House of Lords, *Hansard*, Lords, Vol. 186, col. 651, 24 March, 1954.
271. 'Mankind, if it is to survive': Leader, 'A Hideous Reality', *The Times*, 26 Mar. 1954.
273. 'really monstrous to allow so many decent': Cherwell–Sir Ian Jacob, 21 June 1957, Cherwell Papers.
273. 'that the British tests constitute a danger': quoted A. Pirie (General Ed.), *Fall Out: Radiation Hazards from Nuclear Explosions*, p. 151, (afterwards referred to as 'Pirie').
273. 'On this evidence an appreciable lifetime radiation dose': 'Fall-Out from Nuclear Weapons', 'Annotations', *The Lancet*, 14 July 1956, pp. 81–2, here p. 82.
274. 'There is here a fundamental difficulty': quoted Pirie, p. 107.
275. 'Sir John noted that radiation protection measures': United Kingdom Atomic Energy Authority Press Office, 14 Apr. 1978.
275. 'like many arguments used in discussion': Pirie, p. 17.
276. 'are nearly always harmful ones': Dr Grimmett, Medical Research Council, quoted Gowing (1964), p. 384.
276. 'sufficient quantities of radioactive material': Dr Stafford Warren reported by Colonel Zimmerman, Acting Chief, Advanced Study Group, P. and O., quoted Eisenhower Papers, Vol. VIII, p. 1686.
277. 'quite beyond [Britain's] means': quoted Gowing (1974), Vol. I, p. 439.

277. 'would be an intolerable strain': quoted Gowing (1974), Vol. II, p. 474.
277. 'It would seem that the hydrogen bomb': Julian Amery, House of Commons, *Hansard*, Vol. 549, col. 1091, 28 Feb. 1956.
277. 'Unless we can make a contribution of our own': Churchill, House of Commons *Hansard*, Vol. 537, col. 1897, 1 Mar. 1955.
277. 'One consequence of the evolution from the atomic': Anthony Eden, *Memoirs: Full Circle*, p. 368.
279. 'Man's Peril': Bertrand Russell, B.B.C. Broadcast, 23 Dec. 1954, printed 'Man's Peril from the Hydrogen Bomb', *Listener*, Vol. 52, No. 1348, 30 Dec. 1954, pp. 1135–6.
280. 'Now that Britain has told the world': J. B. Priestley, 'Britain and the Nuclear Bombs', *The New Statesman*, 2 Nov. 1957, pp. 554–6, here p. 555.
282. 'The coming Sunday in an unguarded moment': Georg von Hevesy, quoted Glasstone, p. 666.
284. 'affect the distribution of industry': 'Nuclear Energy as a Source of Power', Note by Messrs. I.C.I., n.d., AB 1/37, P.R.O.
285. 'so sweet that you could not argue': U.S.A.E.C., *In the Matter of J. Robert Oppenheimer*, p. 251.
285. 'most of the labour now in coal-mines': Paper for Cabinet, 2 Feb. 1946, CAB 130/8, P.R.O.
285. 'could quite safely be placed': Report of GEN 112 ad hoc meeting, 28 Jan. 1946, CAB 130/8, P.R.O.
285. 'It did not seem likely': Report of GEN 112 ad hoc meeting, 14 Dec. 1945, CAB 130/8, P.R.O.
287. 'it was built primarily': U.K.A.E.A., *Atomic Energy in Britain, 1939–1975*, p. 11.
288. 'conviction, and one that I shared fully': Lilienthal, p. 109.
289. 'Much of this was an utterly meaningless and wasteful operation': Lilienthal, p. 111.
290. 'One day there may be': Richard Knox, 'Comment', *Nuclear Energy International*, May 1979, p. 15.
291. 'Questions of environmental protection': *Nuclear Energy International*, Dec. 1978, p. 10.
292. 'a thousand ages in Thy sight': *Hymns Ancient and Modern*.
292. 'If one prefers the estimates': Fred Hoyle, *Energy or Extinction: The Case for Nuclear Energy*, p. 76.
294. 'during the four-week period': Richard Curtis and Elizabeth Hogan, *Perils of the Peaceful Atom: The Myth of Safe Nuclear Power Plants*, p. 3.
294. 'Like a breath of fresh air': L. G. Brookes, review of Dr Petr Beckmann, 'The Health Hazards of NOT Going Nuclear', in *Atom*, No. 244, Feb. 1977, pp. 18–22, here p. 18.
294. 'What is quite clear is': 'The Report of the Kemeny Commission', *Atom*, No. 280, Feb. 1980, pp. 38–42, here p. 38.
296. 'long-term dangers to the fabric': Flowers Report.
296. 'A new industry has suddenly appeared': W. W. Waymack. Speech at Des Moines, Iowa, 6 Dec. 1948, A.E.C. Information Release, quoted James S. Allen, *Atomic Imperialism: The State, Monopoly and the Bomb*, p. 15.
297. 'Astronomers sometimes observe how a star': 'Frédéric Joliot-Curie, Nobel Prize Address, Stockholm, 12 Dec. 1935.

Index

LEGE OF EDUCATION
LANCASTER